METHODS IN MOLECULAR BIOLOGY™

Series Editor
John M. Walker
School of Life Sciences
University of Hertfordshire
Hatfield, Hertfordshire, AL10 9AB, UK

For other titles published in this series, go to
www.springer.com/series/7651

The Urinary Proteome

Methods and Protocols

Edited by

Alex J. Rai

Department of Pathology & Cell Biology, Columbia University Medical Center, New York, NY, USA

Editor
Alex J. Rai, Ph.D.
Department of Pathology & Cell Biology
Columbia University Medical Center
622 West 168th Street, CHONY 2C-224
New York, NY
USA
alex.rai@columbia.edu

ISSN 1064-3745 e-ISSN 1940-6029
ISBN 978-1-60761-710-5 e-ISBN 978-1-60761-711-2
DOI 10.1007/978-1-60761-711-2
Springer New York Dordrecht Heidelberg London

Library of Congress Control Number: 2010924744

Printed on acid-free paper

Humana Press is part of Springer Science+Business Media (www.springer.com)

Preface

This book is intended for scientific researchers, clinical laboratorians, clinical and translational scientists, and others interested in proteomics and biomarker discovery. Urine is one of the most easily accessible biological samples, and it provides a treasure trove of molecules important in clinical diagnostics. In this book, we review briefly the classical urine tests that are performed in the clinical laboratory and then delve into the state-of-the-art methods for proteomic analysis using urine specimens. The most recent advances are discussed with regard to sample preparation, data analysis, and finally methods and applications. A multitude of examples are provided including procedural details for the identification and characterization of urine biomarkers that hold potential for the diagnosis and treatment of many different disease conditions.

The text is arranged so as to read systematically: introduction, sample preparation methods, applications, and data analysis. However, it does not necessarily require the reader to read it from start to finish. Each chapter is organized such that it can be read individually without requiring knowledge from other chapters.

I would like to thank the many individuals who made this book possible. These include the many authors who contributed to each of the individual chapters, the corresponding authors who took responsibility in providing the complete and finished versions solicited for the peer review process, and the many scientific reviewers who provided their valuable input and guidance.

I would also like to thank my wife Shilpa and son Aseem who put up with me being at work late for many nights to get this book completed. Finally, I am grateful to Professor John Walker and his colleagues, Patrick Marton and David Casey, at Humana Press for giving me the opportunity and also for keeping things on track. Without them, this edition would not have been possible in its current form.

New York, NY *Alex J. Rai, PhD*

Contents

Contributors

KEITH A. BAGGERLY • *Department of Bioinformatics and Computational Biology, The University of Texas M. D. Anderson Cancer Center, Houston, TX, USA*
REBECCA E. CAFFREY • *Consultant, Richmond, VA, USA*
UTA CEGLAREK • *Institute of Laboratory Medicine, Clinical Chemistry and Molecular Diagnostics, University Hospital Leipzig, Medical Faculty, University Leipzig, Leipzig, Germany*
SHUI-TEIN CHEN • *Institute of Biological Chemistry and Genomics Research Center, Academia Sinica, Taipei, Taiwan; Institute of Biochemical Sciences, College of Life Science, National Taiwan University, Taipei, Taiwan*
WILLIAM CLARKE • *Department of Pathology, Johns Hopkins Medical Institutions, Baltimore, MD, USA*
KEVIN R. COOMBES • *Department of Bioinformatics and Computational Biology, The University of Texas M. D. Anderson Cancer Center, Houston, TX, USA*
DAVID R. CRAFT • *BD Diagnostic, Franklin Lakes, NJ, USA*
GERMÁN ECHEVERRY • *Mount Sinai School of Medicine, New York, NY, USA*
GEORG MARTIN FIEDLER • *Institute of Laboratory Medicine, Clinical Chemistry and Molecular Diagnostics, University Hospital Leipzig, Medical Faculty, University Leipzig, Leipzig, Germany*
AUDE L. FOUCHER • *BD Diagnostic, Franklin Lakes, NJ, USA*
KRIS E. GASTON • *Department of Urology, The University of Texas M.D. Anderson Cancer Center, Houston, TX, USA*
CRAIG A. GELFAND • *BD Diagnostic, Franklin Lakes, NJ, USA*
PATRICIA A. GONZALES • *Laboratory of Kidney and Electrolyte Metabolism, NHLBI, National Institutes of Health, Bethesda, MD, USA*
Department of Chemical and Biomolecular Engineering, University of Maryland, College Park, MD, USA
DAVID GOZAL • *University of Louisville, Louisville, KY, USA*
H. BARTON GROSSMAN • *Department of Urology, The University of Texas M.D. Anderson Cancer Center, Houston, TX, USA*
HOWARD B. GUTSTEIN • *Department of Anesthesiology and Perioperative Medicine, The University of Texas M. D. Anderson Cancer Center, Houston, TX, USA*
WALTER HOFMANN • *Institut für Klinische Chemie, Städtisches Klinikum München GmbH, München, Germany*
GLEN L. HORTIN • *Departments of Pathology, Immunology, and Laboratory Medicine, University of Florida, Gainesville, FL, USA*
MIROSLAV IVANDIĆ • *Synlab GmbH & Ko.KG, München, Germany*
SAEED A. JORTANI • *University of Louisville, Louisville, KY, USA*
MARK A. KNEPPER • *Laboratory of Kidney and Electrolyte Metabolism, NHLBI, National Institutes of Health, Bethesda, MD, USA*

ALEXANDER LEICHTLE • *Institute of Laboratory Medicine, Clinical Chemistry and Molecular Diagnostics, University Hospital Leipzig, Medical Faculty, University Leipzig, Leipzig, Germany*
WENLIN (WENDY) LI • *Pfizer Global Research and Development, Michigan Laboratories, Ann Arbor, MI, USA*
SHU-LING LIANG • *Department of Pathology, Johns Hopkins Medical Institutions, Baltimore, MD, USA*
PETER B. LUPPA • *Institut für Klinische Chemie, Krankenhaus München Rechts der Isar, München, Germany*
DAWN MATTOON • *Protein Array Center, Invitrogen Corporation, Branford, CT, USA*
LIHAO MENG • *Protein Array Center, Invitrogen Corporation, Branford, CT, USA*
JOCHEN METZGER • *Institut für Klinische Chemie, Krankenhaus München Rechts der Isar, München, Germany*
JEFFREY S. MORRIS • *Department of Biostatistics, The University of Texas M. D. Anderson Cancer Center, Houston, TX, USA*
OLGA NEMIROVSKIY • *Pfizer Global Research and Development, St. Louis Laboratories, St. Louis, MO, USA*
BRIANNE OLIVIERI • *Memorial Sloan-Kettering Cancer Center, New York, NY, USA*
LAURENCE D. PARNELL • *Nutrition and Genomics Laboratory, Jean Mayer Human Nutrition Research Center on Aging at Tufts University, US Department of Agriculture, Boston, MA, USA*
SUREE PHUTRAKUL • *Department of Chemistry, Faculty of Science, Chiang Mai University, Chiang Mai, Thailand*
TRAIRAK PISITKUN • *Laboratory of Kidney and Electrolyte Metabolism, NHLBI, National Institutes of Health, Bethesda, MD, USA*
ALEX J. RAI • *Department of Pathology and Cell Biology, Columbia University Medical Center, New York, NY, USA*
SUVI RAVELA • *Department of Clinical Medicine, Division of Clinical Chemistry, Biomedicum, University of Helsinki, Helsinki, Finland*
DAGMAR RUTH • *Siemens Healthcare Diagnostics, Marburg, Germany*
HEIKE SCHNEIDER • *Institut für Klinische Chemie, Krankenhaus München Rechts der Isar, München, Germany*
CHRISTINE M.E. SCHUELLER • *Biomax Informatics AG, Cambridge, MA, USA*
BARRY SCHWEITZER • *Protein Array Center, Invitrogen Corporation, Branford, CT, USA*
CORNELIA SEDLMEIR-HOFMANN • *Unterhaching, Germany*
MARTIN SHAW • *Argutus Medical Inc., Dublin, Ireland*
SUPACHOK SINCHAIKUL • *Institute of Biological Chemistry and Genomics Research Center, Academia Sinica, Taipei, Taiwan*
AYELET SNOW • *University of Louisville, Louisville, KY, USA*
KIMIA SOBHANI • *Department of Clinical Pathology, Cleveland Clinic, Cleveland, OH, USA*
SUPAWADEE SRIYAM • *Institute of Biological Chemistry and Genomics Research Center, Academia Sinica, Taipei, Taiwan; Department of Chemistry, Faculty of Science, Chiang Mai University, Chiang Mai, Thailand*

ROBERT A. STAR • *Renal Diagnostics and Therapeutics Unit, NIDDK, National Institutes of Health, Bethesda, MD, USA*
ULF-HÅKAN STENMAN • *Department of Clinical Medicine, Division of Clinical Chemistry, Biomedicum, University of Helsinki, Helsinki, Finland*
GABRIELLA SZEKELY-KLEPSER • *Allergan Inc., Irvine, CA, USA*
PAYUNGSAK TANTIPAIBOONWONG • *Institute of Biological Chemistry, Department of Chemistry, Faculty of Science, Chiang Mai University, Chiang Mai, Thailand*
JOACHIM THIERY • *Institute of Laboratory Medicine, Clinical Chemistry and Molecular Diagnostics, University Hospital Leipzig, Medical Faculty, University Leipzig, Leipzig, Germany*
CHING TZAO • *Division of Thoracic Surgery, Tri-Service General Hospital, National Defense Medical Center, Taipei, Taiwan*
ROLAND VALDES JR. • *University of Louisville, Louisville, KY, USA*
LEENA VALMU • *Department of Clinical Medicine, Division of Clinical Chemistry, Biomedicum, University of Helsinki, Helsinki, Finland; Finnish Red Cross Blood Service, Helsinki, Finland*
NAM SUN WANG • *Department of Chemical and Biomolecular Engineering, University of Maryland, College Park, MD, USA*
PETER S.T. YUEN • *Renal Diagnostics and Therapeutics Unit, NIDDK, National Institutes of Health, Bethesda, MD, USA*
HUA ZHOU • *Renal Diagnostics a nd Therapeutics Unit, NIDDK, National Institutes of Health, Bethesda, MD, USA*

Chapter 1

Introduction to Urinalysis: Historical Perspectives and Clinical Application

Germán Echeverry, Glen L. Hortin, and Alex J. Rai

Abstract

Urinalysis was the first laboratory test performed in medicine and has been used for several thousand years. Today urinalysis continues to be a powerful tool in obtaining crucial information for diagnostic purposes in medicine. Urine is an unstable fluid, and changes to its composition begin to take place as soon as it is voided. As such, collection, storage, and handling are important issues in maintaining the integrity of this specimen. In the laboratory, urine can be characterized by physical appearance, chemical composition, and microscopically. Physical examination of urine includes description of color, odor, clarity, volume, and specific gravity. Chemical examination of urine includes the identification of protein, blood cells, glucose, pH, bilirubin, urobilinogen, ketone bodies, nitrites, and leukocyte esterase. Finally, microscopic examination entails the detection of crystals, cells, casts, and microorganisms.

Key words: Urinalysis, Physical examination, Chemical examination, Urine microscopy

1. History of Urinalysis

Urinalysis was the first laboratory test performed in medicine, and has been used for more than 6,000 years by countless civilizations. It is such an intricate part of medicine that the urine collection flask was once a symbol of the profession much like the caduceus is today. Babylonian and Sumerian physicians were well aware of the physical changes that occurred to urine in different disease states, and used it in conjunction with other parts of physical examination to diagnose and treat disease. Egyptians were also aware of urine's importance, often recording its frequency and retention. They believed the kidneys to be sacred organs, leaving them in situ during mummification practices, though it is argued that its physiologic role in urine production eluded them.

Alex J. Rai (ed.), *The Urinary Proteome: Methods and Protocols*, Methods in Molecular Biology, vol. 641, DOI 10.1007/978-1-60761-711-2_1, © Springer Science+Business Media, LLC 2010

Hindus associated the sweet taste of urine with diabetes mellitus (DM), noting that black ants were attracted to it.

Many publications throughout history have arisen instructing physicians how to interpret different findings, but it was not until alchemists began their quest to purify its different components that chemical evaluation of urine truly began. Urea was discovered in 1771 by Roulle, and the proof that the sweetness in urine in DM was due to glucose came from Matthew Dobson in the eighteenth century (1). Hippocrates reflected "One can obtain considerable information concerning the general trends by examining the urine," and correctly described urine as a filtrate of blood, one of the four humors of the body along with yellow bile, black bile, and phlegm. More recently, in the early-mid 1800s, an English physician named Richard Bright pioneered the field of kidney research and became known as the "father of nephrology (2)." In addition, Henry Bence Jones systematically studied the chemical composition of both healthy and diseased urine, and later characterized the first tumor marker-kappa and lambda light chains, in the urine of a multiple myeloma patient (3). Today urinalysis continues to be a powerful tool in obtaining crucial information for diagnostic purposes in medicine (4).

2. Collection/Transportation/Storage

Urine is an unstable fluid, and changes to its composition begin to take place as soon as it is voided. It is, therefore, important to properly collect, store, and transport specimens to minimize the risk of altered test results. Often the first void or "morning urine" is collected because it is most concentrated, and trace amounts of substances can be detected. Furthermore, it provides an insight to the kidneys' concentrating abilities.

When collecting the specimens, clear instructions should be provided to the patients in order to minimize the chance of contamination. The male patient should be instructed to retract the foreskin (if present), and clean the glans penis prior to urinating. The female patient should clean the labia and urethral meatus in a similar manner. The patient should begin voiding, placing the container mid stream in the flow of urine to collect the specimen. Once the container is full, the patient can finish urinating in the bed pan or toilet and cap the container taking care not to contaminate its lid or rim.

Urine samples should be evaluated soon after their collection due to the unstable nature of its composition, particularly the formed elements (cells, casts, and crystals). These are more labile than urine's chemical constituents, which are more stable as long as bacterial growth is inhibited, such as by refrigeration. Urine should be stored exposed to minimal light or in darkness, and for no longer than 2 h at room temperature. If longer storage time is required, storing at 4°C for up to 24 h will slow, but not prevent, the process of decomposition. No specimen older than 24 h should be used for urinalysis (4–6).

3. Physical Examination of Urine

There are few instances when the color, odor, or clarity of urine is of clinical significance, but any abnormal finding should be noted on the report.

3.1. Color

Urine is usually of a yellow color, due to a natural pigment urobilin, formerly called urochrome, which is linear tetrapyrrole derived from heme breakdown. Additional yellow coloring may be due to riboflavin, in cases of vitamin supplementation, with alpha-1-microglobulin also contributing a yellow/brown tint. Urine color varies between individuals and is dependent on hydration state, diet, medication, and other factors. Red urine can be observed in the presence of hemoglobin, erythrocytes, myoglobin, porphyrin, fuscin, aniline dye, beets, or menstrual contamination. Black urine is seen in the presence of homogentisic acid, melanin, or methemoglobin. Green urine can be seen with *Pseudomonas* infections, or in the presence of indicans or chlorophyll. Finally, the presence of excessive porphyrins in urine will give it a blue-purple color.

3.2. Clarity

Urine should be translucent, through turbid is not necessarily a sign of disease. Urine can be turbid after refrigeration from precipitation of calcium phosphate, calcium oxalate, uric acidm or other salts. Urine can also appear turbid if it contains white cells and casts vaginal secretions, sperm, bacteria, blood clots, small calculi, or fecal material as a result of a fistula between the colon or rectum and the bladder. Chylomicrons in the urine, termed chyluria, is seldom observed, only when there is lymphatic obstruction by cancer or late-stage filariasis.

3.3. Odor

Depending on its pH, urine often has an ammoniacal odor related to high content of ammonium ion; however, odor varies with diet and medications. Foods such as onions, garlic, and asparagus impart characteristic odors. Strong urine smells are associated with infection, or with specimens that have been left out for too long. Other important odors to recognize are the fruity odor of acetone in diabetic ketoacidosis, "burnt sugar" from maple syrup urine disease, and honey from DM.

3.4. Volume

It is sometimes important in clinical diagnosis to determine the volume of urine produced over time. Typically, adults excrete 1,500–1,600 mL of urine in 24 h with a normal range between 600 and 2,000 mL. The production of less than 500 mL of urine is termed oliguria and is often indicative of dehydration due to prolonged vomiting, diarrhea, or excessive sweating with inadequate hydration resulting in hemoconcentration. Oliguria and anuria can also occur with renal ischemia secondary to heart failure or hypotension, renal disease or renal obstruction. Polyuria is defined as excreting more than 2,000 mL of urine in a 24 h period and occurs with increased water intake, saline or glucose intravenous therapy, alcohol or caffeine consumption, and certain diuretic drugs. Pathologically it is associated with diabetes mellitus and diabetes insipidus.

3.5. Specific Gravity/ Osmolality

Specific gravity and osmolality tests are performed to evaluate the kidney's ability to dilute or concentrate urine in order to maintain homeostasis. Osmolality measures the total number of dissolved particles in urine and serves as a more reliable test than specific gravity for evaluating kidney function. Urine osmolality can range from 50-1200 mosmol/kg, compared with serum osmolality which has a narrow range of about 280–300 mosmol/kg. Specific gravity is the ratio of the mass of solution compared to the mass of an equal volume of water. As this is a measure of weight, it does not measure the exact number of solute particles. Normal urine specific gravity ranges from 1.016 to 1.022 during a 24 h period. Isosthenuria (production of urine with the same osmolality as serum and a specific gravity of about 1.010) may be indicative of end stage renal failure. Methods used to test specific gravity include the hydrometer, refractometer, dipstick reagent pad, and harmonic oscillation (4–6).

4. Chemical Examination of Urine

Knowing certain chemical characteristics of urine is often of importance in clinical medicine. Several semiquantitative and qualitative tests can be performed on reagent strips and tablets, and quantitative methods can be used to measure protein, electrolytes, and porphyrins.

A reagent strip is a narrow band of plastic 4–6 mm wide and 11–12 cm long with a linear series of absorbent pads attached to it. Each pad contains reagents for a separate reaction, so several tests can be carried out simultaneously. Reagent strip methodology encompasses multiple complex chemical reactions. A color change on the pad indicates a reaction which then can be compared to a color chart provided by the manufacturer to interpret the result. This can be done visually, though photometric methods using instruments provided by the manufacturer are preferred as color perception varies among individuals. When using reagent strips it is important to test the urine promptly, understand the advantages and limitations of each test, and use controls. Tablets and chemical tests are used to confirm results obtained by dipstick methods when there are differences in sensitivity and specificity, or to avoid interferences when pigments in the specimens mask the colors obtained on the reagent pads (4–8).

4.1. Protein

Urinary proteins are normally present only in trace amounts, up to about 150 mg/24 h or 10 mg/dL, and originate in the plasma and urinary tract. Albumin is about a third of the total, while the remaining plasma proteins are small globulins. Small plasma proteins usually cross the glomerulus and a major fraction of the filtered protein is reabsorbed by a receptor-mediated process, leaving small amounts in urine. The small amounts of albumin crossing the glomerulus are handled similarly, with only a small proportion excreted in urine. Tamm–Horsfall glycoprotein is secreted by the tubular cells and makes up about a third of the total protein normally lost.

Proteinuria can be separated into a glomerular pattern, a tubular pattern, and an overflow pattern. The glomerular pattern occurs when proteins that are usually retained in the plasma, such as albumin and transferrin, cross the glomerulus in increased amounts. Severe glomerular leakage results in low serum albumin, which can lead to generalized edema and elevated serum lipids. Protein excretion greater than 3.0–3.5 g/dL is observed. Small serum proteins are still reabsorbed because tubular function is normal, therefore no change in their concentration in urine is observed. The tubular pattern is characterized by the increased excretion of low-molecular weight proteins due to the tubular cells' inability to reabsorb them. This pattern occurs in several tubular diseases such as Fanconi's syndrome, Wilson's disease, and pyelonephritis. In tubular pattern the proteinuria is not as high compared to the glomerular pattern, approximately 1–2 g/day. Finally, overflow proteinuria is seen when very high concentrations of small proteins are present in the plasma and crossing the glomerulus, overwhelming the ability of the tubular cells to reabsorb them. This not only leads to the excretion of these proteins in urine, but is also damaging to the tubular cells, and tubular cell casts can be seen in the sediment of these patients. Bence Jones proteins are immunoglobulin light chains and are excreted by up to 80% of multiple myeloma patients. These proteins may not be detected by reagent strip methodology because the method is most sensitive to albumin; therefore, electrophoresis or immunoelectrophoresis is required. Due to the overwhelming amount of work, the tubular cells degenerate, forming inclusions and shedding. The damaged kidney in these patients is sometimes called the "myeloma kidney."

Urine albumin excretion usually is less than 30 mg/day in healthy subjects. Increases of urine albumin excretion to 30–300 mg/day are often termed microalbuminuria, and appear to serve as an early indicator of glomerular injury and risk of progression of renal disease, particularly in diabetes mellitus. Urine albumin excretion also can occur under certain physiological conditions including strenuous exercise, fever, hypothermia, emotional distress, congestive heart failure, and dehydration. This condition is termed "functional proteinuria" and is transitory, resolving with time and supportive care.

The dipstick test for proteinurea takes advantage of the fact that at fixed pH proteins function as hydrogen ion acceptors and cause certain dyes to change colors. The sensitivity of this method ranges from 6 to 30 mg/dL, depending on the manufacturer. This test is most sensitive to albumin; therefore, it is important to do a confirmatory test such as the sulfosalicylic acid (SSA) precipitation method to detect the presence of other proteins. False positive tests can be the result of highly buffered alkaline urine.

In the SSA method, the SSA reagent and urine supernatant are added together, mixed by inversion, and allowed to stand for 10 min at room temperature. After 10 min, the specimen is mixed again by inversion and observed under direct light. The precipitation reaction can be graded in a negative, trace, 1+, 2+, 3+, and 4+ scale according to various protocols, or with concentration values of milligrams per deciliter corresponding to a series of protein standards. The sensitivity is approximately 5–20 mg/dL. False positive SSA results can occur when nonproteins are precipitated by the acid (e.g., radiographic dyes and certain drugs). False negative results can occur with highly alkaline urine which neutralizes the SSA.

4.2. Blood

The basis of detecting the presence of hemoglobin or red blood cells in urine is the peroxidase activity of hemoglobin. In the presence of a peroxide, hemoglobin catalyzes the oxidation of a chromogen to produce a blue-colored product. Myoglobin is capable of performing the same reaction; therefore, it is important to differentiate between the two when in doubt through electrophoresis, or immunochemical tests. False negative results can be seen in the presence of ascorbic acid, as well as in highly concentrated or acidic urine which prevents thorough lysing of the RBCs.

Normal urine may contain as many as five RBCs per microliter; however, this is not detectable using dipstick methodology. Hematuria can be seen as a result of bleeding anywhere from the glomerulus to the urethra, and it can be caused by a variety of factors including bladder and kidney tumors, trauma to the kidney, glomerulonephritis, pyelonephritis, renal calculi, and bleeding disorders related to anticoagulant use. In menstruating women, positive results are the rule rather than the exception. Hemoglobinuria is far less common and can be seen in trauma or transfusion reactions, severe burns, or poisoning. It can also be a result of RBC lysis within the urinary tract. Myoglobinuria is associated with crush injuries and muscle trauma.

4.3. Glucose

Small amounts of glucose may be present in urine, especially after a meal containing a high concentration of carbohydrates. Glucose is freely filtered through the glomerular basement membrane and is mostly reabsorbed in the tubules, so under normal circumstances little is found in the urine. Glycosuria is normally detected when blood glucose levels rise above 180 mg/dL, which exceeds the tubular capacity for reabsorption. Diabetes mellitus is a common clinical cause for glycosuria. In these patients, the absence of insulin leads to an increase in blood glucose concentration, which is reflected in the glomerular ultrafiltrate. This overwhelms the reabsorption capacity of the tubular cells and leads to its excretion. Glycosuria can also be seen in conditions that affect the central nervous system, kidneys, endocrine system, and liver or general metabolic problems such as starvation and obesity. Diuretics and birth control pills can also cause glycosuria. The dipstick method uses a two-step enzymatic reaction in which the enzyme glucose oxidase converts glucose into gluconic acid and Hydrogen peroxide, which then oxidizes chromogen leading to a color change. Results are presented in a 1+ to a 4+ range depending on the amount of glucose in the specimen, which can be between 100 and 2000 mg/dL. There is also a copper reaction available from Bayer Diagnostics in a tablet form that will also detect other reducing sugars such as galactose, lactose, and pentose, but it is both less specific and less sensitive than the reagent strip method. Galactosemia, an inherited metabolic disorder in the newborn that prevents it from turning galactose into glucose, must be screened for using the copper sulfate methodology to prevent the onset of cataract formation, hepatic dysfunction, and mental retardation due to the accumulation of toxic intermediates in these patients.

4.4. pH

The kidneys, lungs, and blood buffers are the primary regulators of acid-base balance in the body. Urine pH may vary from 4.5 to 8; however, most people's urine is slightly acidic due to the predominant acid formation. This is not the case during the "alkaline tide" after a meal when the parietal cells of the stomach secrete hydrochloric acid for digestion and bicarbonate into the interstitial fluid and blood to be excreted by the kidneys. Acidic urine can be seen in patients with diets rich in protein and certain fruits, as well as in metabolic or respiratory acidosis, or taking certain medications. Alkaline urine is observed in patients with vegetarian diets, metabolic or respiratory alkalosis, renal tubular acidosis, or taking certain medications. Urine left at room temperature will become more alkaline with time due to growth of bacteria and the breakdown of urea releasing ammonia.

4.5. Bilirubin

Bilirubin is a byproduct of RBC senescence in the reticuloendothelial system. Hemoglobin is converted to unconjugated bilirubin, which is then released into the blood stream by the reticuloendothelial cells. It then becomes reversibly bound to albumin, which carries it to the liver where hepatocytes remove it by a carrier-mediated active transport. In the hepatocytes, bilirubin is conjugated with glucuronic acid to produce a water soluble compound. The liver excretes this conjugated bilirubin as a constituent of bile and passes to the small intestine. In the intestinal tract, bilirubin is converted back to its unconjugated form and reduced by the flora to urobilinogen. Most of this urobilinogen is then reduced to stercobilinogen, and in the large intestine both compounds are oxidized to urobilins and stercobilins. About 20% of the urobilinogen is reabsorbed into the enterohepatic circulation and taken to the liver, which reexcretes it into the gall bladder. A small amount remains in the bloodstream and is carried to the kidney to be excreted from the body. Small quantities of conjugated bilirubin are normally present in urine (~0.02 mg/dL), but are not detectable with dipsticks. Bilirubinuria is usually indicative of liver dysfunction and intrinsic or extrinsic bilary obstruction (e.g., gallstones, carcinomas at the head of the pancreas) as well as with Dubin–Johnson and Rotor types of congenital hyperbiliruninemia.

Dipstick methodologies are based on a coupling reaction of bilirubin with a diazotized aniline dye to generate a purplish azobilirubin compound. False positive results can occur with certain medications such as pyridium that have a similar color at the low pH of the reagent pad and mask the results. False negative results can occur when urine is not tested promptly or is not stored correctly, since light or air will transform bilirubin to unreactive biliverdin. There is also a tablet method using similar dyes which are more sensitive and can be used to confirm the results.

4.6. Urobilinogen

As noted above, urobilinogen is a product of bilirubin metabolism by bacteria in the intestine. Though most of it is excreted in feces, some of it is reabsorbed and carried by enterohepatic circulation to the liver where the majority is reexcreted in bile. The remaining urobilinogen is then filtered by the kidneys and appears in urine if increased in amount. Normal individuals can excrete up to 4 mg daily, random urine samples containing 0.1–1 mg/dL. Patients with severe liver disease such as cirrhosis or hepatitis are unable to reuptake urobilinogen and can present with elevated serum and urine urobilinogen levels. Patients with hemolytic disorders can have a similar presentation.

The reagent strip method for the detection of urobilinogen uses the Ehrlich reaction. Urobilinogen is mixed with dimethylaminobenzaldehyde in an acid buffer, forming a tan to orange-colored compound. False positive results can occur in the presence of some compounds (i.e., porphobilinogen, sulfonamides, and aminosalicylic acid). False negative results can occur if urine is left exposed to daylight causing urobilinogen to break down. Formaldehyde and high levels of nitrites can also produce a false negative result.

A new and more specific method for testing urobilinogen uses a diazonium salt in an acid buffer. With the Chemstrip dipstick (Roche Diagnostics), a red azo dye color is produced only when urobilinogen is present in excess. Compounds other than urobilinogen do not produce a color change on this pad.

4.7. Ketone Bodies

Ketone bodies are a byproduct of incomplete fat metabolism indicative of a gluconeogenic metabolic state that can occur with starvation, low-carbohydrate diets, febrile illness, or uncontrolled diabetes mellitus. These compounds, acetoacetic acid, acetone, and beta-hydroxy-butyric acid, are released into the blood by the liver and are used by other tissues of the body as an energy source; however, some are also filtered by the kidneys and excreted in urine (ketonuria). The reagent strip test uses the nitroprusside reaction, in which sodium nitroferricyanide and glycine react with acetoacetate and acetone at alkaline pH to yield a violet-colored product. The Multistix system uses a similar reaction, though only testing for acetoacetic acid and producing a pink-maroon colored compound. False positive results can occur in the presence of phthaleins or large amounts of phenyl ketones, 8-hydroxyquinoline, or L-dopa metabolites, along with certain hypertension medications.

4.8. Nitrites

Most common pathogens of the urinary tract, such as *E. coli*, *Proteus*, *Enterobacter*, and *Klebsiella* can have the ability to reduce nitrate to nitrite via the enzyme nitrate reductase. Humans usually excrete nitrate in urine; however, in the presence of these organisms, the nitrate is reduced to nitrite before voiding. The nitrite test takes advantage of this to provide a quick and inexpensive method to test for urinary tract infections (UTIs); however, it should be no substitute for a proper microbiologic workup. The test consists of the diazotization of nitrite with an aromatic amine to produce a diazonium salt, which then undergoes an azo-coupling reaction with an aromatic chromogen, changing the color of the pad from white to pink. False positives can occur in the presence of pigmented materials in urine, as well as contamination of the specimen due to poor storage and collection techniques.

4.9. Leukocyte Esterase

The presence of white blood cells (WBCs) in urine is indicative of an inflammatory process that can be caused by a UTI. Leukocyte esterase is an enzyme present in certain WBCs (neutrophils, basophils, eosinophils, and monocytes) which is used in urinalysis to indirectly test for UTIs. A clean catch or catheter collected specimen is used to prevent unreliable results, since positive test results is the norm when vaginal secretions are present in the sample.

The reagent pad test consists of an ester which is converted into an alcohol by leukocyte esterase. The alcohol subsequently reacts with a diazonium salt to form an azodye giving the color change. False positive results occur in the presence of pigmented substances including medications and foods, while false negative results occur with increased protein, glucose, specific gravity, and certain drugs such as gentamicin and cephalosporin (4–11).

5. Microscopic Examination of Urine

Microscopic examination of the urinary sediment is the third part of routine urinalysis. All nonsoluble components of urine are isolated through centrifugation and are observed under the microscope. Although it has been argued that microscopic examination should be performed on every specimen, most laboratorians agree that it is an essential component in the evaluation of symptomatic patients. Though several microscopic procedures are available for sediment examination, standardized bright-field microscopy is still the most common technique used today. Use of automated particle analyzers to evaluate unspun urine is coming into increased use as a preliminary screen for formed elements.

5.1. Crystals

Due to the kidneys role in metabolic excretion and maintenance of homeostasis, end products of metabolism are found in high concentrations and can precipitate forming crystals, especially if significant time is allowed to pass before evaluating. While finding crystals in urine sediment is not necessarily associated with pathological states, several types of abnormal crystals can be indicative of disease. Leucine and Tyrosine crystals are associated with severe liver disease, while cholesterol crystals are observed in nephrotic syndrome and polycystic renal disease. Uric acid, calcium oxalate, and amorphous phosphate crystals are commonly seen in normal specimens.

5.2. RBCs

Hematuria is commonly observed in many clinical conditions; however, it is most commonly observed when contamination with menstrual blood has occurred. Exercise can also be a causative agent, usually self-resolving after 2 days. RBCs that enter the genitourinary (GU) system through the glomerulus become deformed while passing through the tubules and are termed dysmorphic RBCs. Hematuria originating in the kidneys characteristically is accompanied by dysmorphic RBCs, making them an important indicator of intrarenal hemorrhage.

5.3. WBCs

The presence of an increased number of WBCs in the urine is indicative of an inflammatory state, commonly due to a UTI. Though most of the cells observed are neutrophils, other WBCs may also be present.

5.4. Epithelial Cells

The three types of epithelial cells most often observed in urine sediment are renal tubular, transitional, and squamous. The epithelium of the renal tubules constantly undergoes a turnover process, and as a result it is normal to find a few of these cells in the urinary sediment. Increased numbers of these cells, however, is indicative of renal tubular distress as a result of renal tubular necrosis, heavy metal poisoning, cytomegalovirus infection, renal transplant, or renal vein thrombosis. Transitional epithelium lines a large portion of the urinary tract, and also has a high turnover rate. Increased numbers of these cells is usually seen in certain inflammatory conditions affecting the urinary tract, most importantly bladder carcinomas. Low numbers of squamous epithelial cells should be seen in a properly collected specimen, since contamination with vaginal contents is the most common reason for elevated values.

5.5. Casts

Casts are formed when contents inside the renal tubules solidify and are passed into the urine. This solidification process can be a result of increased concentration or increased filtration of certain proteins, as well as the abnormal presence of different types of cells or compounds. Casts take the shape of the renal tubule in which they are formed; therefore, careful examination can provide valuable information as to the state of the kidney. Two types of casts appear normally in the urinary sediment: hyaline casts and finely granular casts. Erythrocytes, leukocytes, blood, renal tubular epithelial cells, bacteria, and fungi can also form casts under different pathological conditions. Casts can originate also from fibrin, lipids, and bile.

5.6. Microorganisms

In a properly collected specimen, the presence of microorganisms in the urinary sediment is of clinical significance. Specimens containing bacteria, fungi, parasites, or virally infected cells should be followed up by microbiology for confirmation and classification of the pathogen (4, 5, 8, 9).

References

1. Dobson, M. (1776) Nature of the urine in diabetes. Medical Observations and Inquiries 5: 298–310.
2. Bright, R. (1836). Cases and observations illustrative of renal disease accompanied with the secretion of albuminous urine. Guy's Hosp. Rep. 1, 338–379.
3. Bence Jones, H. (1848). On a new substance occurring in the urine of a patient with mollities ossium. Philos. Trans. R. Soc. London 138:55–62.
4. McBride, L.J. (1998). Textbook of Urinalysis and Body Fluids. Philadelphia: Lippincott-Raven.
5. Liang, S.L., Clark, W. (2006). Professional Practice in Clinical Chemistry. Washington, DC: AAAC Press.
6. Simerville, J.A., Maxted, W.C., Pahira, J.J. (2005). Urinalysis: A Comprehensive Review. Am Fam Physician, 71:1153–1162.
7. Kaplan, L.A., Pesce, A.J., Kazmierczak, S. (2003). Clinical Chemistry: Theory, Analysis,

Correlation, 4th edn. St. Louis: Mosby, 1092–1109.
8. McClatchey, K.D., ed. (2002) Clinical Laboratory Medicine, 2nd edn. Philadelphia: Lippincott Williams & Wilkins, pp. 519–551.
9. Fuller, C.E., Threatte, G.A., Henry, J.B. (2001). Basic Examination of Urine. In: Henry JB, ed. Clinical Diagnosis and Management by Laboratory Methods, 20th edn. Philadelphia: WB Saunders, pp. 367–402.
10. Clarke, W., Palmer-Toy, D.E. (2002). Clinical chemistry of urine, pleural effusions, and ascites. In: Lewandrowski, K.D., ed. Clinical Chemistry: Laboratory Management and Clinical Correlations. Philadelphia: Lippincott Williams & Wilkins, pp. 849–864.
11. Hanno, P.M., Wein, A.J., Malkowicz, S.B. (2001). Clinical Manual of Urology, 3rd edn. New York: McGraw-Hill.

Additional References and Further Reading

1. LabTestsOnline, http://www.labtestsonline.org/understanding/analytes/urinalysis/test.html

Chapter 2

A Primer on Clinical Applications and Assays Using Urine: Focus on Analysis of Plasma Cell Dyscrasias Using Automated Electrophoresis and Immunofixation

Brianne Olivieri and Alex J. Rai

Abstract

Urine is a noninvasive sample that is ideal for screening, because it is easy to collect, cost-effective, and can provide a wealth of information on a patient's health status. We provide a brief discussion on the anatomy and physiology of the kidney, a concise overview on B and T cells as key mediators of the immune system, and then delve into the various B-cell neoplasms. This discussion details Waldenstrom's macroglobulinemia, plasmacytoma, heavy chain disease, amyloidosis, and multiple myeloma. Of primary clinical importance from a technical perspective, two commonly applied techniques for the separation and characterization of urine proteins include urine protein electrophoresis and urine protein immunofixation. Procedural details for both techniques are provided herein.

Key words: Urine analysis, Blood cancers, Plasma cell dyscrasia, Urine protein electrophoresis, Immunofixation

1. Introduction

1.1. Urinalysis

Urinalysis is the chemical and physical evaluation of urine and includes a broad spectrum of clinical assays looking for renal, urinary tract, and other disorders (1). It has a long rooted history and has been a method of health evaluation since ancient times (2).

In collecting urine it is important to preserve the composition, as it degrades very quickly upon excretion. The most common means of preservation is refrigeration, although this can cause precipitation of crystalline substances and proteins. In collecting urine, the easiest sample is a spot collection in which the patient collects an excretion at any random time. Spot collection is not as reliable as a timed collection due to the number of variables the

Alex J. Rai (ed.), *The Urinary Proteome: Methods and Protocols*, Methods in Molecular Biology, vol. 641, DOI 10.1007/978-1-60761-711-2_2,

urine composition is contingent upon, including hydration, diet, and/or whether it is the first morning's excretion. The latter is best for a spot collection because of the length of time (approximately 8 h) it has been collecting in the bladder, and thus the concentration of analytes is highest. A spot collection sample is useful in routine analysis, but for more specific tests, a timed collection is preferred. Timed collections are most commonly 24 h of urine, and depending on the tests to be performed (i.e., urine metanephrines or urine catecholamines), preservatives, such as 6N HCl, may be required. In the timed collection the first excretion of the first day is discarded and every excretion following is collected up to and including the first excretion of the next day. This collection method affords an accurate assessment of the patient's renal function and, subsequently, pathology. In a routine urinalysis physicians test for glucose, bilirubin, conjugated bilirubin, ketones, proteins, red blood cells, and the physical properties of the sample (1).

Analysis of urine is usually performed as an initial screen which is subsequently followed up with blood serum and/or other analyses. Urine is an ideal sample for screening since it is easy to collect, cost-effective, and can provide a wealth of information. The collection of the urine sample itself is a noninvasive process making it a better test for an initial or follow-up screen. The results from an urine analysis can include information regarding proteins, casts and monoclonal free light chains, enzymes and cell types present. Clinicians can perform a differential diagnosis or can narrow the investigation of diseases affecting a patient based on the combination of abnormalities detected in the aforementioned components of a urine sample. However, diagnosis of any condition cannot rely solely on urine analysis as the definitive determinant. Confirmatory tests of equal or greater specificity are used once the urine analysis has focused the investigation into a subset of conditions from which a patient may be suffering. For example, when looking at the protein content of a urine sample, there are some tests more commonly used than others. Although a physical evaluation can be done by visual inspection, it is only effective when the protein content of the urine is sufficiently elevated to be seen by the naked eye. Dipstick analysis can also provide qualitative and, sometimes, quantitative information on proteins and pH (3). The different reaction strips on a dipstick provide a more specific analysis than a visual evaluation. To further narrow an investigation, there are additional tests that are protein specific. These analyses include urine immunofixation (UIF) and urine protein electrophoresis (UPE), which can provide specific qualitative and quantitative results regarding free kappa and free lambda light chains. There are many disease states that are detectable in urine including diabetic nephropathy, glomerular disease and other renal disorders, but for this chapter we will review various aspects of urine analysis with particular regard to the diagnosis of patients with malignancies.

1.2. The Kidney

The kidney is comprised of units called nephrons with each nephron comprised of several parts. The renal corpuscle is made up of the glomerulus and Bowman's capsule. The glomerulus is the site of protein filtration from the blood, whereas all other filtrates (small, low molecular weight proteins) enter Bowman's capsule. The filtration process in the glomerulus is based on size and charge with cations more apt to be filtered than anions. The glomerulus is capable of filtering 125 mL/min and catabolizing up to 30 mg of proteins/day. Resorption of necessary blood components into the bloodstream occurs through the semipermeable membranes of the tubules and also secretion of unnecessary blood particles into the tubule. If there is an excess of certain minerals beyond normal levels the tubules resorptive capacity is surpassed and such blood components as bilirubin, glucose, ketones can be deposited in the collection duct. All these materials, to be excreted from the body, are then deposited in the collection duct and leave the body as urine.

In a healthy adult human, a 24-h urine collection should contain <150 mg of protein/day. In the event that there is significantly more protein present in such a urine collection, renal problems should be suspected. In a normal urine sample there are detectable levels of creatinine and urea, but in the background of diseases such as myeloma or proteinuria, there are also detectable levels of proteins such as kappa and lambda free light chains.

In a disease state, renal function is interrupted in several ways, including loss of charge on the glomerulus allowing more proteins to filter through the glomerulus than would normally occur, or renal casts being formed in the tubules. In order for significant amounts of protein to be seen in a urine sample, the renal threshold must be exceeded. This threshold is defined as the point where tubular absorptive capacity is surpassed.

Normal production of free light chains is approximately 500 mg/day and the absorptive capacity in the glomerulus and tubules is 10–30 g/day. When blood serum concentration of monoclonal kappa free light chains is 6–7× the upper limit of the reference range, and in the case of monoclonal lambda free light chains, when the level reaches about 12–18× the upper limit of the reference range, the proteins are significantly detectable in urine. This condition is termed overflow proteinuria, and is often accompanied by genetic mutations causing defective antibody overproduction with kappa and lambda free light chains exhibiting extended half-lives. Overflow proteinuria is a sign of high serum protein levels and can be indicative of myeloma or a related disease. The excess protein filtering thought the kidney adds strain to the glomerulus and organ as a whole. As the kidney begins to fail due to stress from formed casts and excess filtrates, disease becomes more apparent on UIF and UPE gels. Thus, a urine sample presents signs of disease, contingent upon the health of the kidneys, making it a useful indicator of disease state and diagnosis. The types of proteins of particular interest in urine analysis of cancer patients are referred to as Bence Jones proteins – these are monoclonal kappa and lambda free light chains, named after Dr. Henry Bence Jones.

Although Dr. Bence Jones was not the first to discover kappa and lambda light chains, he was the first to describe their importance in diagnosing myeloma (4). Thomas Watson and William Macintyre were the first to discover such proteins as a precipitate in their patient, Thomas Alexander McBean's, urine sample. Not knowing the identity of the precipitate that was formed upon warming, and disappeared when warmed further, they sent a sample of McBean's urine to Dr Henry Bence Jones.

Upon examination of McBean's urine sample, Bence Jones concluded that the precipitate was an "oxide of albumen" further described as "hydrated deutoxide of albumen". He then described a connection between McBean's autopsy results, which included such findings as soft/brittle bones filled with gelatinous substance and an abundance of plasma cells, and the protein in the urine. He is noted as saying "I need hardly remark on the importance of seeking for this oxide of albumen in other cases of mollities ossium". Bence Jones correctly posited that the urine precipitate was related to McBean's autopsy results, which would now indicate that McBean suffered from multiple myeloma. The connection Bence Jones made has proved crucial to testing and analysis used today in evaluating myeloma patients. Bence Jones' oxide of albumen is now known as monoclonal (kappa and lambda) free light chains. The conclusions made by Bence Jones allow physicians today the ability to use detection of kappa and lambda free light chains as a clinical diagnostic tool. The presence of free light chains in urine is ultimately the result of a process of an immune response that has gone awry.

1.3. B and T Lymphocytes

B and T lymphocytes are key mediators of the immune system. B cells are involved in the humoral immune response, where they produce antibodies, which "tag" antigens for destruction by macrophages or natural killer cells.

T cells are involved in the cell-mediated immune response, where an antigen presenting B cell activates the T-cell receptor causing cytokine release. Cytokines are environmental cues used by T cells to direct the immune response – they can cause proliferation of more T cells, attract macrophages, or cause T cells to differentiate into cytotoxic cells.

Defects in B and T cell differentiation and response are the cause of many cancers where cell proliferation occurs at an uncontrolled rate, referred to as B- and T-lymphocyte neoplasms.

B lymphocytes are white blood cells from bone marrow that become plasma cells when stimulated by the appropriate proteins from bacteria or viruses entering the body. Gram-negative bacteria have surface proteins and other molecules that govern the reactions of a cell to the environment. Components such as lipopolysaccharide (LPS) stimulate B cells. LPS is comprised of several moieties – the lipid component, Lipid A, is the toxic member controlling physiological response, and a separate polysaccharide component guides immunogenicity. LPS elicits B lymphocyte differentiation into plasma cells, ultimately causing production of plasma cells commensurate with the levels of the presented antigen.

B and T lymphocytes conspire to mount an immune response to foreign antigens. Once the B cell has engulfed the antigen, digested it, and is presenting fragments of the digested antigen on the MHC, a mature T cell is attracted. B cells produce antibodies against the specific antigen present in the same process in which T cells are activated and give off cytokines to direct the immune response in other capacities. Some cytokines cause maturation of additional T cells, some make the existing T cells cytotoxic, and others attract macrophages to engulf infected cells.

Antibodies are produced by the B lymphocytes to bind and mark each antigen specifically. Once the antibodies have marked the antigen, the foreign component is degraded by phagocytosis and is processed in the spleen or liver. The T-cell response addresses the already infected cells by producing cytotoxic T cells and attracting macrophages. In this coordinated effort, B and T lymphocytes rid the body of the antigen and infected cells to restore a normal state.

In the case of lymphomas, massive overproduction of abnormal plasma cells occurs upon stimulation of the B lymphocytes by a specific antigen. The stimulated B cells produce abnormal plasma cells or myeloma cells, resulting from infection by a virus or through oncogenic mutations they have acquired. An overproduction of myeloma cells leads to overproduction of antibodies, but like the plasma cells from which they arise, they are not healthy antibodies. The abnormal antibodies are overproduced with weak disulfide bonds holding the light and heavy chains together. The inability of the heavy and light chains to stay bound by the disulfide bonds lead to excess monoclonal kappa and lambda free light chains and free heavy chains (usually of IgG subclass, but may also be IgM, IgA, IgE, or IgD). Kappa monoclonal free light chains exist as monomers and lambda monoclonal free light chains exist as dimers. Kappa free light chain is produced at a rate roughly two times that of lambda free light chain. In a UIF it is more likely to see free kappa banding than free lambda.

Accumulation of these cells leads to decreased available space for healthy white blood cells, red blood cells, and platelets in bone marrow. Excess free heavy and light chains lead to tumors of the bones and soft tissue. These tumors are representative of increase in monoclonal (M) protein produced by the B cells. Each B cell responds to a specific antigen and produces a specific M protein to coincide with the antigen. Genetic mutations in the associate signaling pathway can cause plasma cells to continually produce the defective M proteins.

B-lymphocyte neoplasms include a long list of conditions from multiple myeloma to various leukemias and lymphomas. These are described further below. A paucity of healthy cells and growth of tumors weakening the hard bone is indicative of the condition referred to as hypercalcaemia. With this condition, there is increased levels of Ca^{+} in the blood and affects many organs including the kidney, GI, heart, and also the nerves and muscles. Polyuria will occur as the kidney is affected by the increase in Ca^{+} and other components in blood. The glomerulus and tubules of the nephrons suffer in an attempt to filter out excess proteins and Ca^{+}. Eventually the stress on the kidney from the excess proteins causes nephron failure and ultimately renal failure. The blood containing excess protein and Ca^{+} has similar effects on other organs.

1.4. Plasma Cell Disorders

1.4.1. Waldenstrom's Macroglobulinaemia

Waldenstrom's macroglobulinaemia is associated with any detectable IgM monoclonal gammopathy, specific surface proteins, and invasion of bone marrow, spleen, and lymph nodes by plasmacytoma prone lymphocytes. It tends to affect older men more than women and can be accompanied by secondary amyloidosis. Although it is possible to test for Waldenstrom's using urine, it is not the most accurate form of analysis due to polymerization of IgM and glomerular filtration of large molecular weight proteins. Bence Jones proteinuria does occur (often showing elevated free Kappa light chain) in about half the patient population, but is not indicative necessarily of the pathophysiology of the disease.

1.4.2. Plasmacytoma

Waldenstrom's macroglobulinaemia is not the only cancer characterized by plasmacytoid lymphocyte infiltrations into the bone marrow. Plasmacytoma is another form of cancer that can exist in two forms: solitary bone plasmacytoma (SBP) and extramedullary plasmacytoma (EMP). SBP is analagous to a solid tumor in myeloma dyscrasias since it is a single lytic bone lesion invaded by myeloma cells. It can be detected in about two-third of cases in UPE, but in one-third of cases both UIF and UPE should be used. It is otherwise characterized by localized bone pain caused by bone destruction usually in the axial skeleton (i.e., vertebrae) and an absence of symptoms of systemic MM. Tracking of urine protein is not the best way to monitor SBP however. Urine analysis is best used as a diagnostic tool in this case. As for EMP, UIF or UPE are not ideally suited for its detection. It is an invasion of soft tissue by monoclonal plasma cells, most often in the head and neck areas.

1.4.3. Heavy Chain Disease

Heavy chain disease is identified as malignant plasma cells producing only incomplete monoclonal immunoglobulins without light chains being produced. There are three types of heavy chains considered in this disease, IgA, IgG, and IgM.

IgA heavy chain disease occurs primarily in a geographically localized area, the Middle East. It is thought to be caused by a parasite or other microorganism. IgA heavy chain disease occurs in individuals between ages 10 and 30 and is not detectable in urine. It is best identified in intestinal fluids.

IgG heavy chain disease can be asymptomatic and benign but is more often exhibited in malignant lymphoma. This heavy chain disease however can be detected in urine as it displays proteinuria as >1 g protein/24-h collection. Additionally, amyloidosis can develop as a secondary disease.

IgM heavy chain disease occurs in older adults and displays free kappa Bence Jones proteinuria in approximately 1% of patients. Serum analysis is normal or displays hypogammaglobulineamia. Death usually results, caused by uncontrolled proliferation of chronic lymphocytic leukemia (CLL) cells.

1.4.4. Amyloidosis: Primary Amyloidosis (AL), Secondary Amyloidosis (AA)

A detectable B-lymphocyte dyscrasia using UIF and UPE is AL amyloidosis. AL amyloidosis is caused by M protein (amyloid fibrils) deposits in specific organs (localized) or spread throughout the body (systemic). This conditions often leads to organ failure due to tissue damage caused by the amyloid deposits. In diagnosing AL amyloidosis UIF is a more sensitive assay to detect FLC initially and then serum analysis is best to quantify the findings. It is commonly found that MM patients develop AL amyloidosis resulting in renal casts derived from kappa free light chains. These dense protein deposits eventually cause organ failure and death.

Secondary amyloidosis (AA) is similar to AL, except that is it known to accompany other systemic infections and is not characterized as an independent disease. In contrast to AL, in AA the deposited proteins are secondary proteins whereas in AL, the deposited proteins are immunoglobulins. AA usually occurs when there is inflammation or tissue damage in the body. Upon inflammation or tissue damage acute phase reactants are activated to minimize the effects, but when the acute phase reactant serum amyloid A is degraded, it leaves AA proteins to make deposits in the body. This is not detectable in urine, but accompanies diseases such as MM, which are detectable in urine.

1.4.5. Multiple Myeloma: Non-secretory Multiple Myeloma

Multiple myeloma is characterized by an accumulation of myeloma cells in bone marrow, which ultimately leads to tumor growth in bones throughout the body. Multiple myeloma is identified as a monoclonal gammopathy in serum and/or urine. Additional identifiers include osteolytic lesions on the axial skeleton, plasmacytosis, anemia, renal failure, and hypercalcaemia. These types of symptoms are also indicative, potentially, of subclasses of multiple myeloma. For example, nonsecretory multiple myeloma is characterized by the same symptoms as multiple myeloma with the exception of a monoclonal gammopathy. This characteristic means that with such tests as UIF or UPE it would not be possible to detect any free light chains or heavy chains because the myeloma cells are not secreting immunoglobulins. This condition exists with two types of myeloma cells, "producer" type and "nonproducer" type, where producers make immunoglobulins that never pass through the cell membrane for an as of yet undetermined reason and nonproducers do not produce immunoglobulins at all. In such a situation, a clinician would use tests in combination with urinalysis in order to determine the disease, and would rule out standard multiple myeloma. Subsequent work-up (following urinalysis) may include bone marrow aspiration or immunoperoxidase staining to diagnose nonsecretory multiple myeloma.

1.4.6. Multiple Myeloma: Smoldering Multiple Myeloma/MGUS

Another subset of MM is smoldering multiple myeloma, which can best be discussed in combination with monoclonal gammopathies of undetermined significance (MGUS). MGUS is characterized as an intact monoclonal immunoglobulin in a patient showing no other signs of B-cell dyscrasias, <10% clonal plasma cells in bone marrow, and <30 g/L of myeloma cells. This is a very common plasma cell dyscrasia in the population over 50 years of age. It is not a life threatening condition, but is monitored closely due to the possibility of progressing to MM. About 1% of those with MGUS progress to malignant monoclonal gammopathies and most of those cases have IgM or IgA monoclonal gammopathies. In the case of MGUS, Bence Jones proteinuria can be an indication of progression to malignancy if the urine protein exceeds 50 mg/day. From MGUS, disease may progress to SMM and eventually develop to MM. SMM is characterized as 10% or greater clonal plasma cells in bone marrow and >30 g/L of myeloma cells. These patients do not have progressive MM; their disease is stable for an extended period of time during which treatment is not necessary, but abnormal laboratory results are constant. Urine Bence Jones proteins greater than 150 mg/day are a likely indication of SMM. Of the patients with IgA or IgM monoclonal gammopathies in MGUS, those with IgA are more likely to progress to Waldenstrom's macroglobulinaemia or lymphoma.

1.5. Analysis of Urine

1.5.1. Urine Immunofixation Electrophoresis

UIF is a method of resolving proteins of a concentrated urine sample in an effort to identify abnormalities or problems with kidney function. UIF is adapted from a serum IFE in-vitro diagnostic procedure, which is used to help identify such malignancies as MM or liver disease. In serum IF, the goal is to identify any monoclonal, polyclonal, or oligoclonal gammopathies; in UIF the goal is to detect the presence of kappa and lambda light chains in urine (see Figure 1).

Urine samples are only tested for presence of free light chains and serum is tested for the presence of both heavy and light chains. Urine results are indicative of renal failure and excess antibodies present in the blood stream, both of which are signs of oncological malignancies. Appearance of proteins such as monoclonal free light chains in a urine sample is indicative of the protein threshold in the blood being exceeded. The results are read from the gel in a qualitative manner and supplemented by serum Free Light Chain quantitation provided by the Beckman IMMAGE automated system (Beckman Coulter, USA). UIF can be performed regardless of the urine total protein result. Often, UIF is ordered in combination with a UPE. Interpretation of UIF is subjective as it is relative to a set of controls producing banding of fixed intensity.

1.5.2. Urine Protein Electrophoresis

UPE is performed similarly to UIF in that it is also a concentrated urine sample electrophoresed on similar agarose gels, but there is one lane per patient and no antisera are applied, the sample is simply run to resolve the proteins based on charge (see Figure 2). The gel is then scanned and the various protein bands are

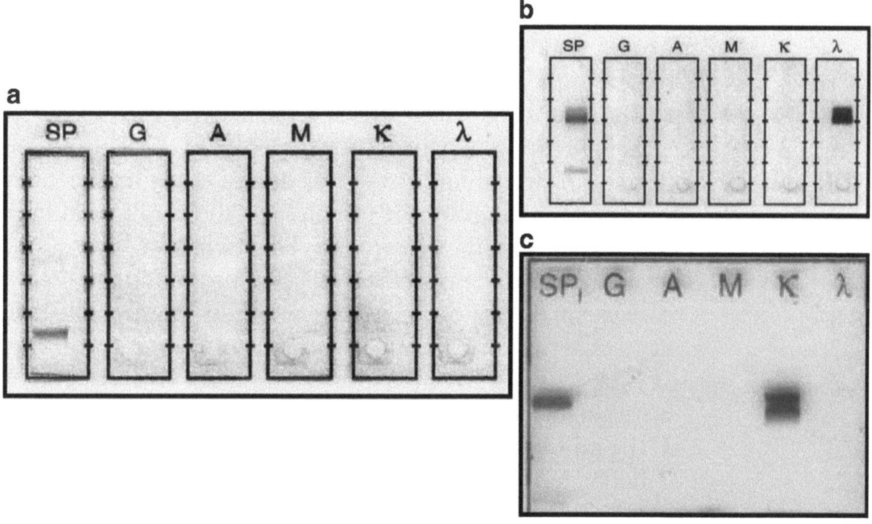

Fig. 1. Urine immunofixation electrophoresis. (**a**) normal pattern from patient urine. (**b**) free lambda light chain. (**c**) free kappa light chain.

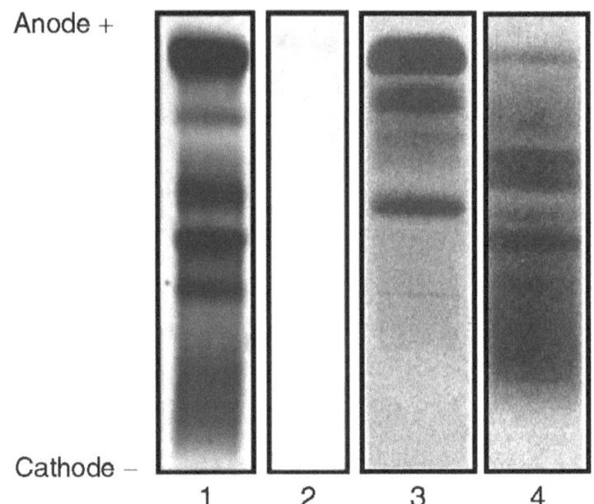

Fig. 2. Urine protein electrophoresis patterns relative to serum proteins. Lane 1, serum proteins; lane 2, urine proteins from normal patient; lane 3, urine proteins from patient with glomerular proteinuria; lane 4, urine proteins from patient with tubular proteinuria. Note that in normal patient, very little protein is detected. Furthermore, the composition of proteins in glomerular (where high molecular weight proteins dominate) is very different from that of tubular proteinuria (low molecular weight proteins dominate).

quantified based on the total protein. If the total protein of a urine sample is less <10 mg/dL, UPE is not performed as the pattern is unable to be differentiated.

Interpretation of PE/UPE results is done after scanning and quantitation is added based on the urine total protein. In the case of a UPE the bands are as follows from left to right: Albumin, Fraction 2, Fraction 3, Fraction 4, and Fraction 5. Abnormal banding patterns in combination with abnormal quantitation can indicate malignancies. Most commonly abnormal banding occurs in the F4 and F5 regions.

From these diseases, the importance of urine analysis in diagnosing and assessing treatment response for several plasma cell dyscrasias is evident. Differential diagnosis entails narrowing the realm of possible afflictions so clinicians can decide on the next step in treating a patient. As a secondary tool urine analysis is helpful in determining stage of disease as in multiple myeloma. Many times the result of urinalysis is the result of kidney status, which is strongly affected by B-lymphocyte dyscrasias. Urine analysis is useful also as a follow up test to monitor FLC levels and disease progression. Although is not highly sensitive, it is cost-effective and a good screening procedure for new patients. It is best used in combination with serum IF and serum PE, and is invaluable because of its ease in sample acquisition and analysis. UIF and UPE have proven to be useful tools in monitoring many cancers, but as the demand for more sensitive and specific tests increases, urine analysis may ultimately become an adjunct procedure.

2. Materials

1. SPIFE 3000 (Helena Laboratories, Beaumont, TX).
2. QuickScan 2000 (Helena Laboratories, Beaumont, TX).
3. SPIFE SPE gels (catalog # 3422, Helena Laboratories, Beaumont, TX).
4. SPIFE Immunofix gels (catalog # 3409, Helena Laboratories, Beaumont, TX).
5. Acid Violet Stain (Helena Laboratories, Beaumont, TX); dissolve powder in 1 L of 10% acetic acid.
6. Acid Blue Stain (Helena Laboratories, Beaumont, TX); dissolve powder in 1 L of 5% acetic acid and stir 30 min prior to use.

3. Methods

3.1. IFE – Electrophoresis Chamber

1. Remove gel (IFE 9 or IFE 6) from the packaging and place on the urine IFE template board. Blot off excess preservative with SPIFE Blotter A. Remove and discard blotter.
2. Select appropriate kit for gel size, 9 or 6 samples. Place three red disposable application templates over the gel and on the pegs on either side of the template board. The corner with the hole should be on the lower left position of the applicator template and the wells should line up with the lanes of the gel.
3. Apply 3 μL of concentrated urine to the kappa and lambda lanes of each sample block. Use the first gel block for Free Kappa and Free Lambda controls. (i.e., each IFE 9 has one control block and eight patient blocks and IFE 6 has one control block followed by five patient blocks).
4. Time sample application for 5 min; blot with blotter A-plus.
5. Remove disposable application templates.
6. Pipette 2 mL of REP-Prep buffer onto the electrophoresis surface of the SPIFE 3000. Place the round hole of the gel onto the round peg at the left of the chamber and lay over the REP-Prep, making sure no bubbles are under the gel as they will result in problems with electrophoresis. Blot any excess REP-Prep from the edges of the gel with a lint free cloth.
7. Place the carbon electrodes on the gel blocks and touching the outside of the magnetic pegs. Close the lid to the electrophoresis chamber.

8. Select SERUM IFE from the test list by hitting Test Select on both the electrophoresis chamber and the staining chamber. On the electrophoresis side hit Start/Stop followed by Test Select/Continue until the screen reads ELECTROPHORESIS; hit Start/Stop to begin gel electrophoresis.
9. The SPIFE 3000 will indicate when electrophoresis is complete and the antisera need to be applied. Remove carbon electrodes and gel blocks.
10. Choose the appropriate antisera template (IFE 9 or 6) place the round hole over the round peg on the left of the chamber. Press down gently to ensure a good seal for each lane over the gel. Using a pipette set to 50 μL, manually dispense free kappa and free lambda antisera into the hole at the right side of the appropriate lanes.
11. Close the chamber lid and press Test Select/Continue to begin antisera application.
12. The next indication beep will signify need to blot excess antisera using the blotter combs for Blot 1 and Blotter D for Blot 2. After placing each blotter in the antisera template and under the antisera template respectively hit Test Select/Continue to time each blot.
13. Following Blot 2, remove the antisera template and replace the electrodes on the outside of the magnetic pegs to predry the gel.

3.2. IFE – Staining Chamber Preparation

1. Select SERUM IFE from the test list. Press Start/Stop to begin chamber preparation approximately 3 min before completion of electrophoresis. Press Start/Stop again followed by Test Select/Continue indicating that the wash container is full and the chamber can be prepared properly.
2. When the chamber is prepared the machine will indicate completion and the screen will read WASH 2. Remove the gel holder until the gel itself is ready to be put in for staining.

3.3. IFE – Staining Chamber

1. With electrophoresis complete and the staining chamber prepared attach the gel to the Gel Holder by the round hole on the peg of the left arm of the gel holder, the middle of the gel under the middle arm and oval hole on the peg of the right arm. Make sure the gel is facing away from you and the middle arm of the gel holder is touching the back of gel.
2. Insert the gel, still facing away from you, into the staining chamber. Press Start/Stop to continue in WASH 2 and through staining, destaining and drying.

3. Upon completion in the staining chamber the gel holder and gel can be removed and the gel scanned using the QuickScan 2000.

3.4. UPE – Electrophoresis Chamber

1. Remove gel (IFE 9 or IFE 6) from the packaging and place on the urine PE template board. Blot off excess preservative with SPIFE Blotter A. Remove and discard blotter.
2. Select appropriate kit for Urine Protein. Place the red disposable application template over the gel and on the pegs on either side of the template board. The corner with the hole should be on the lower left position of the applicator template and the wells should line up with the numbered lanes of the gel.
3. Apply 2 μL of concentrated urine to each well of the template. Use the first two lanes for normal and abnormal controls (Level 1 and Level 2). The controls are run at a 2× dilution. Time sample application for 5 min; blot with blotter A-plus.
4. Remove disposable application template.
5. Pipette approximately 2 mL of REP-Prep onto the electrophoresis surface of the SPIFE 3000. Place the round hole of the gel onto the round peg at the left of the chamber and lay over the REP-Prep, making sure no bubbles are under the gel as they will result in problems with electrophoresis. Blot any excess REP-Prep from the edges of the gel with a lint free cloth.
6. Place the carbon electrodes on the gel blocks and touching the outside of the magnetic pegs. Close the lid to the electrophoresis chamber.
7. Press Test Select/Continue until SERUM PROTEIN appears on the screen. Press Start/Stop to begin electrophoresis.
8. After electrophoresis, remove gel blocks and replace electrodes for predry.

3.5. UPE – Staining Chamber

1. Remove the Gel Holder and attach the round hole of the gel on the left arm peg and the oval hole onto the peg of the right arm making sure the middle arm of the Gel Holder is touching the back of the gel. With the gel facing away, insert the gel and gel holder into the staining chamber. No preparation is necessary for PE/UPE.
2. Press Test Select/Continue until the screen reads SERUM PROTEIN.
3. Press Start/Stop once. At the next prompt press Start/Stop again to begin staining.
4. Upon completion in the staining chamber the gel holder and gel can be removed and the gel scanned using the QuickScan 2000.

4. Notes

1. Unlike serum samples, urine samples may be dilute and should be concentrated using microconcentrators (Amicon) prior to analysis.
2. Urine samples can be concentrated according to the following stipulations:

Protein (mg/dL)	Concentration factor
<50	100×
50–100	50×
100–300	25×
>600	5×

3. If the urine sample is cloudy, it should be first centrifuged at 2,000×*g* for 5 min, to remove any solid precipitates that would interfere with concentration and analysis.
4. It is common to run a single electrophoresis gel with both serum and urine samples, with the different sample types segregated in different rows.

References

1. Simerville, J.A., Maxted, W.C., Pahira, J.J. (2005) Urinalysis: a comprehensive review. Am Fam Physician **71**:1153–1162.
2. Ibn Habib, Abdul Malik d.862CE/283AH "Kitaab Tib Al'Arab" (The Book of Arabian Medicine), Published by Dar Ibn Hazm, Beirut, Lebanon 2007(Arabic).
3. LabTestsOnline, http://www.labtestsonline.org/understanding/analytes/urine_protein/test.html.
4. Bence Jones, H. (1848) On a new substance occurring in the urine of a patient with mollities ossium. Philos Trans R Soc Lond **138**: 55–62.

Additional Reading

1. Sirohi, B., Powles, R. (2004) Multiple myeloma. Lancet **363**:875–887.
2. Rosenfeld, L. (1987) Henry Bence Jones (1813–1873): the best "chemical doctor" in London. Clin Chem **33**:1687–1692.

Chapter 3

Application of Free Flow Electrophoresis to the Analysis of the Urine Proteome

Aude L. Foucher, David R. Craft, and Craig A. Gelfand

Abstract

Urine is a complex fluid, which is thought to contain valuable diagnostic information regarding general health. In particular, there is great diagnostic potential in the peptide and/or protein content of urine, but the information is present in low abundance. Most traditional proteomic techniques lack sufficient sensitivity/dynamic range, especially for dilute and/or complex samples. However, orthogonal separation methods can be applied prior to protein/peptide analysis to increase the success rate of urine proteomic studies and access this potentially valuable information. In this chapter, we describe isoelectric focusing (IEF) of intact urine proteins, via free flow electrophoresis (FFE), prior to typical peptide-based mass spectrometry analysis, facilitating the deep analysis of urine protein detection and identification, for biomarker discovery. Our work demonstrates that such an approach can be used as a preprocessing step and can be integrated into a workflow for the successful identification of protein components (biomarkers) from urine.

Key words: Urine, Proteomics, Free flow electrophoresis

1. Introduction

Urine is the product of blood dialysis within the kidney. As such, urine may contain the same information about the state of the body as blood, but in a different format (mainly large polypeptides or small proteins). In a ddition, urine also contains kidney derived proteins, which reflect the kidney tubular physiology (1). Chemically, urine is defined as an aqueous solution of sodium chloride, urea, and uric acid (MedicineNet.com). The main advantages of urine, as a proteomic sample, are that it can be collected noninvasively and can be collected in relatively large quantities (it is generated at the rate of 1.5–1.8 L/day under normal physiological conditions (2)). Unfortunately, the large volume

Alex J. Rai (ed.), *The Urinary Proteome: Methods and Protocols*, Methods in Molecular Biology, vol. 641,
DOI 10.1007/978-1-60761-711-2_3, © Springer Science+Business Media, LLC 2010

simultaneously carries the hindrance of generally low protein/peptide concentration, estimated at less than 150 mg/day, with high salt content that further complicates most protein analytical processes.

Typical proteomics approaches (i.e., 2D gel) on unfractionated urine provide only limited information about the protein content of the sample, generally limited to data about the most abundant proteins. To increase the depth of the urine proteome coverage, a variety of prefractionation techniques can be used (3). Recently, the use of concanavalin A affinity purification was employed in association with 1D gel electrophoresis and LC–MS/MS for the analysis of *N*-glycoproteins in urine (4). Concentration of the urine proteome using beads coated with a hexameric peptide ligand library was successfully used in association with LTQ-FT mass spectrometry for the discovery of 383 unique proteins (5).

In the present study, we concentrated the urine proteome with ultrafiltration and used IEF–FFE to prefractionate the proteome, in solution, into approximately 50 fractions across a pH gradient between 3 and 8. Each fraction can then be subjected to a typical "bottom-up" proteomics analysis (trypsin digestion, followed by reversed phase liquid chromatography and peptide mass spectrometry (LC/MS)) (6, 7). The combination of IEF–FFE and LC/MS resulted in an extensive list of identified proteins, based on analysis of just three of the FFE fractions, providing depth in protein identification consistent with the Urine Proteome Database, but with advantages in handling and processing provided by use of FFE.

2. Materials

2.1. Sample Collection and Preparation

1. BD Vacutainer® urine collection cup (Becton Dickinson, Franklin Lakes, NJ).
2. Protease inhibitor cocktail, general use (Sigma, St Louis, MO) is dissolved in water and stored in aliquots at –20°C.
3. Centriplus YM-10 centrifugal filter unit (Millipore, Billerica, MA).
4. Vivaspin 4 mL concentrator, 5-kDa filter centrifugation device (Sartorius, Edgewood, NY).
5. Urea (Serva Electrophoresis GmbH, Heidelberg, Germany).
6. Thiourea (Sigma, St. Louis, MO).
7. Mannitol (98%; Sigma, St. Louis, MO).
8. 2D Quant Kit (GE Healthcare, Piscataway, NJ).

2.2. Urine Proteome Fractionation by IEF–FFE

2.2.1. Sample Fractionation by IEF–FFE

1. BD™ Free Flow Electrophoresis System with 0.4 mm spacer (Becton Dickinson, Franklin Lakes, NJ).
2. BD™ FFE–IEF Kit (cat# 441118; Becton Dickinson, Franklin Lakes, NJ), a pH 3–10 gradient kit, which includes hydroxypropyl methyl cellulose (HPMC) powder, pI mix, and buffers 1, 2, and 3 for the pH 3–10 gradient.
3. BD™ FFE–IEF Kit (Becton Dickinson, Franklin Lakes, NJ), a pH 3–8 gradient kit.
4. Urea (Serva Electrophoresis GmbH, Heidelberg, Germany).
5. Mannitol (98%; Sigma, St. Louis, MO).
6. Glycerol (>99.5%; Sigma, St. Louis, MO).
7. Isopropanol (100%; Fisher Scientific, Fair Lawn, NJ).
8. Sodium hydroxide (ACS grade; Sigma, St. Louis, MO).
9. Sulfuric acid (ACS grade; Sigma, St. Louis, MO).
10. Microtiter plate, flat bottom (Becton Dickinson, Franklin Lakes, NJ).
11. *N*-acetylglycine (99.0%; Sigma, St. Louis, MO).
12. Taurine (99.5%; Sigma, St. Louis, MO).
13. Betaine Anhydrous (98.0%; Fluka).
14. TAPS (99.5%; Sigma, St. Louis, MO.
15. AMPSO (99.0%; Sigma, St. Louis, MO).
16. HEPES (99.0%; Sigma, St. Louis, MO.
17. Triethanolamine (99.0%; Sigma, St. Louis, MO).
18. Thiourea (Sigma, St. Louis, MO).

2.2.2. SDS Gel Electrophoresis Analysis

1. NUPAGE® Novex 4–12% Bis–Tris gel, 1.0 mm (Invitrogen, Carlsbad, CA).
2. NUPAGE® MES SDS running buffer (Invitrogen, Carlsbad, CA).
3. NUPAGE® LDS sample buffer (Invitrogen, Carlsbad, CA).
4. Beta mercaptoethanol (Sigma, St. Louis, MO).
5. SeeBlue® Plus 2 Prestained standard (Invitrogen, Carlsbad, CA).
6. SilverQuest™ Silver Staining kit (Invitrogen, Carlsbad, CA).
7. Methanol Chromasolv® for HPLC (>99.9%; Sigma, St. Louis, MO).
8. Acetic acid, glacial (Sigma, St. Louis, MO).
9. Ethanol, 190 proof, 95.0% ACS spectrometric grade (Sigma, St. Louis, MO).

2.2.3. C5 Reversed Phase Purification of IEF–FFE Fraction

1. Discovery® Bio wide pore C5-5 HPLC column (5 cm, 4 mm, 5 μm; Sigma, St. Louis, MO).
2. HPLC 1200 series (Agilent Technologies, Santa Clara, CA).
3. Acetonitrile, HPLC grade with 0.1% (v/v) TFA (Sigma, St. Louis, MO).
4. Milli-Q Water.

2.3. Mass Spectrometry and Data Analysis

2.3.1. In solution Digestion of C5 Purified IEF–FFE Fractions

1. Ammonium Bicarbonate (Sigma, St. Louis, MO).
2. DTT (Sigma, St. Louis, MO).
3. Iodoacetamide (Bio-Rad, Hercules, CA).
4. Trypsin, Proteomics grade (Sigma, St. Louis, MO).

2.3.2. C18 Sep-Pak Column Purification of Peptides from IEF–FFE Fractions

1. Sep-Pak C18 column (Waters, Milford, MA).
2. Acetonitrile, HPLC grade with 0.1% (v/v) TFA (Sigma, St. Louis, MO).
3. TFA (J.T. Baker, Phillipsburg, NJ).
4. Milli-Q Water.

2.3.3. Mass Spectrometry and Data Analysis

1. α-Cyano-4-Hydroxycinnamic acid (CHCA; Bruker Daltonics, Billerica, MA).
2. Ammonium phosphate, monobasic, ASC grade (Thermo Fischer Scientific, Waltham, MA).
3. Matrix-assisted laser desorption/ionization time-of-flight mass spectrometer (ultraflex II; Bruker Daltonics, Billerica, MA).
4. Electrospray Ionization Q-TOF Micro mass spectrometer (Waters, Milford, MA) coupled with a reversed phase chromatography.
5. MASCOT software (Matrix Science, London, UK).

3. Methods

3.1. Sample Collection and Preparation

1. Collect urine in a BD Vacutainer® urine collection cups, from four healthy individuals. Immediately after collection, add 40 μL of protease inhibitor cocktail per 50 mL of urine collected.
2. Centrifuge the sample 10 min at 3,000 × *g* to remove microorganisms and cell debris.
3. Recover the supernatant and filter through a 10-kDa membrane using the Centriplus YM-10 centrifugal filter unit by centrifuging at 3,000 × *g* until there is only about 4 mL left.

4. Transfer the protein concentrate into a Vivaspin 4 mL column (5 kDa) and centrifuge at 3,000 × *g* until the sample volume is about 400 μL.
5. Prepare a buffer exchange solution.

Urea	3.30 g
Thiourea	1.20 g
Mannitol	0.35 g
Milli-Q water	5.18 g

6. Add 3.5 mL of buffer exchange solution in the Vivaspin 4 mL column and centrifuge at 3,000 × *g* for 30 min or until the sample volume is about 400 μL.
7. Repeat step 6 to have a final buffer exchange over 100-fold.
8. Recover the protein from the Vivaspin tube according to the manufacturer's directions.
9. Measure the protein concentration using the 2D Quant kit using the manufacturer's directions.
10. Store the sample at −80°C until ready to use for either 2D gel or FFE separation.

3.2. Urine Proteome Fractionation by IEF–FFE

3.2.1. Sample Fractionation by IEF–FFE

IEF–FFE is a powerful technique, which can accommodate the fractionation of complex protein samples from either small or large sample volumes (8, 9). The entire separation process is performed in solution, without interaction with gels or solid phases, which permits a more thorough separation of complex samples containing, for example, membrane or hydrophobic proteins. The flexibility of the technique is shown in the capacity for modulation of the separation conditions (either native or denaturing condition for example) and/or the pH gradient, which allows for the best fractionation of the sample.

Based on urine protein separation on standard 2D gel electrophoresis (http://www-lecb.ncifcrf.gov/2dwgDB/2DWG.html), we decided to perform isoelectric focusing, using FFE, across a linear pH 3–8 gradient under denaturing conditions.

3.2.1.1. Buffer Preparation

To ensure accuracy, all reagents were weighed out as follows and the buffers were used the same day.

1. Prepare 1 M sodium hydroxide.

Sodium hydroxide	64.0 g
Milli-Q water	to 400 g

2. Prepare 5 M sulfuric acid.

Sulfuric acid	49.0 g
Milli-Q water	73.3 g

3. Counterflow buffer.

Urea	378.0 g
Thiourea	138.0 g
Mannitol	13.6 g
Milli-Q water	600.0 g

4. Prepare anodic stabilization media.

5 M Sulfuric acid	4.0 g
Counterflow buffer	196.0 g
N-Acetylglycine	0.703 g
Taurine	5.00 g
Betaine anhydrous	1.17 g

5. Prepare separation buffer.

Urea	126.0 g
Thiourea	46.0 g
Mannitol	13.6 g
IEF prolytes buffer 3–8	52.5 g
Milli-Q water	147.5 g

6. Prepare cathodic stabilization buffer.

4 M Sodium hydroxide	11.30 g
Counterflow buffer	288.70 g
TAPS	5.47 g
AMPSO	10.23 g
HEPES	2.15 g
Triethanolamine	1.37 g

7. Prepare electrolyte anode buffer.

1 M Sulfuric acid	40 g
Milli-Q water	to 400 g

8. Prepare electrolyte cathode buffer.

1 M Sodium hydroxide	40 g
Milli-Q water	to 400 g

9. Prepare HPMC 0.8% solution.

HPMC powder	8.0 g
Milli-Q water	1,000 g
Add the water to a large glass beaker containing a large stirrer bar	
Stir the water with a strong vortex and gradually add the HPMC powder over a 10-min period	
Mix overnight until the HPMC is completely dissolved. Store at 4°C	

10. Prepare HPMC 0.1% solution.

0.8% HPMC solution	105.0 g
Milli-Q water	525.0 g

3.2.1.2. Establish the pH 3–8 IEF–FFE Gradient

1. Assemble the FFE system (with 0.4 mm spacer) and fill the separation chamber with H_2O according to the manufacturer's instructions.
2. Coat the chamber with the 0.1% HPMC solution by placing all the media tubes and counterflow tubes into the 0.1% HPMC solution and set the media flow rate to 170 mL/h for 5 min (Note 1).
3. Stop the media pump, transfer the media tubes and counterflow tubes into a beaker containing Milli-Q water. Turn on the pump at 170 mL/h for 5 min.
4. Stop the media pump and place the FFE media tubes and counterflow tubes in the appropriate buffer:

Media tubes 1 and 2	Anodic stabilization buffer
Media tubes 3 and 4	Separation buffer
Media tubes 5, 6, and 7	Cathodic stabilization buffer
Counterflow tubes 1, 2, and 3	Counterflow buffer

5. Run the media pump at 70 mL/h.
6. Rinse the sample tube by uncapping the tube and allowing the media flowing through the separation chamber to passively flow through the tube.
7. Place the anode tubes in the anode electrolyte and the cathode tubes in the cathode electrolyte. Place the electrode contact lid on the electrode pump chamber and switch on the electrode pump.
8. Place the electrode contact lid onto the separation chamber.
9. Set the separation parameters as follows:

Voltage (V_{max})	400 V
Current (I_{max})	50 mA
Power (P_{max})	60 W

10. Turn on the voltage and allow the current to equilibrate for about 20 min, until stable at approximately 20 mA (Note 2).

3.2.1.3. Fractionation of the Urine Proteome by IEF–FFE

1. Dilute the sample one-half in FFE separation buffer (1.5 mg of protein/mL) and add 1% (v/v) of SPADNS.
2. Inject the sample into the FFE separation chamber through the middle inlet at a flow rate of 1 mL/h. Continue sample injection until the 6 mL of sample have been loaded. Stop the sample pump.
3. Collect one microtiter plate of FFE fractions (200 μL per fraction) for running the NUPAGE gel and analysing the pH gradient.
4. Collect the rest of the sample in 2 mL deep 96-well plates (one plate per hour) containing 150 μL of 10% TFA/well (to prevent carbamidomethylation of proteins).
5. To determine the efficiency of the fractionation, run 10 μL of FFE fractions onto a NUPAGE gel (see Subheading 3.3). The urine proteome fractionated using IEF–FFE is shown in Fig. 2.

3.2.2. SDS Gel Electrophoresis Analysis

1. Determine the efficiency of the IEF–FFE fractionation by SDS gel electrophoresis.
2. Mix 10 μL of FFE fractions with 3 μL of the NUPAGE LDS sample buffer, and 3 μL of total urine with 3 μL of the NUPAGE LDS sample buffer, and incubate for 10 min at 70°C.
3. Load the samples onto the gel and run for 35 min at 200 V.
4. Following separation, remove the gel from the plastic cassette according to the manufacturer's instructions.
5. Prepare the fixing solution.

Methanol	40 mL
Acetic acid	10 mL
Milli-Q water	50 mL

6. Transfer the gel into the fixing solution and proceed to staining using the SilverQuest silver stain kit following the manufacturer's instructions.
7. The complexity of a urine sample, as seen on NUPAGE gel, is shown in Figs. 1 and 2.

3.2.3. C5 Reversed Phase Purification of IEF–FFE Fractions

This step is mainly a desalting and concentrating step for the protein from the FFE fractions. Removing the urea from the FFE fractions can significantly reduce the potential for protein carbamylation that might otherwise be induced during prolonged exposure to urea that breaks down into isocyanic acid.

1. Transfer the FFE fractions of interest into a HPLC vial (Note 3).

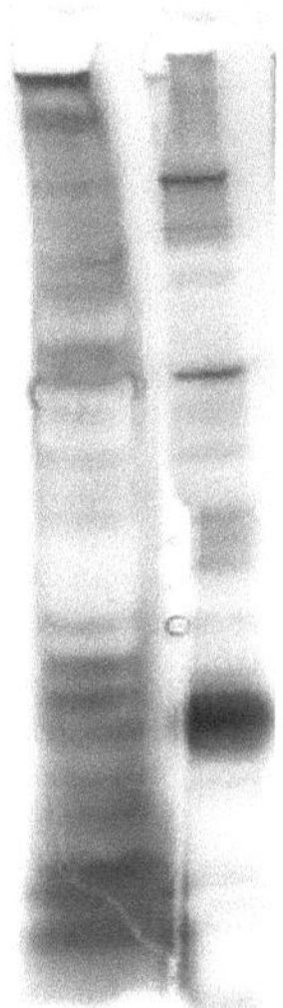

Fig. 1. Analysis of the total urine sample by SDS-PAGE. *Lane 1*: total urine sample, *Lane 2*: molecular weight marker. Silver staining reveals some abundant proteins (negatively stained) in the mass region of 21–66 kDa, which cannot be resolved into sharp separated bands on the gel.

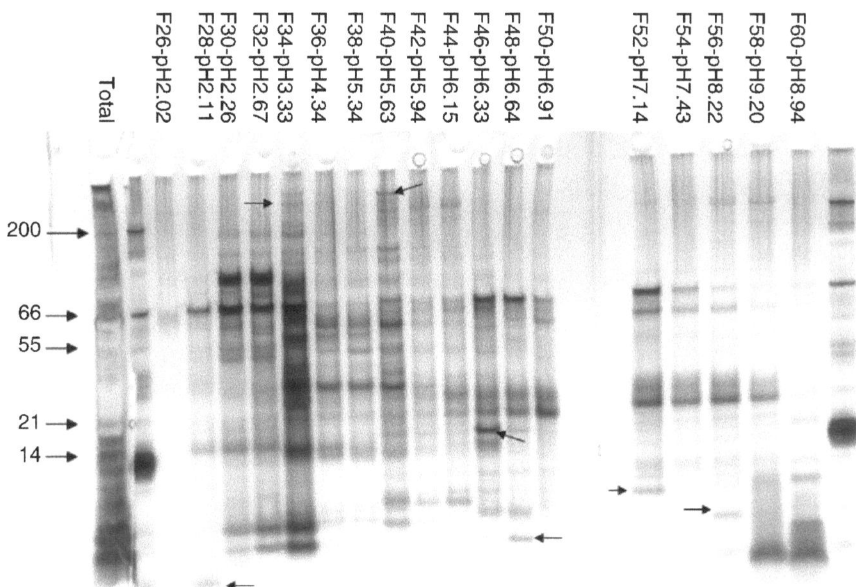

Fig. 2. Evaluation of the fractionation of the urine proteome by IEF–FFE. The FFE instrument elutes fractions into a standard 96-well microtiter plate format. Fractions 26–60 were separated by standard 1D SDS-PAGE. Since FFE provides the IEF dimension, this image is analogous to the results of a typical 2D gel. The corresponding measured pH of each fraction is shown. The extent of fractionation and precision of the isoelectric focusing (indicated by *arrows*) can be seen in the unique staining patters of each fraction on the 1D gels.

2. Prepare the following solution:

(a) Solvent A	
Acetonitrile	20 mL
TFA	1 mL
Milli-Q Water	979 mL
(b) Solvent B (Note 4)	
Acetonitrile	900 mL
Milli-Q Water	100 mL

3. Run the following program on the HPLC to clean up and concentrate the proteins:

Injection	3 × 900 μL
Gradient	0–5 min: 100% solvent A, flow rate: 3 mL/min
	5.01–5.02 min: 50% solvent B, flow rate: 1 mL/min
	5.02–5.50 min: 90% solvent B, flow rate: 1 mL/min
	5.50–6.00 min: 100% solvent B, flow rate: 1 mL/min
	6.00–9.00 min: 100% solvent B, flow rate: 1 mL/min
	9.00–20.00 min: 100% solvent A, flow rate: 3 mL/min
Collection	From 7.2 to 8.6 min, corresponding to 1.4 mL

4. Vacuum centrifuge the collected fractions, to remove the acetonitrile, to a final volume of 500 μL and store at 4°C until ready for the *in solution* trypsin digestion.

3.3. Mass Spectrometry and Data Analysis

3.3.1. In solution Digestion of C5 Purified IEF–FFE Fractions

1. Prepare the following solution:

(a) 500 mM Ammonium bicarbonate	
Ammonium bicarbonate	0.39 g
Milli-Q water	10 mL
(b) 100 mM DTT	
DTT	15 mg
Milli-Q water	1 mL
(c) 1 M Iodoacetamide	
Iodoacetamide	184 mg
Milli-Q water	1 mL

2. Add 100 μL of 500 mM ammonium bicarbonate to the 500 μL FFE fractions that were purified using the C5 column.
3. Add 60 μL of 100 mM DTT. Incubate 30 min at 56°C.
4. Add 66 μL of 1 M iodoacetamide. Incubate 30 min at room temperature in the dark.

5. Add 60 μL of 100 mM DTT to quench the iodoacetamide.
6. Add 10 μg of trypsin. Incubate overnight at 37°C.
7. Desalt and concentrate the peptides using a Sep-Pak C18 column.

3.3.2. C18 Sep-Pak Column Purification of Peptides from IEF–FFE Fractions

1. Add 5 μL of 10%TFA per sample to acidify the sample.
2. Prepare the following solution:

(a) Solution 1	
Acetonitrile	6 mL
TFA	0.5 mL
Milli-Q water	3.5 mL
(b) Solution 2	
Acetonitrile	0.5 mL
TFA	10 μL
Milli-Q water	9.49 mL
(c) Solution 3	
Acetonitrile	6 mL
TFA	0.6 μL
Milli-Q water	4 mL

3. Prewet the C18 Sep-Pak column with 1 mL of solution 1.
4. Equilibrate the C18 Sep-Pak column with 1 mL of solution 2.
5. Bind the peptides to the C18 Sep-Pak column (Note 5).
6. Wash with 1 mL of solution 2.
7. Elution 1: 2 mL of solution 3.
8. Dry to 200 μL using the speed vacuum centrifuge.
9. Prepare the matrix solution:

CHCA	5 mg
20 mM ammonium phosphate	0.5 mL
Acetonitrile	0.5 mL
TFA	1 μL

10. Mix the sample 1:1 with the matrix solution and spot on a ground steel MALDI target.
11. Air dry and analyze using the MALDI-TOF mass spectrometer.

3.3.3. Mass Spectrometry and Data Analysis of the IEF–FFE Fractions

1. Perform direct MALDI on the eluted peptides from the Sep-Pak column to determine the complexity of the FFE fraction analyzed (Fig. 3).

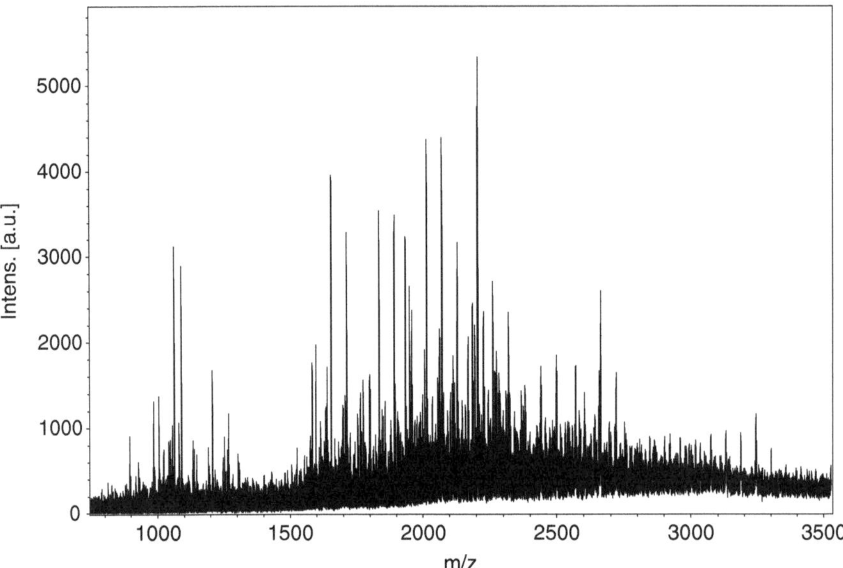

Fig. 3. Direct MALDI spectrum of FFE fraction 34 after trypsin digestion and Sep-Pak C18 clean up. The spectrum, showing mass to charge ratio versus peptide intensity, demonstrates the complexity in protein content of a single FFE fraction.

2. Run the peptides onto C18 reversed phase chromatography system coupled to the Q-TOF. Elute at a flow rate of 200 nL/min into the Q-TOF mass spectrometer using the following conditions:
   (a) 0–6 min: 5% acetonitrile.
   (b) 6–66 min: linear gradient from 5 to 60% acetonitrile.
   (c) 66–76 min: 95% acetonitrile.
3. Select the following automatic parameters for the Q-TOF analysis:
   (a) MS scan in 200–1,990 *m/z*.
   (b) MS/MS analysis on the three most abundant peaks, using dynamic exclusion.
   (c) MS/MS scan in 50–1,990 *m/z*.
4. Search the IPI database using the MASCOT search engine with the following parameters:
   (a) Database: IPI human database.
   (b) Global modifications: Carbamidomethyl [C].
   (c) Variable modifications: Oxidation [M], Carbamyl [K], and [N-term].
   (d) Missed cleavage: 1.
   (e) Mass tolerance: 100 ppm for MS, 0.2 Da for MS/MS.

5. Identification of a protein was only accepted if the protein responded to the two following criteria:
   (a) At least one peptide has a MASCOT score >39.
   (b) And at least one peptide was ranked first by MASCOT.
6. The Q-TOF analysis resulted in the identification of 22 proteins in FFE fraction 30, 32 proteins in fraction 34, and 25 proteins in fraction 60. Proteins identified with Q-TOF analysis are listed in Table 1.
7. Cross-reference of the proteins identified in this study with the Urine Proteome Database (http://proteome.biochem.mpg.de/urine/) (10) reveals that the majority of the proteins identified in these three FFE fractions have been previously identified in urine.
8. No carbamylated peptides were identified in the three FFE fractions used meaning that:
   (a) The urine proteins identified in this study were not modified by urea in the body.
   (b) Urine proteins identified in this study were not modified by urea during the denaturing IEF–FFE fractionation process.
9. Abundant proteins, such as serum albumin and AMBP, are identified across a broad pH range, represented by the three fractions analyzed (Fig. 4), suggesting that these proteins are either degraded or modified, or both in urine. Protein degradation is an expected and inherent condition of the urine *in vivo*, as filtration at the glomeruli is based on the size and charge of the serum proteins. Posttranslational modifications are a well-known characteristic of blood proteins and thus commonly observed in urine. By comparison, apart from these abundant proteins, there is very little overlap in proteins identified in the various FFE fractions, demonstrating the efficiency of the IEF–FFE to fractionate the urine proteome.
10. The urine proteome is enriched with lysosomal, extracellular, and membrane proteins (5). Some of these proteins are difficult to analyze by typical proteomic methods mainly due to their size and hydrophobicity. However, they are easily accessible by IEF–FFE (Table 1) as all FFE separation are performed in solution.
11. As exemplified by fraction 60, even the FFE fraction that appears to have a low complexity, as seen on 1D SDS-PAGE (Fig. 2), can have a surprising number of proteins identified by typical "bottom-up" approaches.
12. One-dimensional SDS gel analysis of fractionated urine (Fig. 2) shows that 34 FFE fractions contain some amount of

Table 1
List of proteins from FFE fraction 30, 34, and 60 identified by Q-TOF MS

FFE Fraction	IPI number	Protein name	MASCOT score	Peptide ID	MW	% Coverage	Ref literature
F30	00022426	AMBP protein precursor	1,183	35	39,886	23	Yes
	00022418	Fibrinonectin precursor	662	15	266,034	6	Yes
	00745872	Serum albumin precursor	383	18	71,317	25	Yes
	00021841	Apolipoprotein A-I precursor	345	13	30,759	26	Yes
	00395488	Vasorin precursor	183	5	72,751	7	?
	00553177	Alpha 1 antitrypsin precursor	149	4	46,878	10	Yes
	00218875	Isoform C of osteopontin precursor	132	2	32,507	5	Yes
	00297160	CD44 antigen precursor	122	6	39,904	8	Yes
	00478493	Haptoglobin precursor	106	6	38,941	17	Yes
	00298388	Kringle domain containing protein HGFL precursor	97	1	28,686	5	Yes
	00154742	IGLα protein	89	3	25,119	14	Yes
	00383032	Hepatitis A virus cellular receptor 2 precursor	81	2	16,594	9	No
	00419424	IGKV1-5 protein	75	1	26,503	6	Yes
	00061977	IGHA1 protein	68	5	55,203	14	Yes
	00382606	Factor VII active site mutant immunoconjugate	65	1	77,386	2	Yes
	00017601	Ceruloplasmin precursor	52	2	122,983	2	Yes
	00164623	Complement C3 precursor	49	2	188,725	1	Yes
	00022429	Alpha 1 acid glycoprotein 1 precursor	47	2	23,725	9	Yes
	00011140	Protein NOV homolog precursor	46	1	41,473	3	Yes
	00023858	Fc gamma receptor IIIb	41	1	30,321	5	Yes
	00022488	Hemopexin precursor	40	1	52,385	4	Yes

(continued)

Table 1 (continued)

FFE Fraction	IPI number	Protein name	MASCOT score	Peptide ID	MW	% Coverage	Ref literature
	00103636	WAP four disulfide core domain protein 2 precursor	40	1	8,571	24	Yes
F34	00745872	Serum albumin precursor	656	36	71,317	33	Yes
	000395488	Vasorin precursor	462	12	72,751	14	?
	00216914	Vitelline membrane outer layer protein 1 homolog precursor	347	4	22,034	17	Yes
	00797833	Kininogen 1	340	15	48,954	29	Yes
	00022463	Serotransferrin precursor	312	11	79,280	12	Yes
	00019568	Prothrombin precursor	239	5	71,475	7	Yes
	00006662	Apolipoprotein D precursor	227	11	21,547	26	Yes
	00022426	AMBP protein precursor	216	14	39,886	25	Yes
	00018236	Ganglioside GM2 activator precursor	181	4	21,265	10	Yes
	00328113	Fibrillin 1 precursor	150	3	332,682	0	Yes
	00553177	Alpha 1 antitrypsin precursor	146	4	46,878	5	Yes
	00022418	Fibronectin precursor	109	3	266,034	2	Yes
	00006971	CD248 isoform 1 of endosialin precursor	94	2	82,803	2	Yes
	00011302	CD59 glycoprotein precursor	94	15	14,795	28	Yes
	00021841	Apolipoprotein A-I precursor	92	2	30,759	9	Yes
	00218413	Biotinidase precursor	91	1	62,006	2	Yes
	00419424	IGKV1-5 protein	86	1	26,503	6	Yes
	00022431	Alpha 2 HS glycoprotein precursor	84	3	40,098	14	Yes
	00103871	Roundabout homolog 4 precursor	83	2	108,417	2	Yes
	00304808	Kallikrein 1 precursor	82	2	29,498	4	Yes
	00291262	Clusterin precursor	72	2	53,031	7	Yes
	00021000	Osteopontin precursor	72	1	35,572	5	Yes

(continued)

Table 1 (continued)

FFE Fraction	IPI number	Protein name	MASCOT score	Peptide ID	MW	% Coverage	Ref literature
	00022417	Leucine-rich alpha 2-glycoprotein precursor	63	3	38,382	14	Yes
	00009030	Lysosomal 2A associated membrane glycoprotein 2	63	1	45,503	1	Yes
	00107731	Osteoclast-associated receptor	61	2	27,897	5	Yes
	00477597	Haptoglobin-related protein precursor	61	2	39,496	2	Yes
	00009276	Endothelial protein C receptor precursor	56	2	31,038	6	Yes
	00296992	AXL receptor tyrosine kinase	54	2	99,528	1	Yes
	00001952	Endonuclease domain containing 1 protein precursor	52	1	55,723	2	Yes
	00218192	Inter alpha trypsin inhibitor heavy chain H4 precursor	49	1	101,488	2	Yes
	00021885	Isoform 1 Fibrinogen alpha chain precursor	48	1	95,656	1	Yes
	00022488	Hemopexin precursor	46	2	52,385	4	Yes
F60	00021841	Apolipoprotein A-I precursor	446	16	30,759	34	Yes
	00745872	Serum albumin precursor	323	22	71,317	31	Yes
	00022418	Fibrinonectin precursor	317	10	266,034	5	Yes
	00553177	Alpha 1 antitrypsin precursor	240	7	46,878	10	Yes
	00641737	Haptoglobin precursor	196	12	47,378	22	Yes
	00430808	IGKC protein	169	6	25,915	24	Yes
	00022431	Alpha 2 HS glycoprotein precursor	130	1	40,098	5	Yes
	00292150	Latent transforming growth factor beta binding protein 2	89	1	204,059	0	Yes

(continued)

Table 1 (continued)

FFE Fraction	IPI number	Protein name	MASCOT score	Peptide ID	MW	% Coverage	Ref literature
	00061977	IGHA1 protein	88	1	55,203	3	?
	00006662	Apolipoprotein D precursor	79	1	21,547	9	Yes
	00305461	Inter alpha trypsin inhibitor heavy chain H4 precursor	78	2	106,826	2	Yes
	00029819	Neurogenic locus notch homolog protein 3 precursor	78	3	256,668	0	Yes
	00022426	AMBP protein precursor	78	2	39,886	3	Yes
	00216304	Isoform alpha Stromal cell derived factor 1 precursor	66	2	10,382	15	Yes
	00032258	Complement C4-A precursor	62	2	194,247	2	Yes
	00032293	Cystatin C precursor	55	6	16,017	32	Yes
	00395488	Vasorin precursor	54	2	72,751	4	Yes
	00005721	Neutrophil defensin 1 precursor	50	2	10,536	9	Yes
	00021085	Peptidoglycan recognition protein precursor	50	1	22,116	8	Yes
	00164623	Complement C3 precursor	50	1	188,728	0	Yes
	00549330	Myosin reactive immunoglobulin light chain variant region	49	2	12,699	18	No
	00220327	Keratin, type II cytoskeletal 1	47	3	66,149	6	Yes
	00022290	Beta Defensin 1 precursor	45	2	7,814	33	Yes
	00291866	Plasma protease C1 inhibitor precursor	44	2	55,347	4	Yes
	00737338	Uromodulin precursor	43	3	57,663	1	Yes

The content by column: IPI accession number, protein name, MASCOT score, number of peptides identified, theoretical molecular weight (MW), the percentage of sequence coverage, and the presence/absence of the IPI number in the Urine Proteome Database (http://proteome.biochem.mpg.de/urine/). "Yes" indicates the presence of the IPI accession number in the database, "No" indicates that the IPI accession number is not present in the database, and "?" indicates that the IPI accession number is present but the name of the protein is different. The majority of the proteins identified in these three FFE fractions have been previously identified in urine using LTQ-FT and LTQ-Orbitrap mass spectrometers

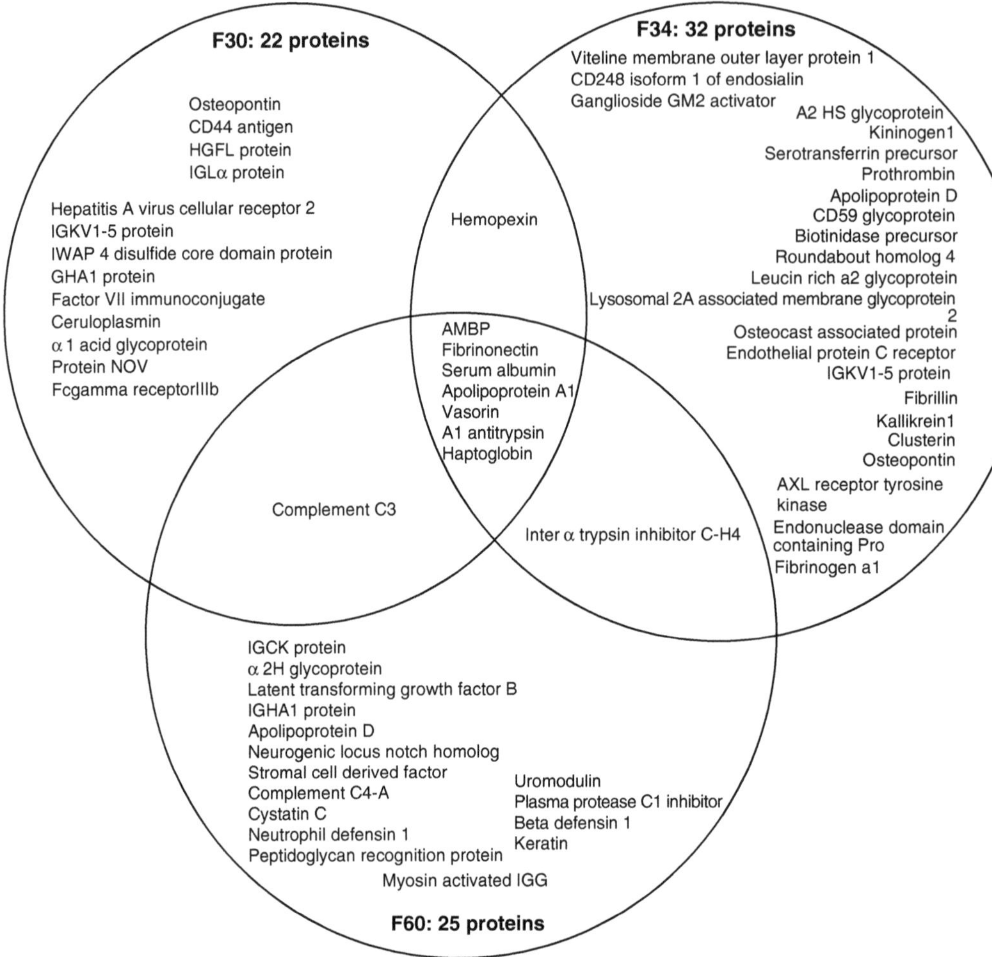

Fig. 4. Uniqueness of the identified protein content between the three FFE fractions. The FFE fractions examined cover the range from pH 2 (fraction 30) to pH 8.5 (fraction 60). Seven proteins are identified in all three fractions, but we attribute this observation to be that these abundant proteins likely exist in the sample in a variety of isoforms and/or naturally proteolytically damaged forms, and thus might well be expected to exist across a diverse pI range. By comparison, only one protein (Hemopexin) overlaps between F30 and F34, demonstrating the resolution of the FFE fractionation method. Complement C3 is identified in both F30 and F60, but not in F34, suggesting that at least two isoforms and/or truncated forms of complement C3 are present in urine. See full protein names and IPI numbers in Table 1.

detectable protein, under these particular FFE running conditions. It is possible that separation conditions can be further optimized to yield higher fractionation resolution, and thus more fractions with protein content. Under the conditions described, by extrapolation of the numbers of proteins identified in three FFE fractions analyzed, we predict identification of approximately 900 or more urine proteins should be enabled, by simply combining IEF–FFE in front of traditional "bottom-up" proteomics protocols.

4. Notes

1. The coating with HPMC is performed to reduce electroendosmosis.
2. Once the current is stable and before loading the protein sample, the system should be tested by conducting a control experiment with the pI marker mix, which demonstrates accurate isoelectric focusing, before committing sample to the system.
3. The FFE fractions were acidified with TFA directly in the FFE fraction collection plate, and so do not need to be reacidified before the C5 clean-up and concentration.
4. The acetonitrile used to make solvent B already contain 0.1% TFA and so there is no need to add more TFA in this solvent.
5. For best binding of the peptide to the C18 Sep-Pak column, it is useful to apply the flow-through of the sample back onto the column.

References

1. Christensen, E., Gburek, J. (2004). Protein reabsorption in renal proximal tubule-function and dysfunction in kidney pathophysiology. *Pediatric Nephrology* **19**: 714–721.
2. Brunzel, N. A. (2004). Fundamental of Urine and Body Fluid Analysis. Philadelphia, PA: Saunders.
3. Mataija-Botelho, D., Murphy, P., Pinto, D. M., MacLellan, D. L., Langlois, C., Doucette, A. A. (2009). A quantitative proteome investigation of the sediment portion of human urine: Implications in the biomarker discovery process. *Proteomics – Clinical Applications* **3**: 95–105.
4. Wang, L., Li, F., Sun, W., Wu, S., Wang, X., Zhang, L., Zheng, D., Wang, J., Gao, Y. (2006). Concavalin A captured glycoproteins in healthy human urine. *Molecular and Cellular Proteomics* **5**: 560–562.
5. Castagna, A., Cecconi, D., Sennels, L., Rappsilber, J., Guerrier, L., Fortis, F., Boschetti, E., Lomas, L., Righetti, P. G. (2005). Exploring the hidden human urinary proteome via ligand library beads. *Journal of Proteome Research* **4**: 1917–1930.
6. Moritz, R. L., Simpson, R. J. (2005). Liquid-based free flow electrophoresis-reversed phase HPLC: a proteomic tool. *Nature Methods* **2**(11): 863–873.
7. Moritz, R. L., Ji, H., Schutz, F., Connolly, L. M., Kapp, E. A., Speed, T. P., Simpson, R. J. (2004). A proteome strategy to fractionate proteins and peptides using continuous free-flow electrophoresis coupled to off-line reversed phase high-performance liquid chromatography. *Analytical Chemistry* **76**: 4811–4824
8. Xie, H., Rhodus, N. L., Griffin, R. J., Carlis, J. V., Griffin, T. J. (2005). *Molecular and Cellular Proteomics* **4**(11): 1026–1030.
9. Cho, S. Y., Lee, J. S., Kim, H. Y., Park, J. M., Kwon, M. S., Park, Y. K., Lee, H. J., Kang, M. J., Kim, J. Y., Yoo, J. S., Cho, J. W., Kim, H. S., Paik, Y. K. (2005). Efficient prefractionation of low abundance proteins in human plasma and construction of a two-dimensional map. *Proteomics* **5**: 3386–3396.
10. Adachi, A., Kumar, C., Zhang, Y., Olsen, J. V., Mann, M. (2006). The human urinary proteome contains more than 1500 proteins, including a large proportion of membrane proteins. *Genome Biology* 7: R80, 1–16.

Chapter 4

Standardized Preprocessing of Urine for Proteome Analysis

Georg Martin Fiedler, Uta Ceglarek, Alexander Leichtle, and Joachim Thiery

Abstract

Proteome/peptidome profiling of human urine is a promising tool for the discovery of novel disease-associated biomarkers. However, a wide range of preanalytic variables influence the results of proteome/peptidome analysis regardless of the method used. We present a validated pretreatment protocol, which allows standardization of preanalytic modalities and facilitates reproducible peptidome profiling of human urine by means of magnetic bead (MB) separation in combination with MALDI-TOF MS. Such a procedure is necessary for generating consistent and reliable data from which meaningful results may be obtained for biomarker discovery and general proteomic experiments.

Key words: MALDI-TOF MS, Mass spectrometry, Magnetic bead separation, Preanalytics, Peptidome, Proteome, Proteomics, Renal diseases, Standardization, Urine protein profiling

1. Introduction

A semiquantitative estimate of the total urinary protein content by dipstick assessment and quantitative methods for the determination of the urinary protein content are clinically established tools for the screening of kidney diseases. Quantitative immunoassays of selected urinary proteins such as albumin, α_1-microglubulin, immunoglobulin G, and α_2-macroglobulin are regularly used to characterize the nature of protein excretion. The established protein assays are helpful to classify abnormal proteinuria and to give indices for the underlying disease. However, the diagnostic value of the results is still limited due to the low specificity and sensitivity of the present urinary protein methods.

The rapid developments in mass spectrometry have opened up the challenge for large scale profiling of the protein complement of human urine. First results from urinary proteome/peptidome

Alex J. Rai (ed.), *The Urinary Proteome: Methods and Protocols*, Methods in Molecular Biology, vol. 641,
DOI 10.1007/978-1-60761-711-2_4, © Springer Science+Business Media, LLC 2010

investigations show that the identification of characteristic urinary protein and peptide excretion profiles may enable early detection and classification of disease (1–15). However, a wide range of preanalytic variables must be taken into account for accurate data generation and interpretation independently from the applied method (13, 16–23). Both, exogenous and endogenous factors can markedly influence the results of peptide profiling. Therefore, standardized protocols for sample collecting, handling, and preparation have to be used minimizing the influence of preanalytic variables on the results. We describe a recently validated standard protocol for studies of urinary proteomics/peptidomics using magnetic bead (MB) separation followed by matrix-assisted laser desorption/ionization time-of-flight (MALDI-TOF) mass spectrometry (MS) (17).

2. Materials

2.1. Preanalytical Conditions

1. Collection of urine in 100 mL urine cups Sarstedt (Nümbrecht, Germany), long-term storage of urine aliquots in 450 μL vials (CryoTubes™, Sarstedt, Nümbrecht, Germany).
2. Multivariable urine strip: protein, hemoglobin, leucocytes, and nitrite. For the determination of the total protein concentration we used a commercial turbidimetric test using benzethonium chloride (Roche diagnostics) (see Note 1).
3. Prepare 1 M solutions of acetic acid and ammonium hydroxide. Prepare the solutions daily fresh.

2.2. Magnetic Bead Fractionation

Store the magnetic beads (Bruker Daltonics, Bremen, Germany) at 4°C (see Note 2). The magnetic beads are characterized as follows: particle size, <1 μm; mean pore size, 40 nm; specific surface area, 100 cm^2/g.

1. MB-HIC 8 (*M*agnetic *B*eads based *H*ydrophobic *I*nteraction *C*hromatography).
   The kit contains: Magnetic Beads (MB) C8, MB-HIC Binding Solution (BS), MB-HIC Wash Solution (WS), deionized water.
2. MB-WCX (*M*agnetic *B*eads based *W*eak *C*ation Exchange *C*hromatography).
   The kit contains: Magnetic Beads MB-WCX, MB-WCX Binding Solution (BS), MB-WCX Wash Solution (WS), MB-WCX Elution Solution (ES), MB-WCX Stabilization Solution (SS).
3. MB-IMAC Cu (*M*agnetic *B*eads Cu based on *I*mmobilized *M*etal ion *A*ffinity *C*hromatography).

The kit contains: Magnetic Beads MB-IMAC Cu, MB-IMAC Cu Binding Solution (BS), MB-IMAC Cu Washing Solution (WS), MB-IMAC Cu Elution Solution (ES).

2.2.1. Consumables for Manual Bead Fractionation

1. 0.2 mL tubes (PCR tubes, Biozym, Hess. Oldendorf, Germany).
2. 450 μL CryoTubes™ (Sarstedt, Nümbrecht, Germany).
3. 1.5 mL polypropylene tubes (Eppendorf safe-lock, Eppendorf, Hamburg, Germany).
4. 0.5–20 μL, 20–200 μL, 200–1,000 μL pipette standard tips (Eppendorf).
5. Bottles (Nalgene PP, FEP (Teflon) bottles), Nalgene (Hereford, United Kingdom).
6. Magnetic separator (Bruker Daltonics, Bremen, Germany).

2.2.2. Consumables for Robot-Assisted Magnetic Bead Fractionation

1. 96 MTP well plates (Biozym).
2. 96 TubePlates (Biozym).
3. 8-strip low-profile polypropylene tubes (Abgene, Hamburg, Germany).
4. Modular Reservoir Quarter Modules (Beckman Coulter, Krefeld, Germany).
5. Customized Tips (Bruker Daltonics, Bremen, Germany).

The above listed brands have been proven to be compatible with MALDI-TOF MS analysis (see Note 3).

2.3. MALDI-TOF MS Profiling

2.3.1. MALDI Targets

1. MTP AnchorChip™ 600/384 (384 anchors, 600 μm diameter).
2. MTP 384 target ground steel, MTP target frame (Bruker Daltonics, Bremen, Germany).

2.3.2. Spotting Solvents

1. Trifluoroacetic acid (TFA, ≥99.5% pro analysis).
2. Ethanol.
3. Acetone (HPLC gradient-grade).
4. Deionized water (for HPLC).
5. Trifluoroacetic acid (2%): dilution of 20 μL TFA per mL deionized water. Store at 4°C. The solution is stable for about 2 months.

2.3.3. Cleaning Solvents

1. Methanol.
2. Acetone (HPLC gradient-grade).
3. Deionized water (for HPLC).
4. TFA (≥99.5% pro analysi).

5. TFA (80%): dilution of 800 μL TFA per ml deionized water. Prepare the solution fresh.

2.3.4. MALDI-Matrix: α-Cyano-4-hydroxycinnamic Acid (HCCA)

1. HCCA for AnchorChip 600 μm target: 0.3 g/L HCCA (0.3 mg HCCA per mL ethanol:acetone 2:1 (volume/volume)).
2. HCCA for ground steel target: 3 g/L HCCA (3 mg HCCA per mL acetonitrile: 2% TFA 1:1 (volume/volume)).

All MALDI matrix solutions should be prepared fresh each day (see Note 4).

2.3.5. Mass Calibration

1. Peptide calibration standard I (Bruker Daltonics) angiotensin II, angiotensin I, substance P, bombesin, ACTH clip 1-17, ACTH clip 18-39, somatostatin 28.
2. Protein calibration standard I (Bruker Daltonics, Bremen, Germany), insulin, ubiquitin I, cytochrome C, myoglobin.
3. TFA (0.1%): Dilution of 1 μL TFA per ml deionized water. Store at 4°C. The solution is stable for 2 months.
4. Ammonium acetate (10 mM): 0.77 mg ammonium acetate per mL deionized water. Store at 4°C. The solution is stable for 2 months.

2.4. Data Interpretation

1. ClinProTools bioinformatics software (Bruker Daltonics, v2.0) (24).
2. ASCII converter for the MS data files (Bruker Daltonics, v2.0) (25).

3. Methods

Magnetic bead (MB) separation followed by MALDI-TOF MS enables high-throughput profiling of proteins and peptides in the region of low M_r (1,000–20,000) in human body fluids without laborious sample-pretreatment (13, 24, 26–30). Therefore, MB-MALDI-TOF MS is appropriate for investigations of alterations of the proteome caused by preanalytical conditions.

3.1. Preanalytical Conditions

The preanalytic phase is affected by a wide range of preanalytic variables, both exogenous (instrument settings, urine collection and storage methods, freezing conditions, the number of freeze-thaw cycles) and endogenous (pH, urine concentrations of salts and proteins, blood and bacterial interferences), that can markedly influence the results of peptide profiling (13, 17, 18, 20, 21, 23). The use of a reliable pretreatment protocol for collecting and storing urine samples is the key for valid and reproducible analyses of urinary peptide patterns.

3.1.1. Urine Sampling and Storage

1. Use 20 mL of first or second morning urine of the patient for proteome analysis (see Note 5). The first 5 mL urine fraction should be rejected by the patient before sampling (see Note 6).
2. The urine sample should reach the laboratory 2 h after sampling. Avoid transporting temperatures > 25°C (see Note 7).
3. 5 mL of the fresh urine has to be used for clinical-chemical analysis. The residual urine sample has to be centrifuged at 10,000 × *g* for 10 min at room temperature. Aliquots of the supernatant (450 μL) have to be stored at −80°C or in liquid nitrogen (see Note 8).

3.1.2. Clinical Chemical Urine Characterizations

5 mL fresh urine has to be investigated for blood, leukocytes, and nitrite, using multivariable strips. Additionally, electrolytes (Na^+, K^+, Ca^{2+}) and total protein have to be determined (see Note 9).

3.1.3. pH Normalization

Adjust the urine pH value to pH 7 using 1 M solutions of acetic acid or ammonium hydroxide before MB-WCX or MB-IMAC Cu purification (see Note 10). Control the pH adjustment by use of indicator stripes. After pH adjustment the urine should be centrifuged again at 10,000 × *g* for 10 min.

3.2. Magnetic Bead Fractionation

The preparation process can be performed manually or using the ClinProt™ Liquid handling robot (Fig. 1).

3.2.1. For MB-HIC C8 Preparation of Urine the Following Protocol is Recommended

1. Shake the magnetic bead suspension carefully at least 20 times. Do not vortex or sonicate (see Note 11). Repeat shaking between pipetting steps.
2. Transfer 60 μL BS and 5 μL of the magnetic beads suspension to a standard thin wall PCR-tube. Add 30 μL of urine sample and mix the solutions carefully by pipetting up and down five times. Incubate the mixture for 1 min.
3. Place the tube in a magnetic separator, and move the tubes back and forth 20 times between the adjacent wells. Note the movement of the magnetic beads in the tube. Finally, wait for 20 s to separate the beads from the supernatant.
4. Remove the supernatant carefully using a pipette. Avoid contact of pipette tips with the magnetic beads and take care not to remove the beads.
5. Normalize the protein content by increasing the loaded urine sample volume in a stepwise manner. After the first loading step, remove residual urine and BS and repeat the procedure with a second urine fraction (90 μL maximum per loading step (30 μL urine/60 μL BS)). After sample loading, start the washing procedure.

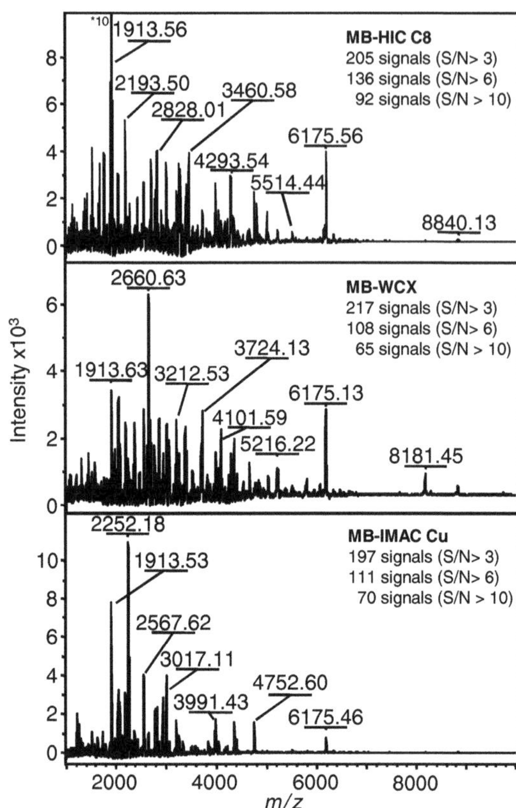

Fig. 1. Characteristic proteome profiles after purification using MB-HIC C8, MB-WCX, and MB-IMAC Cu.

6. Remove the tube from the magnetic separator. To wash, add 100 µL WS.
7. Replace the tube in the magnetic separator, and move the tubes back and forth 20 times between the adjacent wells.
8. Wait 20 s for collecting magnetic beads at the wall of the tubes, and remove the supernatant carefully using a pipette.
9. Repeat steps 6–8 twice.
10. For the elution step remove the tube from the magnetic separator. To elute peptides and proteins, add 5 µL of a gradient-grade acetonitrile:water 1:1 (volume/volume) mixture and mix thoroughly. Wait for 1 min. Avoid evaporation (see Note 12).
11. Place the tube in the magnetic separator, and wait for 30 s to separate the beads from the elution solution at the wall of the tubes.
12. Finally, transfer the elution solution containing purified peptides/ proteins into a fresh tube.

The eluate is now ready for spotting onto MALDI-TOF MS targets and measurement.

3.2.2. For MB-WCX Preparation of Urine the Following Protocol is Recommended

1. Shake the magnetic bead suspension carefully for at least 20 times to get a homogenous suspension. Do not vortex or sonicate (see Note 11). Repeat shaking between pipetting steps if necessary.
2. Transfer 60 μL BS and 5 μL of the magnetic beads suspension to a standard thin wall PCR-tube. Add 30 μL of urine sample, and mix the solutions carefully by pipetting up and down five times. Incubate the mixture for 5 min.
3. Place the tube in a magnetic separator, move the tubes back and forth 20 times between the adjacent wells. Note the movement of the magnetic beads in the tube. Finally, wait for 20 s to separate the beads from the supernatant.
4. Remove the supernatant carefully using a pipette. Avoid contact of pipette tips with the magnetic beads, and take care not to remove the beads.
5. Normalize the protein content by increasing the loaded urine sample volume in a stepwise manner. After the first loading step, remove residual urine and BS, and repeat the procedure with a second urine fraction (90 μL maximum per loading step (30 μL urine/60 μL BS)). After sample loading, start with washing procedure.
6. Remove the tube from the magnetic separator.
7. To wash, add 100 μL WS.
8. Place the tube in the magnetic separator and move the tubes back and forth 20 times between the adjacent wells.
9. Wait 20 s for collecting magnetic beads at the wall of the tubes, and remove the supernatant carefully using a pipette.
10. Repeat steps 6–9 twice.
11. For elution remove the tube from the magnetic separator, add 5 μL ES, and mix thoroughly. Wait for 1 min.
12. Place the tube in the magnetic separator, collect the beads at the tube wall, and transfer the clear supernatant into a fresh tube.
13. Add 5 μL SS to the eluate, and mix intensively by pipetting up and down.
14. The eluate is now ready for spotting onto MALDI-TOF MS targets and measurement.

3.2.3. For MB-IMAC Cu Preparation of Urine the Following Protocol is Recommended

1. Shake the magnetic bead suspension carefully for at least 20 times to get a homogenous suspension. Do not vortex or sonicate (see Note 11). Repeat shaking between pipetting steps if necessary.
2. Pretreat 5 μL of magnetic beads suspension with 50 μL BS. Place the tube in a magnetic separator, and move it there between adjacent wells 10 times. Note the movement of the magnetic beads in the tube.

3. Collect the beads at the tube wall.
4. Discard the supernatant carefully using a pipette.
5. Repeat steps 2–4 twice.
6. Resuspend the magnetic beads in 60 μL BS; add 30 μL sample and mix carefully by pipetting up and down five times. Incubate the mixture for 5 min.
7. Place the tube in a magnetic separator, move the tubes back and forth 20 times between the adjacent wells. Note the movement of the magnetic beads in the tube. Finally, wait until the separation of the beads from the supernatant.
8. Remove the supernatant carefully using a pipette. Avoid contact of pipette tips with the magnetic beads, and take care not to remove the beads.
9. Normalize the protein content by increasing the loaded urine sample volume in a stepwise manner. After the first loading step, remove residual urine and BS, and repeat the procedure with a second urine fraction (90 μL maximum per loading step/30 μL urine/60 μL BS).
10. After sample loading start with washing procedure.
11. Remove the tube from the magnetic separator.
12. To wash, add 100 μL WS.
13. Place the tube in the magnetic separator, and move the tubes back and forth 20 times between the adjacent wells.
14. Wait for collecting magnetic beads at the wall of the tubes, and remove the supernatant carefully using a pipette.
15. Repeat steps 10–13 twice.
16. For elution remove the tube from the magnetic separator. To elute, add 10 μL ES and mix thoroughly. Wait for 5 min.
17. Place the tube in the magnetic separator, and wait for about 20 s to separate the beads from the elution solution at the wall of the tubes.
18. Transfer the supernatant containing the eluted molecules into a fresh tube.

The eluate is now ready for spotting onto MALDI-TOF MS targets and measurement.

3.3. Target Cleaning

Target cleaning is an important process for reproducible MALDI-spotting. The following procedures are recommended.

3.3.1. AnchorChip 600-μm Target

Scour the target intensively under flowing hot tap water, wipe it intensively first with acetone following with deionized water using kimwipe tissues, and then clean it with methanol and let it dry at RT.

3.3.2. Ground Steel Target

If required wipe the target carefully with 80% TFA (see Note 13). Scour the target intensively under flowing hot tap water, wipe it intensively first with acetone following with deionized water using kimwipe tissues, and clean it with methanol. For drying put the target at about 40°C for 20 min.

3.4. Target Preparation

3.4.1. AnchorChip 600-μm Target Preparation

For spotting of purified urine samples prepare a mixture containing 10 μL 0.3 g/L HCCA and 1 μL of the eluate. Prepare at least four spots each with 1 μL per sample purification. Let it dry at RT (see Notes 14 and 15). The samples are ready for MALDI-TOF analysis. The crystallized spots are stable for about 4 h (store in the dark and under dry conditions).

For target preparation of the calibration standard, mix 1 μL of the standard with 10 μL of freshly prepared 0.3 g/L HCCA. Spot 0.5 μL of the mixture on the defined calibrant positions near your crystallized sample spots and let dry at RT. Introduce the target into your instrument, and choose an appropriate acquisition method.

3.4.2. Ground Steel Target Preparation

For spotting of purified urine samples, apply 1 μL of eluate from bead experiment per spot and let it dry at RT for about 10 min (see Note 16). Prepare at least four spots each with 1 μL per sample purification. Apply 1 μL of matrix consisting of 3 g/L HCCA carefully over the whole eluate spot. Let it dry at RT. The samples are ready for MALDI-TOF analysis. The crystallized spots are stable for about 24 h (store in the dark and under dry conditions).

For target preparation of the calibration standard thaw an aliquot of the calibration standard at RT. Mix 1 μL of the standard with 10 μL of freshly prepared 3 g/L HCCA. Spot 1 μL of the mixture on the positions near your crystallized sample spots, and let dry at RT for a few minutes. Introduce the target into your instrument, and choose an appropriate acquisition method.

3.5. Data Acquisition

3.5.1. Preparation of Calibration Standard

1. Solubilize Peptide Calibration Standard I in 125 μL of 0.1% TFA for 5 min at RT and vortex for 1 min.
2. Solubilize Protein Calibration Standard I in 125 μL of 0.1% TFA for 5 min at RT and vortex for 1 min.
3. Mix the standard regarding the following scheme: 5 μL (alternatively 25 μL) Peptide Calibration Standard, 20 μL (alternatively 100 μL) 10 mM ammonium acetate, and 25 μL (alternatively 125 μL) Protein Calibration Standard I.
4. Mix well for 1 min by vortexing. Store it in aliquots at –20°C. The standard is stable for months.

3.5.2. Calibration and Tuning of the Instrument

The calibration of the MALDI-TOF MS has to be performed before each analysis. For evaluation of resolution and intensity over the mass range from 1 to 10 kDa (up to 20 kDa is possible),

Table 1
Resolution values for calibration standard

Substance	Average mass	Resolution
Angiotensin II (M+H)+	1,047.18	360
Angiotensin I (M+H)+	1,297.48	365
Substance P (M+H)+	1,348.64	380
Bombesin (M+H)+	1,620.86	420
ACTH clip 1–17 (M+H)+	2,094.42	475
ACTH clip 18–39 (M+H)+	2,466.68	540
Somatostatin 28 (M+H)+	3,149.57	580
Ubiquitin (M+2H)+	4,283.45	800
Insulin (M+H)+	5,734.56	580
Cytochrome C (M+2H)+	6,181.05	560
Ubiquitin (M+H)+	8,565.89	345

test first different values of the laser power. To eliminate possible impurities in the crystallized spot, use a "matrix blaster". Accumulate about ten laser shots with approximately 10–20% higher laser power as the final power. After matrix blaster sum up 30 shots from the same spot position using a laser power that is slightly above the desorption threshold (see Note 16). Accumulate shots from 5 to 10 different positions to minimize variability of inhomogeneous crystallization surface.

After accumulations of approximately 200 shots (in chunks of 30 shots), calibrate the flexControl method you want to use with the generated spectrum. By tuning the acquisition method parameters, you should reach the above listed resolution values of Table 1 for your calibration spectrum. Try to use all peptides from the standard with the quadratic equation (refer to the flexControl Manual) using the default calibration list. Save the calibration to the flexControl method. The averaged mass deviation should be better than 100 ppm compared to the last calibration.

3.5.3. Acquisition Method Parameters

Linear MALDI-TOF mass spectrometer (Autoflex, Bruker Daltonics, Bremen, Germany), flexControl acquisition software (Bruker Daltonics, v2.0).

Ion source 1:20 kV; ion source 2, 18.50 kV; lens, 9.00 kV; pulsed ion extraction, 120 ns; nitrogen-pressure, 2,500 mbar; Irradiation with a nitrogen laser ($\lambda = 337$ nm) operating at 50 Hz. For matrix suppression we used a high gating factor with signal suppression up to 500 Da. Mass spectra are detected in linear positive mode.

3.5.4. Acquisition of MS Data

After tuning and calibrating the MS instrument, the sample analysis can be performed. Accumulate about 10 laser shots with approximately 10–20% higher laser power as the final power. Accumulate 30 shots from at least 15–20 different position per spot (sum: 450–600 shots) to minimize variations in the crystallization surface. The approximately ten preshots should not be added to the sum spectra.

The data acquisition can be performed manually or using the AutoXecute tool of the FlexControl software (refer to the flexControl Manual).

3.6. Data Interpretation

The ClinPro Tools bioinformatics software is used for proteome pattern analysis. The ASCII converter renders MS data files suitable for analysis with any software capable of importing ASCII raw data (e.g., http://ms proteomics.net).

4. Notes

1. It has to be considered, that different test systems for determination of total protein have different sensitivities to the varying proteins. In addition, most automated assays show high imprecision and poor accuracy for the measurement of urinary protein in the normal range (31). We used an automated benzethonium chloride-turbidimetric assays (Roche diagnostics), which is practical in use.
2. Avoid freezing of magnetic bead material due to the loss of the pore structure.
3. Many plastics are not compatible with matrix preparation on AnchorChip or mass spectrometric analysis, because they may release polymers. Avoid all kind of siliconized tubes. The use of improper plastics typically results in abnormal crystallization and polymer signals. In extreme cases, samples do not crystallize at all and remain liquid even after hours.
4. The matrix solution contains volatile compounds, which can evaporate rapidly. Prepare the matrix solutions daily fresh. Keep the solution during the day in sealed vials or bottles. To prevent photo dissociation, store the matrix solution in the dark.
5. For quantitative analysis of urinary proteins/peptides, 24-h urine collection would be desirable, because it enables the measurement of the daily excretion rate. However, 24-h urine collection is limited by the low patient compliance and preanalytical alterations of the urine. The collection of first or second morning spot urine is an alternative (32). The peptide pattern of first and second morning urine are more appropriate for

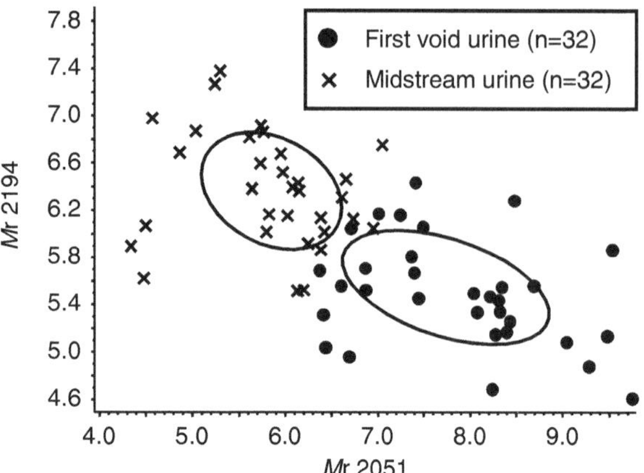

Fig. 2. Two-dimensional mass plot comparing relative peak areas of two characteristic mass signals after MB-HIC C8 purification with respect to first-void and midstream urine samples. The ellipses represent the standard deviations of the selected *Mr*s of each group.

clinical investigations, but they differ significantly due to the varying concentration of the urine components. Therefore, the time of urine collection has to be exactly defined.

6. The proteome pattern of first-void vs. midstream urine samples shows significant differences. Generally, the first 5–10 mL of urine sampling will be considered as first-void urine. Due to possible bacterial contaminations, the first void urine should be rejected (Fig. 2).
7. Storage temperature and time differently influence the stability of the urine peptide pattern. The peptide patterns of once-frozen urine samples are stable up to 24 h at 4°C with no detectable degradation; however, significant changes in peptide patterns at 25°C after 6 h of storage were observed. At 40°C, peptide patterns showed significant variations in signal intensities for numerous signals by 3 h (see Note 17).
8. One-time frozen urine shows a better reproducibility and improved stability against increased temperatures compared with fresh urine. Recurrent freeze-thaw cycles (3×) did not further affect urine peptide patterns (17).
9. The measurement of clinical-chemical parameters is necessary for estimation of potential interfering effects of the proteome analysis. Urea and sodium chloride up to a concentration of 1,000 mmol/L do not interfere with the MB purification. Blood contamination (>0.4 μL blood/per mL urine) interferes with the analysis (Fig. 3). In addition, bacterial contamination of the urine can influence the obtained mass spectra by suppression effects (see Note 17). The analysis of total protein

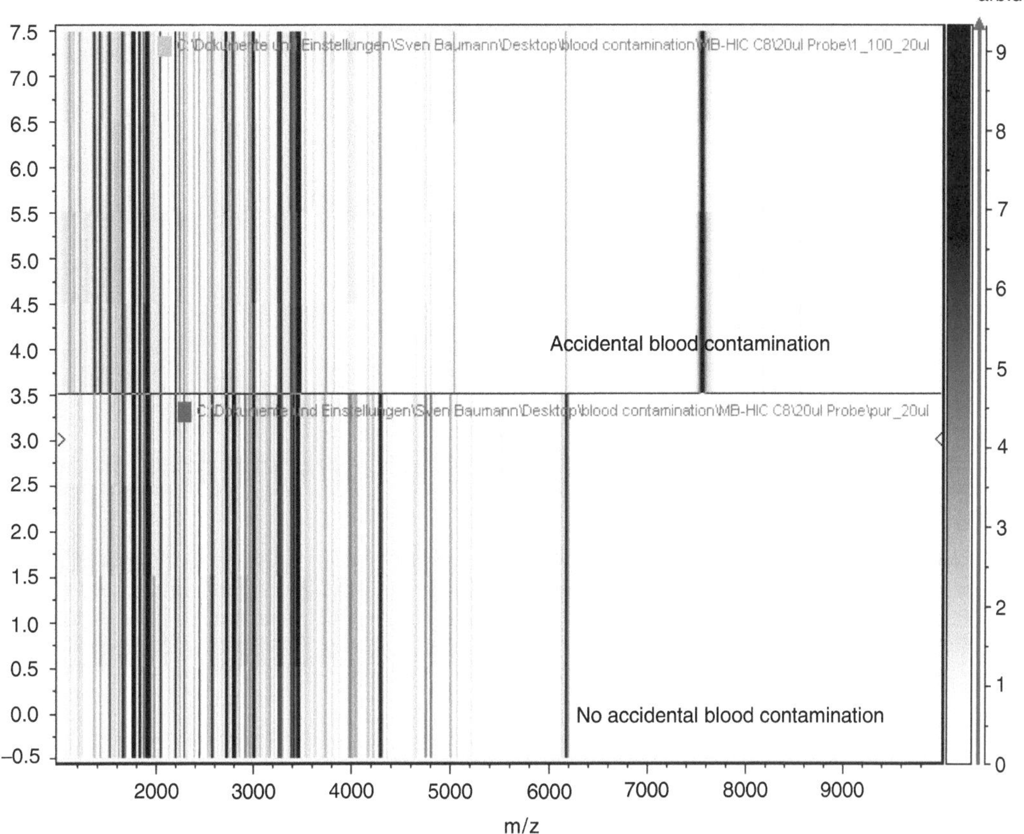

Fig. 3. Densitometric display of the influence of accidental blood on urine proteome profiles. Sample purification was performed using MB-HIC C8.

content is necessary for normalization of the urine before further analysis (Fig. 4).

10. The urinary pH influences the magnetic bead separation with the MB-WCX and MB-IMAC Cu beads. The numbers and intensities of signals with M_rs > 3,500 were decreased above pH 7. MB-HIC C8 beads show no pH-dependent differences in peptide patterns (see Note 17).
11. Avoid vortex and sonication especially of MB-HIC C8 beads due to irreversible decomposition of the magnetic beads and the suspending solution.
12. Acetonitrile is volatile, and evaporation can occur rapidly. Prepare the elution solution daily fresh, and store it for using over the day in a sealed vial or bottle.
13. The repeated application of the ground steel target may lead to a surface impurity despite cleaning with water,

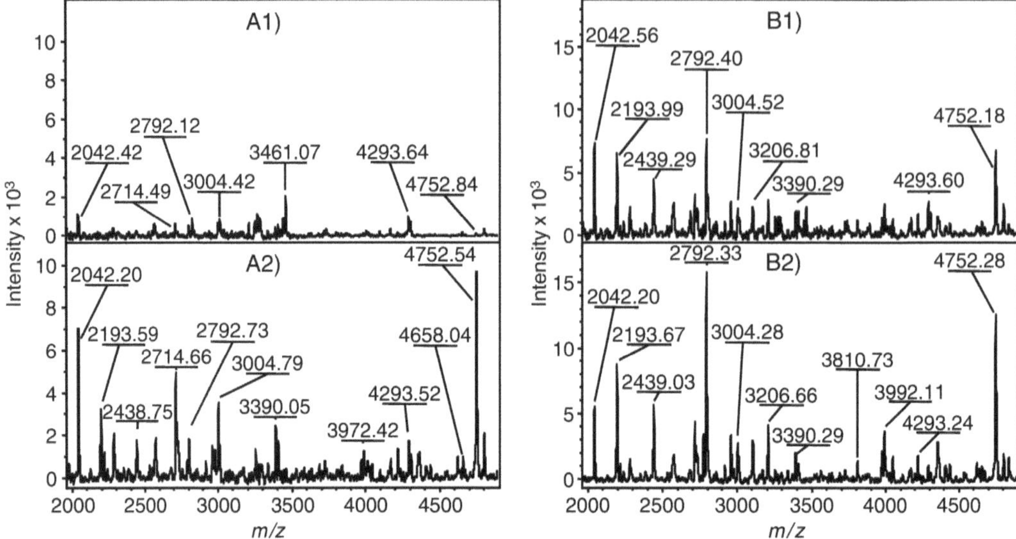

Fig. 4. Normalization effects on MB-HIC C8 urine peptide patterns for two healthy volunteers (**a** and **b**): (**a1**); Without normalization, total protein: 22.8 mg/L, sample volume: 30 μL (represents 0.7 μg absolute) (**b1**); Without normalization, total protein: 110.1 mg/L, sample volume: 30 μL (represents 3.3 μg absolute) (**a2**); Normalized by applied sample volume, total protein: 22.8 mg/L, sample volume: 153 μL (represents 3.5 μg absolute) (**b2**); Normalized by applied sample volume, total protein: 110.1 mg/L, sample volume: 33 μL (represents 3.5 μg adsorbed).

acetone, and methanol. The target surface may develop a milky appearance. In this case, carefully wipe it with 80% TFA. Thereafter, complete the cleaning procedure in accordance to the above-mentioned recommendation.

14. The environmental conditions significantly influence the quality of MALDI target preparation and crystallization process. For reproducible results make sure that the relative humidity ranges between 30 and 60% and the temperature between 18 and 25°C. Prevent direct air condition on the target during the crystallization process.
15. To ensure a homogeneous crystallization and to exploit the function of the hydrophilic anchors, avoid air bubbles during target spotting.
16. The time for drying the eluate spots depends on ambient conditions. The eluate must be completely dried before matrix spotting. To prevent oxidation do not extend standing time before matrix spotting.
17. The desorption threshold of the crystallized analytes underlies day-to-day variations. Try to figure out the desorption

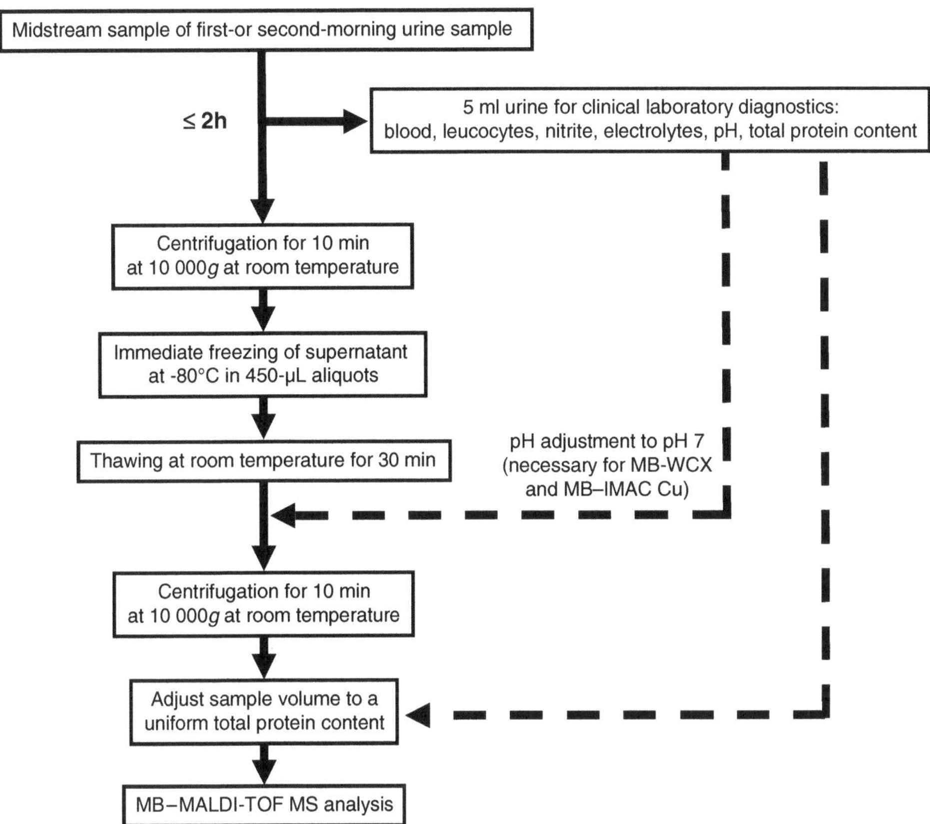

Fig. 5. Recommended sample pretreatment protocol for reproducible MB-based urinary profiling (see Note 17).

threshold before each MS run. The lower the adjusted laser power the higher the resolution of the mass signals. By accumulation of 30 shots, the intensity of the highest peak within the spectrum should be between 1,000 and 3,000 counts. Too high laser intensities will lead to worse resolution of the mass signals.

Proteome/peptidome analysis of urine provides a powerful tool for the discovery of new potential disease related markers, which may improve the understanding of renal physiology/pathophysiology and the diagnostic power. However, standardization of preanalytic and analytic steps is critical. The established pretreatment protocol (Fig. 5) allows standardization of the critical preanalytic phase and enables a reproducible proteome/peptidome profiling of human urine. Preliminary data from our urine proteome study in diabetic patients and other groups let expect that the peptidome profiling of urine may be helpful for early detection and classification of diabetic proteinuria (5).

Acknowledgments

This work was supported by a grant from the Sächsische Aufbaubank, by a "formel-1" grant of the Medical Faculty of the University Leipzig and by a grant of the Deutsche Forschungsgemeinschaft Th 374/2-3. We greatly acknowledge the skilful technical contributions and analytical advice provided by Sven Baumann.

References

1. Kistler, A.D., Mischak, H., Poster, D., Dakna, M., Wuthrich, R.P., Serra, A.L. (2009) Identification of a unique urinary biomarker profile in patients with autosomal dominant polycystic kidney disease. *Kidney Int* **76**(1), 89–96.
2. Quintana, L.F., Sole-Gonzalez, A., Kalko, S.G., Banon-Maneus, E., Sole, M., Diekmann, F., et al. (2009) Urine proteomics to detect biomarkers for chronic allograft dysfunction. *J Am Soc Nephrol* **20**, 428–35.
3. Decramer, S., Gonzalez de Peredo, A., Breuil, B., Mischak, H., Monsarrat, B., Bascands, J.L., Schanstra, J.P. (2008) Urine in clinical proteomics. *Mol Cell Proteomics* 7, 1850–62.
4. Mischak, H., Coon, J.J., Novak, J., Weissinger, E.M., Schanstra, J.P., Dominiczak, A.F. (2009) Capillary electrophoresis-mass spectrometry as a powerful tool in biomarker discovery and clinical diagnosis: An update of recent developments. *Mass Spectrom Rev* **28**(5), 703–24.
5. Rossing, K., Mischak, H., Dakna, M., Zurbig, P., Novak, J., Julian, B.A., et al. (2008) Urinary proteomics in diabetes and CKD. *J Am Soc Nephrol* **19**, 1283–90.
6. Zimmerli, L.U., Schiffer, E., Zurbig, P., Good, D.M., Kellmann, M., Mouls, L., et al. (2008) Urinary proteomic biomarkers in coronary artery disease. *Mol Cell Proteomics* 7, 290–8.
7. Goligorsky, M.S., Addabbo, F., O'Riordan, E. (2007) Diagnostic potential of urine proteome: a broken mirror of renal diseases. *J Am Soc Nephrol* **18**, 2233–9.
8. Gonzalez-Buitrago, J.M., Ferreira, L., Lorenzo, I. (2007) Urinary proteomics. *Clin Chim Acta* **375**, 49–56.
9. Theodorescu, D., Mischak, H. (2007) Mass spectrometry based proteomics in urine biomarker discovery. *World J Urol* **25**, 435–43.
10. Clarke, W. (2006) Proteomic research in renal transplantation. *Ther Drug Monit* **28**, 19–22.
11. Chalmers, M.J., Mackay, C.L., Hendrickson, C.L., Wittke, S., Walden, M., Mischak, H., et al. (2005) Combined top-down and bottom-up mass spectrometric approach to characterization of biomarkers for renal disease. *Anal Chem* 77, 7163–71.
12. Haubitz, M., Wittke, S., Weissinger, E.M., Walden, M., Rupprecht, H.D., Floege, J., et al. (2005) Urine protein patterns can serve as diagnostic tools in patients with IgA nephropathy. *Kidney Int* **67**, 2313–20.
13. Schaub, S., Wilkins, J., Weiler, T., Sangster, K., Rush, D., Nickerson, P. (2004) Urine protein profiling with surface-enhanced laser-desorption/ionization time-of-flight mass spectrometry. *Kidney Int* **65**, 323–32.
14. Thongboonkerd, V. (2008) Urinary proteomics: towards biomarker discovery, diagnostics and prognostics. *Mol Biosyst* **4**, 810–5.
15. Weissinger, E.M., Kaiser, T., Meert, N., De Smet, R., Walden, M., Mischak, H., Vanholder, R.C. (2004) Proteomics: a novel tool to unravel the patho-physiology of uraemia. *Nephrol Dial Transplant* **19**, 3068–77.
16. Lee, R.S., Monigatti, F., Briscoe, A.C., Waldon, Z., Freeman, M.R., Steen, H. (2008) Optimizing sample handling for urinary proteomics. *J Proteome Res* 7, 4022–30.
17. Fiedler, G.M., Baumann, S., Leichtle, A., Oltmann, A., Kase, J., Thiery, J., Ceglarek, U. (2007) Standardized peptidome profiling of human urine by magnetic bead separation and matrix-assisted laser desorption/ionization time-of-flight mass spectrometry. *Clin Chem* **53**, 421–8.
18. Thongboonkerd, V. (2007) Practical points in urinary proteomics. *J Proteome Res* **6**, 3881–90.

19. Khan, A., Packer, N.H. (2006) Simple urinary sample preparation for proteomic analysis. *J Proteome Res* **5**, 2824–38.
20. Traum, A.Z., Wells, M.P., Aivado, M., Libermann, T.A., Ramoni, M.F., Schachter, A.D. (2006) SELDI-TOF MS of quadruplicate urine and serum samples to evaluate changes related to storage conditions. *Proteomics* **6**, 1676–80.
21. Hortin, G.L. (2005) Can mass spectrometric protein profiling meet desired standards of clinical laboratory practice? *Clin Chem* **51**, 3–5.
22. Kumps, A., Duez, P., Mardens, Y. (2002) Metabolic, nutritional, iatrogenic, and artifactual sources of urinary organic acids: a comprehensive table. *Clin Chem* **48**, 708–17.
23. Kouri, T., Harmoinen, A., Laurila, K., Ala-Houhala, I., Koivula, T., Pasternack, A. (2001) Reference intervals for the markers of proteinuria with a standardised bed-rest collection of urine. *Clin Chem Lab Med* **39**, 418–25.
24. Cheng, A.J., Chen, L.C., Chien, K.Y., Chen, Y.J., Chang, J.T., Wang, H.M., et al. (2005) Oral cancer plasma tumor marker identified with bead-based affinity-fractionated proteomic technology. *Clin Chem* **51**, 2236–44.
25. Conrad, T.O.F., Leichtle, A., Hagehülsmann, A., Diederichs, E., Baumann, S., Thiery, J., Schütte, C. (2006) Beating the noise: new statistical methods for detecting signals in MALDI-TOF spectra below noise level, in Proceeding of CompLife (M. R. Berthold RG, and I. Fischer, ed.), Vol. 4216: Springer-Verlag Berlin Heidelberg, pp. 119–28.
26. Hsieh, S.Y., Chen, R.K., Pan, Y.H., Lee, H.L. (2006) Systematical evaluation of the effects of sample collection procedures on low-molecular-weight serum/plasma proteome profiling. *Proteomics* **6**, 3189–98.
27. Baumann, S., Ceglarek, U., Fiedler, G.M., Lembcke, J., Leichtle, A., Thiery, J. (2005) Standardized approach to proteome profiling of human serum based on magnetic bead separation and matrix-assisted laser desorption/ionization time-of-flight mass spectrometry. *Clin Chem* **51**, 973–80.
28. de Noo, M.E., Tollenaar, R.A., Ozalp, A., Kuppen, P.J., Bladergroen, M.R., Eilers, P.H., Deelder, A.M. (2005) Reliability of human serum protein profiles generated with C8 magnetic beads assisted MALDI-TOF mass spectrometry. *Anal Chem* 77, 7232–41.
29. Villanueva, J., Philip, J., Entenberg, D., Chaparro, C.A., Tanwar, M.K., Holland, E.C., Tempst, P. (2004) Serum peptide profiling by magnetic particle-assisted, automated sample processing and MALDI-TOF mass spectrometry. *Anal Chem* **76**, 1560–70.
30. Zhang, X., Leung, S.M., Morris, C.R., Shigenaga, M.K. (2004) Evaluation of a novel, integrated approach using functionalized magnetic beads, bench-top MALDI-TOF-MS with prestructured sample supports, and pattern recognition software for profiling potential biomarkers in human plasma. *J Biomol Tech* **15**, 167–75.
31. Dube, J., Girouard, J., Leclerc, P., Douville, P. (2005) Problems with the estimation of urine protein by automated assays. *Clin Biochem* **38**, 479–85.
32. Bottini, P.V., Ribeiro Alves, M.A., Garlipp, C.R. (2002) Electrophoretic pattern of concentrated urine: comparison between 24-hour collection and random samples. *Am J Kidney Dis* **39**, E2.

Chapter 5

Different Sample Preparation and Detection Methods for Normal and Lung Cancer Urinary Proteome Analysis

Supachok Sinchaikul, Payungsak Tantipaiboonwong, Supawadee Sriyam, Ching Tzao, Suree Phutrakul, and Shui-Tein Chen

Abstract

The urinary proteome is known to be a valuable field of study related to human physiological functions because many components in urine provide an alternative to blood plasma as a potential source of disease biomarkers useful in clinical diagnosis and therapeutic application. Due to the variability and complexity of urine, sample preparation is very important for decreasing the dynamic range of components and isolating specific urinary proteins prior to analysis. We discuss many useful sample preparation methods in this chapter, including those of lung cancer urine samples. In addition, protein detection methods are also crucial in visualizing protein profiles and for quantification of protein content in urine samples from both normal donor and lung cancer patients. This chapter also provides alternative choices of urine sample preparation and detection methods for selective use in urinary proteome analysis and for identifying urinary protein markers in lung cancer and other diseases.

Key words: Urine, Lung cancer, Proteomics, Sample preparation, Detection

1. Introduction

Human urine is a readily obtainable biological fluid that contains various types of biomarkers, and these biomarkers are useful for prognosis, diagnosis, and drug development for many diseases. It is formed in the kidney by ultrafiltration from the plasma to eliminate waste products, for instance urea and metabolites, and excreted less than 1% of ultrafiltrate. Thus, protein concentration in normal donor urine is very low (less than 100 mg/L when urine output is 1.5 L/day), and normal protein excretion is less than 150 mg/day. Diseases which adversely affect the function of the kidneys also cause excessive losses of proteins in the urine, and

Alex J. Rai (ed.), *The Urinary Proteome: Methods and Protocols*, Methods in Molecular Biology, vol. 641,
DOI 10.1007/978-1-60761-711-2_5, © Springer Science+Business Media, LLC 2010

excretion of more than 150 mg/day protein is defined as proteinuria (1–3). Thus, the evaluation of these proteins may lead to an increased understanding of human physiology and possibly provides the differentiation of protein markers in subgroups of many diseases, which are useful for further diagnostic and therapeutic applications.

In the present study, the urinary proteome has been analyzed by many proteomic techniques, such as RP-HPLC (4, 5), CE (6, 7), two-dimensional gel electrophoresis (2-DE) (8–10), mass spectrometry (MS) (11–13), and through a combination of multidimensional separation techniques, e.g., CE-MS (14–16), surface-enhanced laser desorption/ionization time-of-flight (SELDI-TOF)-MS (17, 18). More than 1,500 proteins have been identified in the human urinary proteome (19), and some urinary protein markers have been observed from normal and diseased states, such as inflamed pilonidal abscess (20), kidney/renal diseases (10, 21, 22), and cancers, e.g., lung cancer (23), bladder cancer (24), ovarian cancer (25). Although these proteomic techniques are powerful in being able to analyze the urinary proteome, the complexity of the urinary proteome is great and is not yet fully characterized. Human urine has a very dilute protein concentration with high concentrations of salts, metabolic wastes and small molecules that interfere with proteomic analysis. Moreover, the yield of protein recovery remains unsatisfactory due to losses during sample preparation. It is crucial to use an effective protocol to eliminate interferences and to concentrate urinary proteins. Therefore, preparation of urine samples is an important step in proteomic analysis, and influences further protein separation and data generation. There are many protocols that can be employed to isolate and concentrate urinary proteins, e.g., centrifugal filtration (9, 10, 23, 26), lyophilization (11, 12, 23), precipitation (9, 23, 26, 27), ultracentrifugation (27, 28). Not only is sample preparation crucial for urinary proteome analysis, but the choice of detection method after protein separation is also important for data visualization. It allows for the surveillance of protein expression profiles and is also used for quantification in proteomic research. There are many detection methods used, such as protein gel staining methods, e.g., Coomassie blue (9, 10), silver staining (26, 29), SYPRO® Ruby stain (20, 23), 2-D fluorescence difference in-gel electrophoresis (2-D DIGE) using different colors of CyDyes (30), and gel image analysis (30, 31). However, sample preparation and detection methods need to be optimized and judiciously selected for examining the urinary proteome.

In this chapter, the objectives are to provide simple preparation methods of urinary samples and different detection methods for subsets of urinary proteins, including phosphoproteins and glycoproteins. We provide illustrative examples for each of the following procedures: (1) comparison of urine preparation and detection methods, (2) comparison of urinary protein/phosphoprotein/glycoprotein contents from normal human and lung cancer patients, and (3) investigation of urinary protein markers of lung cancer.

2. Materials

2.1. Equipment

1. Centricon® Centrifugal Filter Devices, cut-off 10 kDa membrane (Millipore, Billerica, MA).
2. PD-10 Desalting column (GE Healthcare, Piscataway, NJ).
3. Microcentrifuge machine (Becker).
4. Filter (45 μm for organic solvent) (Millipore, Billerica, MA).
5. Sonicator bath (Brasonic 52, Branson, Shelton, CT, USA).
6. RP-HPLC column: 4.6×250 mm (7 μm, Nucleosil 7C18, Machery & Nagel, Germany).
7. HPLC machine: L-4250 UV-VIS detector and L-1700 pump (Hitachi, Tokyo, Japan).
8. IPG strips (18 cm, pH 3-10 nonlinear/pH4-7, GE Healthcare).
9. IPGphor IEF (GE Healthcare).
10. PROTEAN II xi Multi-Cells (Bio-Rad, Hercules, CA).
11. Typhoon 9200 image scanner (GE Healthcare).

2.2. Buffer and Solvent for Protein Precipitation

1. Buffer: 10 mM phosphate buffer, pH 7.5.
2. Acetone.
3. Acetonitrile (ACN)/Trifluoroacetic acid (TFA): 100% ACN and 0.1% TFA.
4. Chloroform.
5. Methanol.
6. Trichloroacetic acid (TCA)/Acetone: 20% TCA in acetone. Store at −20°C.

2.3. Buffers for Reverse Phase-High Performance Liquid Chromatography (RP-HPLC)

1. RP-HPLC buffer A (mobile phase A): 95% ddH_2O, 5% ACN, 0.1% TFA. Filter by 45 μm-filter (for organic solvent, Millipore) and degas by sonicator bath.
2. RP-HPLC buffer B (mobile phase B): 5% ddH_2O, 95% ACN, 0.1% TFA. Filter by 45 μm-filter (for organic solvent, Millipore) and degas by sonicator bath.

2.4. SDS-Polyacrylamide Gel Electrophoresis (SDS-PAGE)

1. Thirty percent acrylamide/bis solution (37.5:1 with 2.6% C) (see Note 1) and N,N,N,N′-Tetramethyl-ethylenediamine (TEMED, Bio-Rad, Hercules, CA) (see Note 2).
2. Separating buffer (4×): 1.5 M Tris–HCl, pH 8.8. Store at room temperature.
3. Stacking buffer (4×): 0.5 M Tris–HCl, pH 6.8. Store at room temperature.
4. Ammonium persulfate: prepare 10% solution in water and immediately freeze in single use (200 μL) aliquots at –20°C.
5. Running buffer (10×): 25 mM Tris-HCl, 192 mM glycine, 0.1% (w/v) SDS. Store at room temperature.
6. Sample loading buffer (3×): 50 mM Tris–HCl, pH 6.8, 0.1 M DTT, 10% glycerol, 2% (w/v) SDS, 0.1% (w/v) bromophenol blue. Store at 4°C within 2 weeks or aliquot each 500 μL and store at –20°C until use.

2.5. Two-Dimensional Electrophoresis (2DE)

1. Lysis buffer: 7 M Urea, 2 M Thiourea, 4% CHAPS, 0.002% bromophenol blue, 65 mM DTT.
2. Equilibration buffer: 50 mM Tris–HCl (pH 8.8), 6 M urea, 30% glycerol, 2% SDS, a trace of bromophenol blue.
3. Thirty percent acrylamide/bis solution (37.5:1 with 2.6% C) (see Note 1) and N,N,N,N′-Tetramethyl-ethylenediamine (TEMED, Bio-Rad, Hercules, CA) (see Note 2).
4. Separating buffer for second SDS-PAGE: 1.5 M Tris–HCl, pH 8.8, 0.4% SDS. Store at room temperature.
5. 0.5% agarose: 0.5 g agarose, 100 mL separating buffer, a trace of bromophenol blue. Add all ingredients into a 500 mL Erlenmeyer flask, swirl to disperse, and heat in a microwave oven at a low temperature until the agarose is completely dissolved. Allow the agarose to cool to 40–50°C before sealing on the top of gel.

2.6. Protein Gel Staining

2.6.1. Coomassie Staining

1. Fixing solution: 50% ethanol, 10% acetic acid.
2. Washing solution: 50% methanol, 5% acetic acid.
3. Stain: 0.1% Coomassie brilliant blue R250, 20% methanol, 10% acetic acid. Dissolve 0.4 g of Coomassie blue R250 in 200 mL of 50% (v/v) methanol in water with stirring as needed. Filter the solution to remove any insoluble material. Add 200 mL of 20% (v/v) acetic acid in water. The final concentration is 0.1% (w/v) Coomassie blue R250, 20% (v/v) methanol, and 10% (v/v) acetic acid.
4. Destain solution: 50% methanol, 5% acetic acid.

2.6.2. SYPRO® Ruby Protein Gel Stain

1. Fix solution: 50% methanol, 7% acetic acid.
2. Stain solution: SYPRO® Ruby protein gel stain (Molecular Probes, Invitrogen, Carlsbad, CA).
3. Wash solution: 10% methanol, 7% acetic acid.

2.6.3. Phosphoprotein Staining by Pro-Q® Diamond Phosphoprotein Gel Stain

1. The phosphoprotein gel stain is carried out by a Pro-Q® Diamond Phosphoprotein Gel Stain Kit (Molecular Probes).
2. Fix solution: 50% methanol, 10% acetic acid.
3. Destain solution: 20% acetonitrile, 50 mM sodium acetate pH 4.0.

2.6.4. Glycoprotein Staining by Pro-Q® Emerald 488 Glycoprotein Gel Stain

1. The glycoprotein gel stain is carried out by a Pro-Q® Emerald 488 Glycoprotein Gel Stain Kit (Molecular Probes).
2. Fix solution: 50% methanol, 5% acetic acid.
3. Wash solution: 3% acetic acid in ddH_2O.
4. Oxidizing solution: Add 250 mL of 3% acetic acid to the bottle containing the periodic acid (Component C) and mix until completely dissolved.
5. Pro-Q Emerald 488 stock solution: Add 0.5 mL of DMSO or DMF to one vial of Pro-Q Emerald 488 reagent (Component A) and mix thoroughly. Warm the vial to room temperature before opening. Always use the stock solution within a few hours of preparation.
6. Fresh Pro-Q Emerald 488 staining solution: For SDS-PAGE gel, dilute the Pro-Q Emerald 488 stock solution 50-fold into Pro-Q Emerald 488 staining buffer (Component B). For example, dilute 500 μL of the Pro-Q Emerald 488 stock solution into 25 mL of the staining buffer to make enough staining solution for one 8 × 10 cm gel. Large 2-DE gels require 250 mL of staining solution.

2.6.5. Glycoprotein Staining by Fluorescence-Labeled Lectin Stain

1. Membrane: Polyvinylidene difluoride membrane (PVDF) (0.45 μm pore size from Millipore).
2. Transfer buffer: 25 mM Tris-Base, 190 mM glycine, 20% methanol, 0.05% SDS
3. Fluorescein isothiocyanate (FITC)-labeled lectins: ConA (concanavalin A), WGA (Wheat germ agglutinin); PNA (Peanut agglutinin), AAL (Aleuria aurantia lectin).
4. PBS buffer: 10 mM phosphate buffer, pH 7.4, 150 mM NaCl.
5. PSBT buffer: 10 mM phosphate buffer, pH 7.4, 150 mM NaCl, 0.05% Tween 20.
6. Blocking buffer: 5% (w/v) nonfat dried milk in PBST buffer.

2.7. Two-Dimensional Difference In-Gel Electrophoresis (2-D DIGE)

1. PBS buffer: 10 mM phosphate buffer, pH 7.4, 150 mM NaCl.
2. CyDye™ monofunctional meleimide dyes: Cy3 and Cy5 maleimide dyes (GE Healthcare) (see Note 3).

3. Methods

3.1. Preparation of Human Urine Samples

1. First morning urine samples are collected into the polypropylene containers containing 0.2% sodium azide.
2. The containers are covered with caps and stored in the refrigerator during the collection period.
3. Each urine sample is centrifuged at 10,000 × *g* for 30 min at 4°C and the supernatant is collected into fresh container.
4. The containers containing urine samples are labeled (e.g., name, date, time of completion, age, sex, disease, and stage of disease) and stored at –80°C until further processing. Urine samples can also be stored for long periods or for easy transportation by lyophilization to dryness, and storage at –80°C until further processing.

3.2. Desalting Column (PD-10)

1. The bottom tip of the column is cut off, the cap is removed and excess liquid is decanted.
2. The column is placed in the PD-10 Desalting Workmate and equilibrated with approximately 25 mL 10 mM phosphate buffer, pH 7.5. Discard flow-through.
3. The urine sample of a total volume of 2.5 mL is added into the column. If the sample is less than 2.5 mL, then add buffer until the total volume of 2.5 mL is achieved. Discard flow-through.
4. The proteins in column are eluted with 3.5 mL buffer.
5. The fractionated protein solutions are collected (0.5 mL/fraction), lyophilized, and stored at –20°C until further use. An example of separation of urinary proteins by PD-10 desalting column is shown in Fig. 1.

3.3. Centrifugal Filtration (Centricon Tube)

1. Urine samples are loaded into the Centricon tube (mw-cutoff 10 kDa) and centrifuged at 5,000 × *g* for 15 min at 4°C.
2. Distilled water is added to the Centricon tube for desalting and eluting interfering substances. Repeat this step more than three times to make sure that you are removing the high concentrations of salt and small interferences (see Note 4).
3. The urine samples are lyophilized and stored at –20°C until further use.

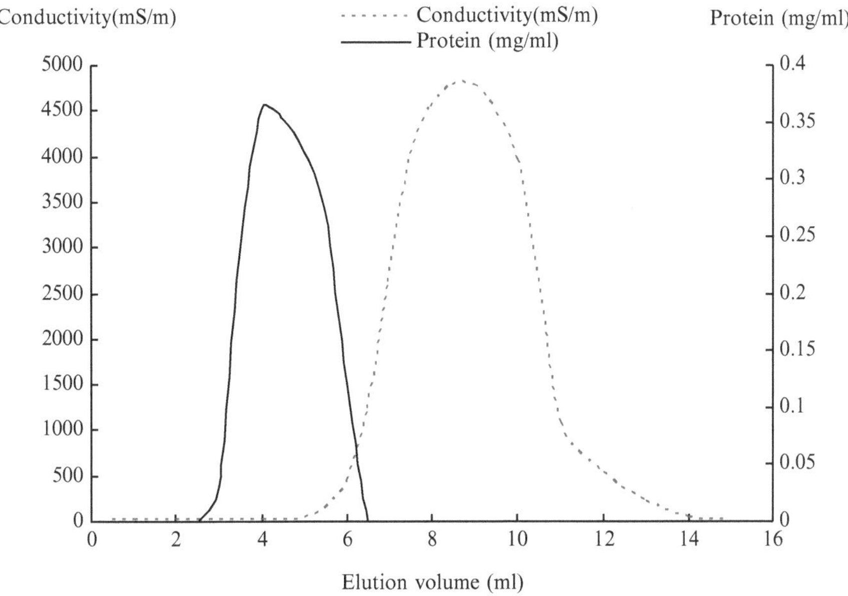

Fig. 1. Separation of urinary proteins by Sephadex gel filtration using PD-10 Desalting column. After 2.5 mL of urine samples were loaded into a column, urinary proteins were eluted with 10 mM phosphate buffer pH 7.5. All eluted protein fractions were collected (0.5 mL/fraction). Dash (*black*) and straight (*black*) lines represent the determined conductivity and protein contents, respectively. The protein fractions no. 3–6 were pooled and concentrated by lyophilization.

4. Examples of separation of urinary proteins prepared by centrifugal filtration using the Centricon tube are shown in Fig. 2 (a1, b1, c1), Fig. 3 (lane 1, 5), and Fig. 4 (a1, b1).

3.4. Acetone Precipitation

1. Two volumes of cold acetone are added to the urine samples and the mixture is stored overnight at −20°C.
2. A pellet is obtained by centrifugation at 10,000 × *g* at 4°C for 15 min and then vacuum dried.
3. Examples of separation of urinary proteins prepared by acetone precipitation are shown in Fig. 2 (a2, b2, c2), Fig. 3 (lane 2, 6), and Fig. 4 (a2, b2).

3.5. ACN/TFA Precipitation

1. Two volumes of cold ACN/0.1%TFA are added to the urine samples.
2. The mixture is vortexed and centrifuged at 10,000 × *g* at 4°C for 15 min.
3. The supernatant is removed and the pellet is vacuum dried.
4. Examples of separation of urinary proteins prepared by acetone precipitation are shown in Fig. 2 (a3, b3, c3), Fig. 3 (lane 3, 7), and Fig. 4 (a3, b3).

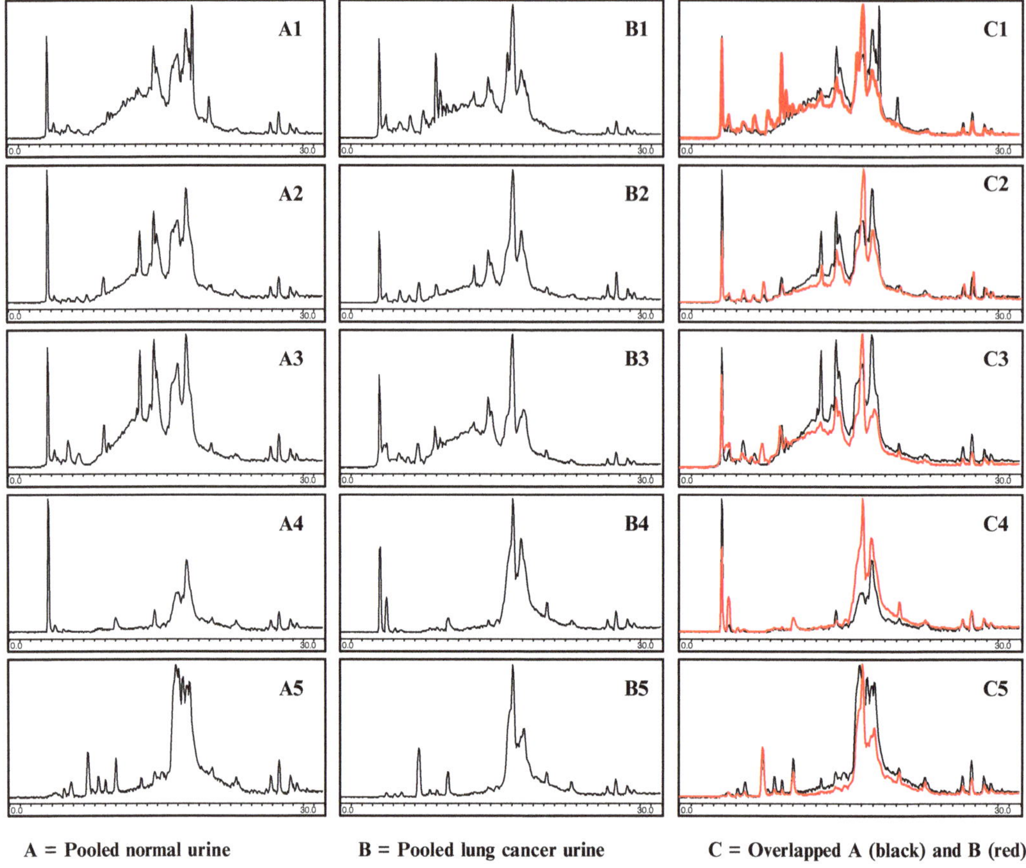

Fig. 2. RP-HPLC chromatograms of pooled normal (**a**) and lung cancer (**b**) urine samples preparing by centrifugal filtration, acetone precipitation, ACN/TFA precipitation, MeOH/$CHCl_3$/water precipitation and TCA in acetone precipitation. The overlapped chromatogram profiles of urine samples (*black line*) from normal and lung cancer (*red line*) patients show in **c1**–**c5**. The samples were analyzed on a Nucleosil 7C18, 4.6×250 mm column using a linear gradient of 0–100% mobile phase B at a flow rate of 1 mL/min for 25 min and washed with 100% mobile phase B until the running time reached to 30 min. Mobile phase A was 5% v/v ACN, 0.1% v/v TFA and mobile phase B was 95% v/v ACN, 0.1% v/v TFA. Protein elution was monitored at absorbance 280 nm. Label: 1, centrifugal filtration; 2, acetone precipitation; 3, ACN/TFA precipitation; 4, MeOH/$CHCl_3$/water precipitation; 5, TCA/acetone precipitation.

3.6. Methanol/ Chloroform Precipitation

1. 0.4 mL of methanol is added to the urine sample (0.1 mL).
2. The mixture is vortexed and centrifuged at 10,000×*g* for 10 s.
3. 0.1 mL of chloroform is added to the mixture, vortexed, and centrifuged again at 10,000×*g* for 10 s.
4. Water (0.3 mL) is added and the sample is vortexed vigorously and centrifuged at 10,000×*g* for 1 min. The upper phase is carefully removed and discarded.

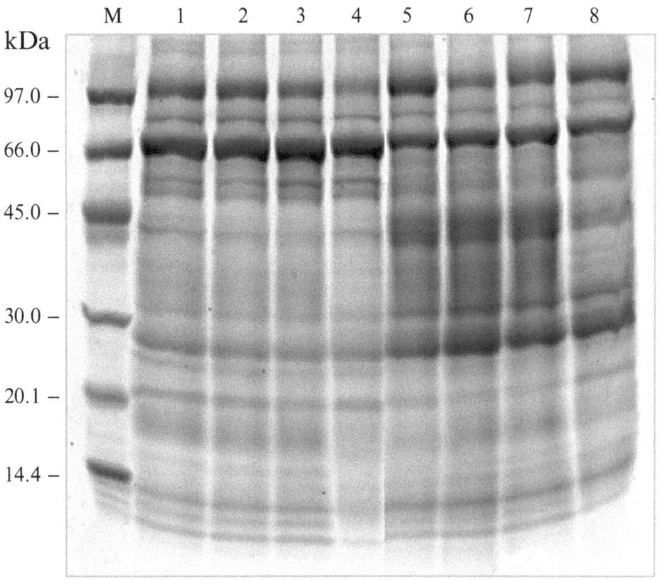

Fig. 3. SDS-PAGE of pooled urine samples of normal and lung cancer patients, prepared by different methods. SDS-PAGE was carried out in a mini-gel 4–20% gradient gel with 10 μg protein/well. Lane M is the protein standard markers, lanes 1–4 are pooled normal urine samples prepared using centrifugal filtration, acetone precipitation, ACN/TFA precipitation and $MeOH/CHCl_3/H_2O$ precipitation, respectively. Lanes 5–8 are pooled lung cancer urine samples prepared using ultrafiltration, acetone precipitation, ACN/TFA precipitation and $MeOH/CHCl_3/H_2O$ precipitation, respectively.

5. A further 0.3 mL of methanol is added into the rest of the chloroform phase and the interphase with the precipitated protein.
6. The sample is mixed and centrifuged again at 10,000 × *g* for 2 min.
7. The supernatant is removed and the protein pellet is vacuum dried.
8. Examples of separation of urinary proteins prepared by methanol/chloroform precipitation are shown in Fig. 2 (a4, b4, c4), Fig. 3 (lane 4, 8), and Fig. 4 (a4, b4).

3.7. TCA/Acetone Precipitation

1. To the urine sample is added two volumes of cold 20% TCA in acetone (–20°C) and the mixture is stored overnight at –20°C.
2. The pellet is obtained by centrifuging at 10,000 × *g*, 4°C for 15 min.
3. The pellet is washed twice: first with cold acetone containing 20 mM DTT and then with cold acetone without DTT, and centrifuged at 10,000 × *g*, 4°C for 15 min.

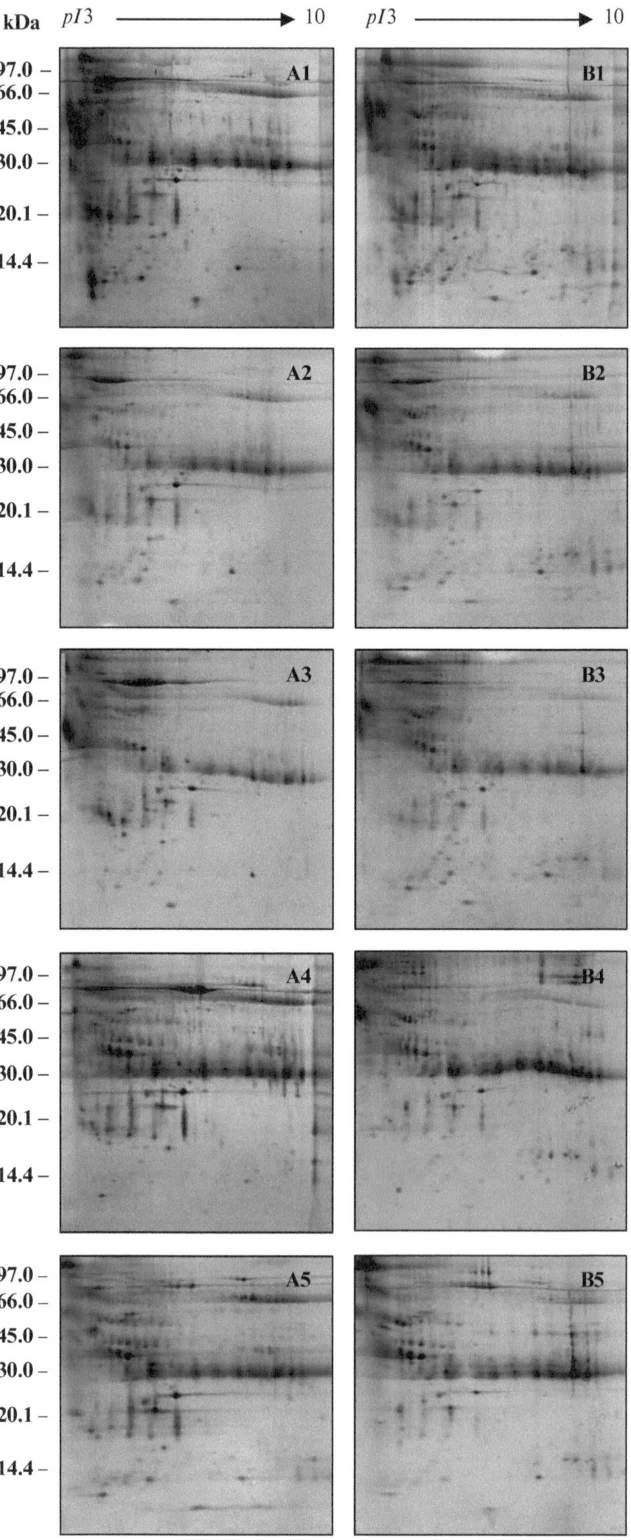

Fig. 4. 2-DE gel images of pooled normal urine samples (**a1**–**a5**) and lung cancer urine sample (**b1**–**b5**), prepared by different methods. Gels **a1** and **b1** represent urine

4. The supernatant is removed and the pellet is vacuum dried.
5. Examples of separation of urinary proteins prepared by methanol/chloroform precipitation are shown in Fig. 2 (a5, b5, c5) and Fig. 4 (a5, b5).

3.8. RP-HPLC

1. Analysis of urine samples are performed using a reverse-phase HPLC column (4.6 × 250 mm, Nucleosil 7C18) attached to a L-4250 UV-VIS detector and an L-7100 pump with a 20 μL sample injection loop.
2. The urine sample is resuspended in mobile phase A at a concentration of 1 mg/mL, filtered and injected into the HPLC column with a final volume of 10 μL.
3. A linear gradient of 5% ACN to 95% ACN (0–100% mobile phase B) is used for 25 min within a 30 min running time at a flow rate of 1 mL/min. Protein is monitored by measuring UV absorption at 280 nm.
4. Examples of HPLC separation of urinary proteins prepared by various preparation methods are shown in Fig. 2.

3.9. SDS-PAGE

1. Prepare a 1.5-mm thick, 10% gel by mixing 6.25 mL of 4× separating buffer, with 8.33-mL acrylamide/bis solution, 0.25 mL SDS, 10.03 mL water, 125 μL ammonium persulfate solution, and 15 μL TEMED. Pour the gel, leaving space for a stacking gel, and overlay with absolute ethanol. The gel should polymerize in about 30 min. Alternatively, a 4–20% polyacrylamide slab gel (Invitrogen, Leak, The Netherlands) can also be used.
2. Pour off the ethanol and rinse the top of the gel twice with water.
3. Prepare the stacking gel by mixing 1.25 mL of 4× stacking buffer with 0.67 mL acrylamide/bis solution, 3 mL water, 25 μL ammonium persulfate, and 5 μL TEMED. Use about 0.5 mL of this to quickly rinse the top of the gel and then insert the comb. The stacking gel should polymerize within 30 min.

Fig. 4. (continued) samples prepared using centrifugal filtration of pooled normal and lung cancer patients, respectively. Gels **a2** and **b2** represent urine samples prepared using acetone precipitation of pooled normal and lung cancer patients, respectively. Gels **a3** and **b3** represent urine samples prepared using ACN/TFA precipitation of pooled normal and lung cancer patients, respectively. Gels **a4** and **b4** represent urine samples prepared using MeOH/$CHCl_3$/water precipitation of pooled normal and lung cancer patients, respectively. Gels **a5** and **b5** represent urine samples prepared using TCA in acetone precipitation of pooled normal and lung cancer patients, respectively.

4. Prepare the running buffer by diluting 50 mL of the 10× running buffer with 450 mL of water in a measuring cylinder. Cover with Para-Film and invert to mix.
5. Once the stacking gel has set, carefully remove the comb and wash the wells with running buffer. Add the running buffer to the upper and lower chambers of the gel unit
6. To each sample is added 1/3 volume of 3× sample loading buffer (with maximum volume approximately 25 μL), and placed in a boiling water bath for 5 min. The samples are centrifuged at 12,000 × *g* for 10 min before loading. Load 25 μL of each sample in a well. Include one well for molecular weight markers.
7. Complete the assembly of the gel unit and connect it to a power supply. The gel can pre-run with 25–30 V until the dye front (blue) reaches to the separating gel, and then run at a constant voltage of 100 V until the dye front reaches the bottom of the gel.
8. Examples of SDS-PAGE separation of urinary proteins prepared by various preparation methods are shown in Fig. 3.

3.10. 2-DE

1. To the samples add: lysis buffer (total 1 ml), TBP (5 mM, 25 μL/mL) or DTT (65 mM, 10 mg/mL), and IPG buffer (0.5%, 5 μL/mL, pH 3–10 nonlinear or pH 4–7). Vortex and let stand for 2–4 h at 4°C. The samples are centrifuged at 12,000 × *g* for 30 min at 20°C, and the supernatants are ready for loading.
2. The first dimensional electrophoresis is performed on the IPGphor isoelectric focusing system (GE Healthcare).
3. The 18 cm-strip holder(s) is selected corresponding to the 18 cm-length IPG strip (pH 3–10 nonlinear or pH 4–7) chosen for the experiment and placed on the IPGphor IEF system.
4. The sample solutions with a final protein concentration of 300 μg in 350 μL are slowly applied into each strip holder and the IPG strips are placed in each strip holder, followed by IPG cover fluid or mineral oil.
5. IPGphor IEF is performed under the following condition: IPG strips are rehydrated passively and/or actively for 12 h at 30 V followed by ramping to 100 V for 1 h, 250 V for 1 h, 500 V for 1 h, 1,000 V for 1 h, 2,000 V for 1 h, and focusing at 6,000 V for up to 50,000 Vh.
6. After IEF, the IPG strips are equilibrated in 3 mL equilibration buffer containing 2% DTT for 15 min, and then subsequently alkylated in 3 mL equilibration buffer containing 2.5% iodoacetamide for 15 min.

7. Prepare a 1.5-mm thick, 15% gel (18×18 cm) by mixing 25 mL of 4× separating buffer, with 50-mL acrylamide/bis solution, 1 mL of 1% SDS, 23.5 mL water, 500 μL ammonium persulfate solution, and 33 μL TEMED. This recipe is used for a fixed percentage gel. Pour the gel, leaving space for strip loading and a stacking gel, and overlay with 2 mL absolute ethanol. After allowing a minimum of 1 h for polymerization, the slab gels are rinsed twice with water.
8. The IPG strips are placed on the top of the 15% polyacrylamide gel and covered with 0.5% agarose.
9. The second-dimensional separation is carried out at 40 mA per gel at 15°C until the bromophenol blue dye front reached the bottom of the gel, approximately 5–6 h.
10. Examples of 2-DE separation of urinary proteins prepared by various preparation methods are shown in Fig. 4.

3.11. Different Staining Methods

3.11.1. Coomassie Staining

1. After electrophoresis, the apparatus is disassembled and the gel is washed off the glass plates with 500 mL of the gel-fixing solution and soaked in that solution for 1 h. At the end of this time, remove the solution by aspiration.
2. The gel is added with 500 mL of the gel-washing solution and continue to fix the proteins in the gel by incubating overnight at room temperature with gentle agitation. At the end of this time, the solution is removed by aspiration.
3. 400 mL of the Coomassie stain is added to the gel and incubated at room temperature for 3–4 h with gentle agitation. The Coomassie stain is removed by aspiration after staining.
4. ~250 mL of the destain solution is added to the gel and allow to destain with gentle agitation. The destain solution should be changed several times, removing it at each change by aspiration. Continue the destaining until the protein bands are seen without background staining of the gel.
5. An example of urinary proteins stained by Coomassie blue is shown in Fig. 5b.

3.11.2. SYPRO® Ruby Protein Gel Stain

1. The gels are placed into a clean polypropylene container with 100 mL of fix solution (1 L for 2-DE gel), and the container is agitated on an orbital shaker for 30 min (3 h for 2-DE gel). Repeat once more with fresh fix solution. After fixing, the gels are washed thrice in ddH_2O for 10 min each, before proceeding to the staining step.
2. The gels are incubated in 100 mL SYPRO® Ruby gel stain (350 mL for 2-DE gel) for at least 3 h or overnight with agitation on an orbital shaker (see Notes 5 and 6).

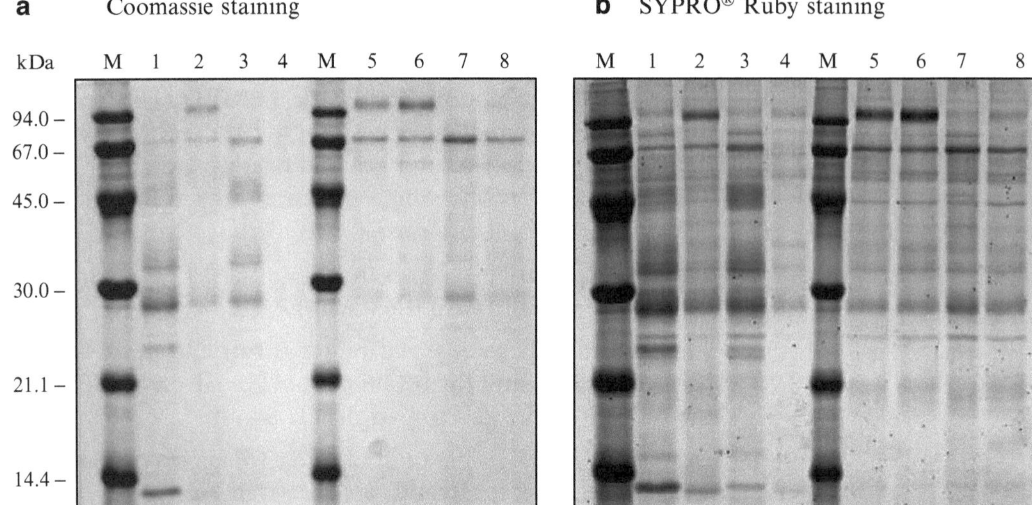

Fig. 5. SDS-PAGE of normal and lung cancer urinary proteins stained by Coomassie blue (**a**) and SYPRO® Ruby gel stain (**b**). Labels: M, protein markers; 1–4, normal urine samples; 5–8, lung cancer urine samples.

3. The gels are transferred to a clean container and washed with 100 mL of wash solution (1 L for 2-DE gel) for 30 min.
4. The gels are rinsed with ddH_2O twice for 5 min and scanned using a Typhoon 9200 image scanner (GE Healthcare) that emit at 450, 473, 488, or 532 nm. It can also be visualized using a UV or blue light source.
5. An example of comparative protein staining methods between Coomassie blue and SYPRO Ruby stain is shown in Fig. 5.5. The SDS-PAGE gel stained by SYPRO® Ruby gel stain gives a better urinary protein profile than the gel stained by Coomassie stain.

3.11.3. Phosphoprotein Staining by Pro-Q® Diamond Phosphoprotein Gel Stain

1. The gels are placed into a clean polypropylene container with 100 mL of fix solution (500 mL for 2-DE gel) and incubated at room temperature with gentle agitation for at least 30 min. Repeat the fixation step once more to ensure that all of the SDS is washed out of the gels. Gels can be left in the fix solution overnight.
2. The gels are incubated in 100 mL of ddH_2O (500 mL for 2-DE gel) with gentle agitation for 15 min.
3. The gels are incubated in a volume of Pro-Q® Diamond phosphoprotein gel stain equivalent to ten times the volume of the gel (e.g., 60 mL for SDS-PAGE mini gel, 350 mL for 2-DE gel), with gentle agitation in the dark for 1–2 h.

b SYPRO® Ruby staining

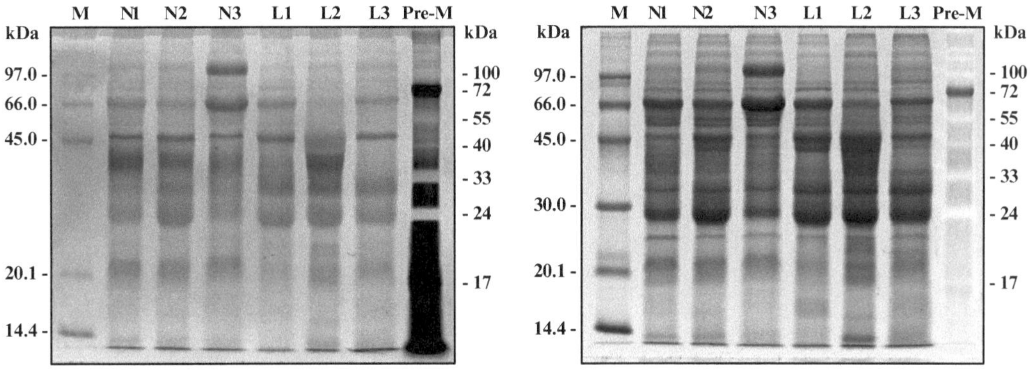

Fig. 6. SDS-PAGE represents urinary phosphoproteins stained by Pro-Q® Diamond phosphoprotein gel stain (**a**) and subsequently to SYPRO® Ruby stain (**b**). Labels: M, protein markers; Pre-M, pre-stained protein markers; N1–N3, normal urine samples; L1–L3, lung cancer urine samples.

4. The gels are incubated in 80–100 mL of destain solution (500 mL for 2-DE gel) with gentle agitation for 30 min at room temperature, protected from light. Repeat this step two more times. The optimal total destaining time is about 1.5 h.
5. The gels are washed twice with 100 mL of ddH$_2$O (500 mL for 2-DE gel) at room temperature for 5 min per wash. If the background is high or irregular, the gel may be left in the second wash for 20–30 min and re-imaged.
6. The gels are scanned using a Typhoon 9200 image scanner (GE Healthcare) with emission maximum at ~580 nm. Stained gels can be visualized using visible-light illumination with wavelengths between 532 and 560 nm (see Note 7).
7. An example of urinary proteins on SDS-PAGE gel stained by Pro-Q® Diamond phosphoprotein gel stain and subsequently by SYPRO® Ruby gel stain is shown in Fig. 6. Different urinary phosphoproteins in normal and lung cancer urine samples can be observed on SDS-PAGE gel and the total protein staining makes it easier to localize a protein to a particular band/spot in the complex protein pattern.

3.11.4. Glycoprotein Staining by Pro-Q® Emerald 488 Glycoprotein Gel Stain

1. The gels are placed into a clean polypropylene container with 100 mL of fix solution (1 L for 2-DE gel) and incubated at room temperature with gentle agitation for at least 1 h. Repeat this step once. For best results with gels, the gels should be incubated overnight the second time to ensure that all of the SDS is washed out of the gel.

2. The gels are incubated in 100 mL of wash solution (1 L for 2-DE gel) with gentle agitation for 10–20 min. Repeat this step once.
3. The gels are incubated in 25 mL of oxidizing solution (500 mL for 2-DE gel) with gentle agitation for 20 min.
4. The gels are washed in 100 mL of wash solution (1 L for 2-DE gel) with gentle agitation for 10–20 min. Repeat this step three times.
5. The gels are incubated in the dark in 25 mL of Pro-Q® Emerald 488 staining solution (250 mL for 2-DE gel) with gentle agitation for 1.5–2 h (2.5 h for 2-DE gel).
6. The gels are incubated in 100 mL of wash solution (1 L for 2-DE gel) at room temperature for 15–30 min. Repeat this wash twice for a total of three washes, using 30–45 min for the second and third washes.
7. The gels are rinsed twice with 100 mL of ddH_2O for 1 min each and scanned using a Typhoon 9200 image scanner (GE Healthcare) with emission maximum at ~520 nm. Stained gels can be visualized using visible-light illumination with wavelengths between 470 and 500 nm (see Note 7).
8. An example of urinary proteins on SDS-PAGE gel stained by Pro-Q® Emerald 488 glycoprotein stain and subsequently by SYPRO® Ruby gel stain is shown in Fig. 7. Different urinary glycoproteins in normal and lung cancer urine samples can be observed on SDS-PAGE gel and the total protein staining makes it easier to localize a protein to a particular band/spot in the complex protein pattern.

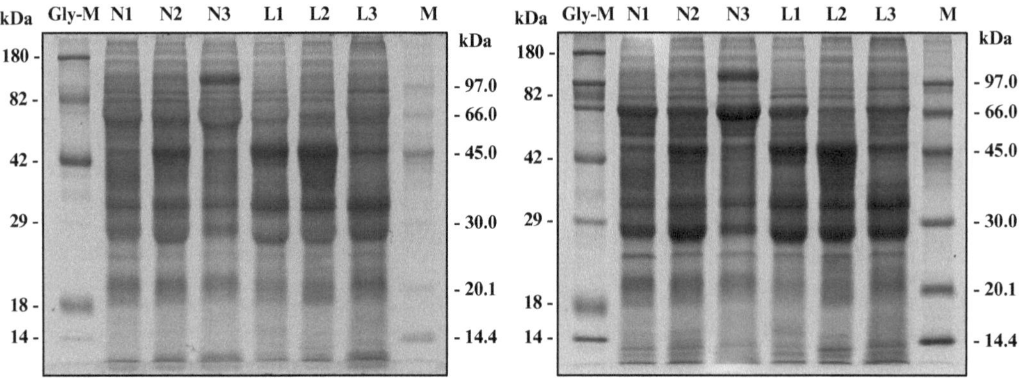

Fig. 7. SDS-PAGE represents urinary glycoproteins stained by Pro-Q® Emerald 488 glycoprotein gel stain (**a**) and subsequently to SYPRO® Ruby stain (**b**). Labels: M, protein markers; Gly-M, CandyCane™ glycoprotein standard markers; N1–N3, normal urine samples; L1–L3, lung cancer urine samples.

3.11.5. Glycoprotein Staining by Fluorescence-Labeled Lectin Staining

1. This fluorescence-labeled lectin stain requires the use of a PVDF membrane.
2. After electrophoresis (12.5% SDS-PAGE gel), the gel is soaked in transfer buffer for 10–20 min.
3. The PVDF membrane is prepared by first soaking in 100% methanol for about 3 s and then rinsed in water for 5 min to remove the methanol (see Note 8). The PDVF membrane and filter papers are then equilibrated with the transfer buffer for 10–20 min.
4. The cassette is assembled for transfer as follows, starting from the negative electrode (Black): five pieces of filter papers (Whatman 3 MM) moistened with transfer buffer, protein gel (remove bubbles between gel and the filter papers), PVDF membrane (remove all bubbles), five pieces of moistened filter paper, respectively. The cassette is inserted into the blotting apparatus, making sure that the membrane is facing the positive electrode and the gel is facing the negative electrode.
5. The protein is transferred over a 2-h period at a constant current of 250 mA in a Nova Blot (semi dry transfer apparatus, GE Healthcare).
6. After transfer, the membrane is washed three times with 100 mL of PBST buffer for 5 min.
7. The membrane is blocked with 100 mL of blocking buffer for 1 h at room temperature or at 4°C overnight and washed three times with PBST.
8. The membrane is then added with an FITC-labeled lectins (ConA, WGA, PNA, and AAL) dissolved at 10 μg/mL in PBS buffer, and then incubated at room temperature for 1 h. This step must avoid the light or the container should be covered with the dark seal.
9. The membrane is washed three times with 100 mL of PBST for 5 min and the wet membrane is scanned using a Typhoon 9200 image scanner (GE Healthcare) with emission maximum at 526 nm. An example of urinary proteins from normal and lung cancer urine samples stained by fluorescence-labeled lectin staining is shown in Fig. 8. The different glycoprotein profiles in normal and lung cancer urine samples stained by FITC-labeled lectins can provide information on carbohydrate content of the glycoproteins in urine samples based upon the lectin specificity; e.g., ConA is specific to mannose, WGA is specific to sialic acid and N-acetylglucosamine, PNA is specific to galactose and N-acetylgalactosamine, and AAL is specific to fucose.

a FITC-labeled ConA staining

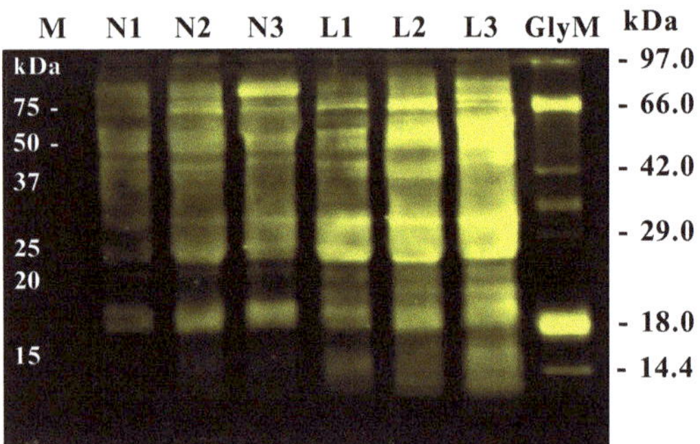

b FITC-labeled WGA staining

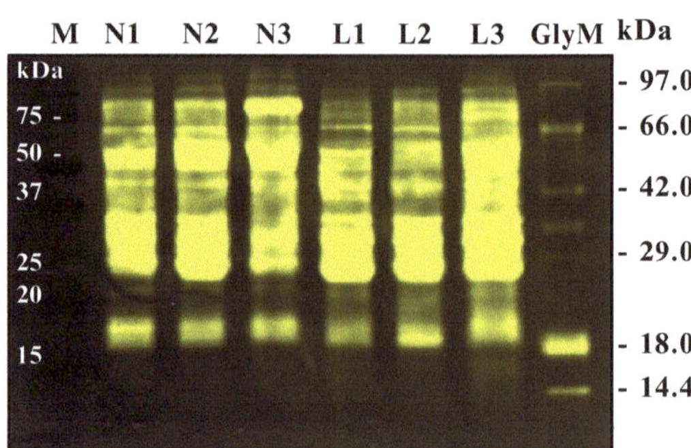

c FITC-labeled PNA staining

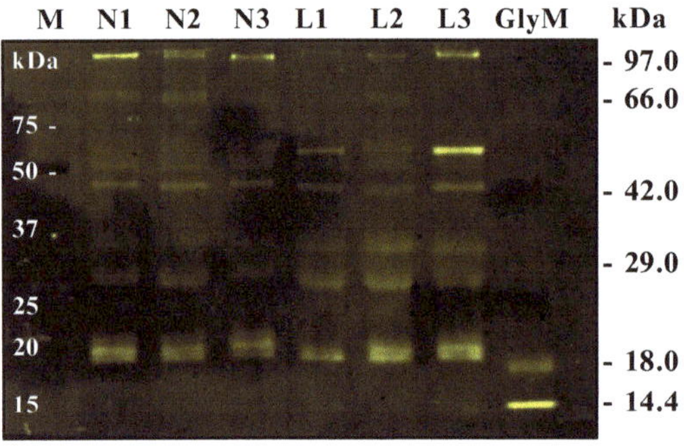

d FITC-labeled AAL staining

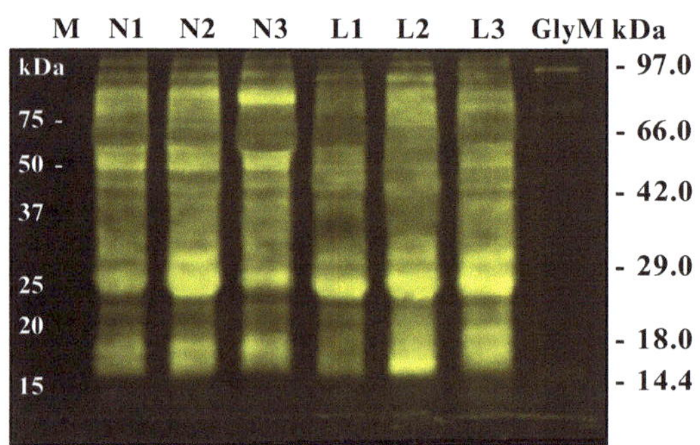

Fig. 8. SDS-PAGE represents urinary glycoproteins stained by fluorescence-labeled lectin staining. Four FITC-labeled lectins used are ConA (**a**), WGA (**b**), PNA (**c**) and AAL (**d**). Labels: M, protein markers; Gly-M, CandyCane™ glycoprotein standard markers; N1–N3, normal urine samples; L1–L3, lung cancer urine samples.

3.12. 2-D DIGE

1. The urine samples are dissolved at 1 mg/mL in degassed PBS buffer and left for 30 min at room temperature.
2. The sample solutions are added with a 100 molar excess of Tris-(2-carboxethyl)phosphine (TCEP; Sigma) (180 μg, 10 μL of a 18 mg/mL TCEP solution in PBS, per 1 mg of protein). The sample solutions in vials are flushed with Nitrogen gas, covered with cap, and mixed thoroughly.
3. The reaction is incubated at room temperature for 10 min. While protein reduction is occuring, prepare a dye solution by adding 50 μL of anhydrous DMF to one pack of dye. The vial is flushed with Nitrogen gas, covered with cap, and mix thoroughly.
4. The dye solution (50 μL) is added to 1 mL of reduced proteins. The normal urinary proteins are labeled with Cy3 (green) and the lung cancer urinary proteins are labeled with Cy5 (red). The vial is flushed with Nitrogen gas, covered with cap, and mix thoroughly. The reaction is incubated at room temperature for 2 h with additional mixing every 30 min, and then left overnight at 2–8°C (see Note 9).

5. After labeling urinary proteins with CyDye, the Cy3-labeled normal urinary protein and Cy5-labeled lung cancer urinary proteins are mixed with an equal volume of samples to get the final protein concentration of 150 μg, and the mixture of Cy3/Cy5-labeled proteins is then analyzed on 2-DE.
6. After 2-DE analysis, the gel is scanned using a Typhoon 9200 image scanner (GE Healthcare) with emission maximum at 580 nm for Cy3 and at 670 nm for Cy5.
7. An example of 2-DIGE of urinary samples between Cy3 labeled normal urine sample and Cy5 labeled lung cancer urine sample is shown in Fig. 9. The different colors of Cy3 (green) and Cy5 (red) dyes represent the different protein expression levels of each urine sample on a single 2-DE gel. In addition, the subsequent staining of 2-D DIGE gel by SYPRO® Ruby gel stain can visualize the whole protein quantity on 2-DE gel and easily detected by visible UV.

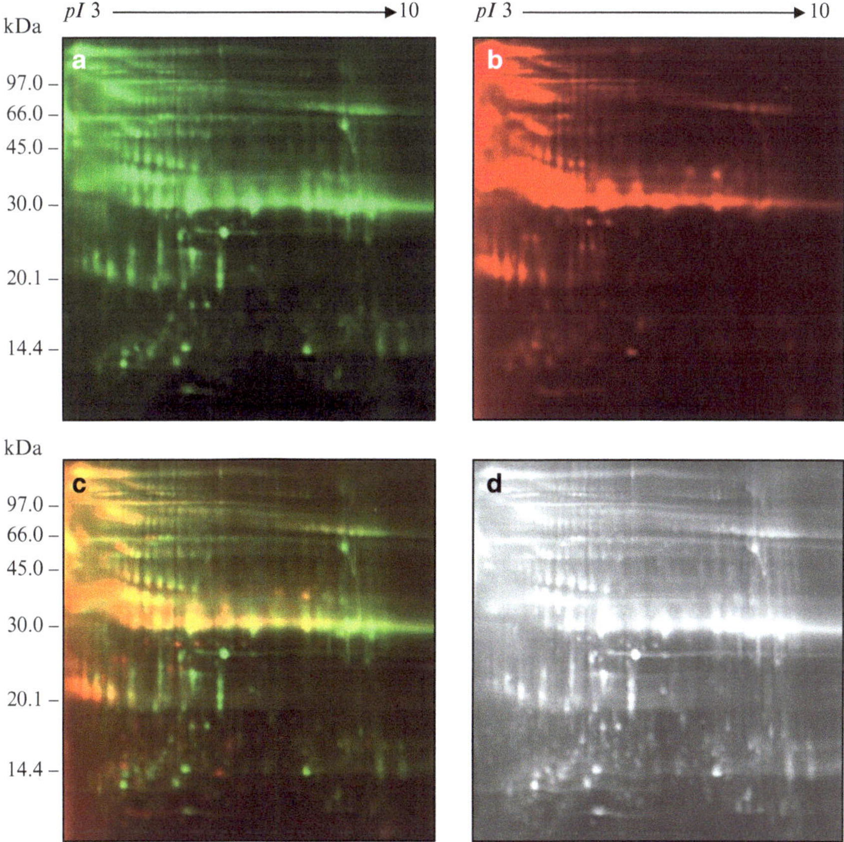

Fig. 9. 2-D DIGE of urinary proteins from normal and lung cancer patients. Labels: **a**, Cy3-normal urinary proteins; **b**, Cy5-lung cancer urinary proteins; **c**, mixture of Cy3 and Cy5; **d**, Cy3/Cy5 DIGE gel subsequently stained with SYPRO® Ruby gel stain.

3.13. 2-DE Gel Image Analysis

1. Examples of image analysis software used for comparative urinary protein patterns include PDQuest™ 2-D software version 7.11 or higher version (BioRad) and ImageMaster™ 2D Platinum software version 5.0 (GE Healthcare).
2. The instruction manual of the appropriate 2-D image analysis software should be consulted for analysis methods. Examples of comparative urinary protein profiles from normal urine sample and lung cancer urine sample by PDQuest™ and ImageMaster™ 2D Platinum software are shown in Figs. 10 and 11, respectively.

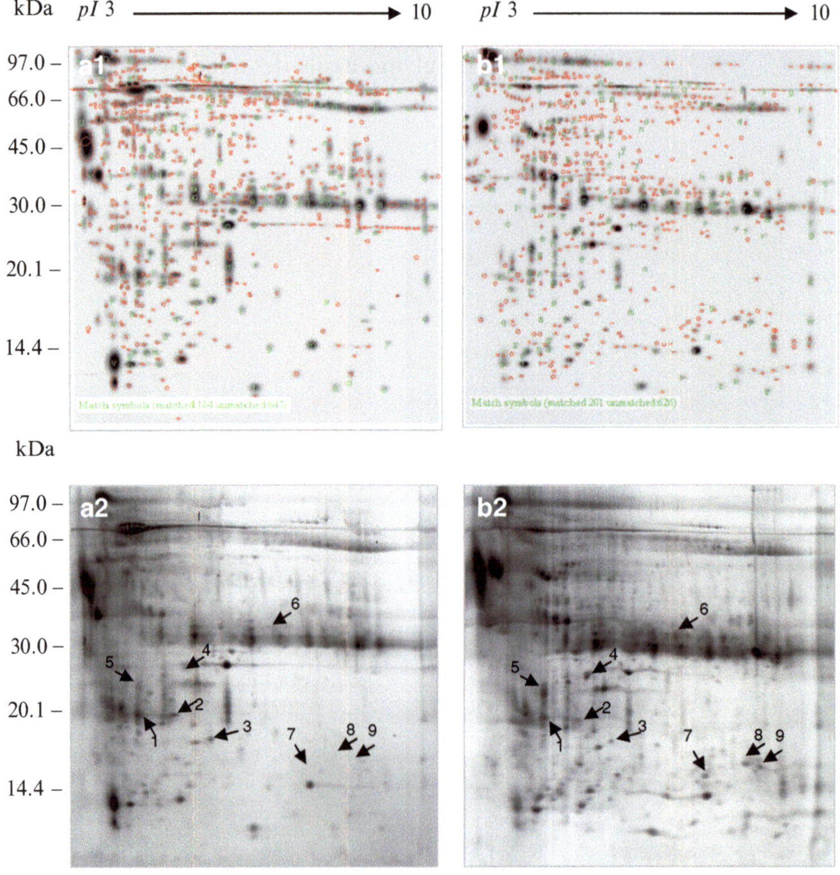

Fig. 10. Comparison of 2-D gel images of pooled normal and lung cancer urine samples prepared by centrifugal filtration, which analyzed by PDQuest™ 2-D analysis software version 7.1.1. Gels: (**a1**) normal urine as reference gel comparing with lung cancer urine; (**b1**) lung cancer urine as reference comparing with normal urine; (**a2**) normal urinary proteins; (**b2**) lung cancer urinary proteins. *Green* labeled spots represent matched spots and *red* labeled spots represent unmatched spots. The *black arrows* show the different proteins that are excised and analyzed by MALDI Q-TOF MS analysis. Nine identified proteins indicating with arrows: 1, CD59 glycoprotein; 2, activator of cAMP responsive element modulator; 3, transthyretin; 4, plasma retinol-binding protein; 5, GM2 activator protein (GM2AP); 6, Ig lambda light chain; 7–9, Ig kappa chain C region (23).

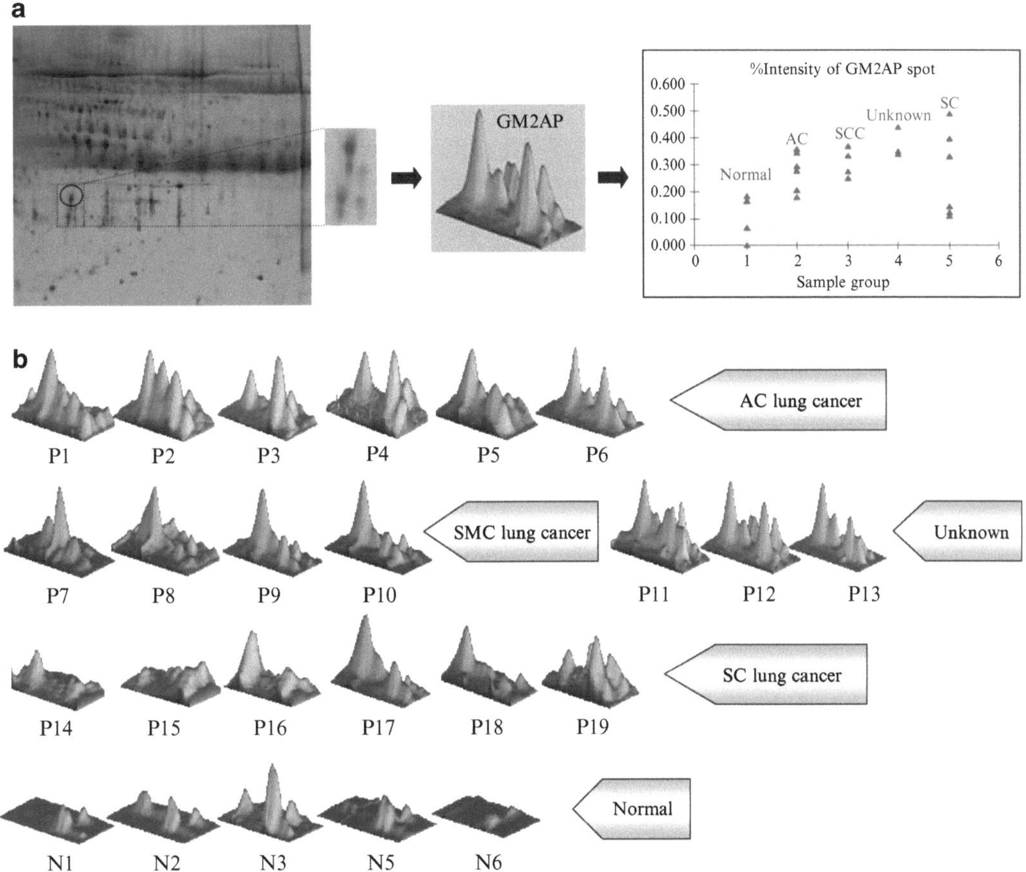

Fig. 11. 3-D image views of GM2AP marker in lung cancer urine samples by ImageMaster™ 2D Platinum software version 5.0. Labels: (**a**) example of 3-D view of GM2AP in lung cancer urine sample and the comparative intensities of GM2AP in various types of lung cancer; (**b**) 3-D views of GM2AP in normal urine sample and lung cancer urine samples (AC-adenocarcinoma, SMC-squamous cell carcinoma, SC-small cell carcinoma, unknown).

4. Notes

1. Polyacrylamide is a neurotoxin when unpolymerized, and so care should be exercised.
2. TEMED is best stored at room temperature in a desiccator. Buy small bottles as its quality may decline (gels will take longer to polymerize) after opening.
3. CyDye™ monofunctional maleimide dyes used in this procedure are designed to label the thiol group on cysteine of proteins.
4. When desalting or purifying small volumes (i.e., 50–250 μL), the sample should be diluted to 2 mL before spinning. Often this will reduce salts or interferences to an acceptable level in

a single spin. For example, if a 100 μL sample is diluted to 2 mL, then concentrated to 25 μL, over 98% of the salt will be removed.

5. SYPRO® Ruby protein gel stain is characterized as an irritant due entirely to the solvent system and buffer salts in the product. It should be handled with care, consistent with good laboratory practices, and using powder-free gloves. The staining must be carried out in a dark container to protect the stain from light.
6. SYPRO® Ruby stain can be used subsequent to staining with other gel stains such as Pro-Q® Diamond phosphoprotein gel stain, or Pro-Q® Emerald 300/488 glycoprotein gel stain. SYPRO® Ruby stain should always be used last because its bright fluorescent signal tends to dominate over signal from other stains. SYPRO® Ruby stain does not work well as a post-stain for colorimetric stains such as Coomassie and silver stains.
7. After staining with Pro-Q® Diamond stain or Pro-Q® Emerald 488 stain, the gel can be stained with a total-protein stain, such as SYPRO® Ruby protein gel stain. The gel should be rinsed in 500 mL ddH_2O two times for 5 min prior to staining with SYPRO® Ruby protein gel stain.
8. The porous PVDF membrane is hydrophobic, and it is difficult to get water or buffer into the pores. Be sure to hydrate or "wet" it properly by soaking the entire membrane in 100% methanol and then rinsing in ddH_2O. If it does become dry, then it must be re-wetted with methanol before it can be used again.
9. The CyDye-labeled proteins can be separated from the excess or unconjugated dyes by dialysis, centrifugal filtration, or PD-10 Desalting column. Dialysis does not give as efficient and rapid a separation as gel filtration. We, therefore, recommend that protein purification by centrifugal filter or PD-10 desalting column be used. In addition, many CyDye-labeled proteins can be stored at 2–8°C for a few days or at –20°C for 1–2 weeks without further manipulation.

Acknowledgments

We gratefully thank Academia Sinica and the National Taiwan Science Council, Taiwan for supporting a grant of lung cancer research. This work is part of the Royal Golden Jubilee Ph.D. project of Ms. Supawadee Sriyam supported by the Thailand Research Fund (TRF), Thailand. We also gratefully acknowledge the provision of urine samples from Lampang Regional Cancer Center (Thailand).

References

1. Hall, C. L., and Hardwicke, J. (1979) Low molecular weight proteinuria. *Ann. Rev. Med.* **30**, 199–211.
2. Lillehoj, E. P., and Poulik, M. D. (1986) Normal and abnormal aspects of proteinuria. Part I: Mechanisms, characteristics and analyses of urinary protein.Part II: Clinical considerations. *Exp. Pathol.* **29**, 1–28.
3. Waller, K. V., Ward, K. M., Mahan, J. D., and Wismatt, D. K. (1989) Current concepts in proteinuria. *Clin. Chem.* **35**, 755–765.
4. Xu, G., Di Stefano, C., Liebich, H. M., Zhang, Y., and Lu, P. (1999) Reversed-phase high-performance liquid chromatographic investigation of urinary normal and modified nucleosides of cancer patients. *J. Chromatogr. B* **732**, 307–313.
5. Yang, J., Xu, G., Kong, H., Zheng, Y., Pang, T., and Yang, Q. (2002) Artificial neural network classification based on high-performance liquid chromatography of urinary and serum nucleosides for the clinical diagnosis of cancer. *J. Chromatogr. B* **780**, 27–33.
6. Liebich, H. M., Xu, G., Di Stefano, C., and Lehmann, R. (1998) Capillary electrophoresis of urinary normal and modified nucleosides of cancer patients. *J. Chromatogr. A* **793**, 341–347.
7. Ma, Y., Liu, G., Du, M., and Stayton, I. (2004) Recent developments in the determination of urinary cancer biomarkers by capillary electrophoresis. *Electrophoresis* **25**, 1473–1484.
8. Marshall, T., and Williams, K. M. (1998) High resolution two-dimensional electrophoresis of human urinary proteins. *Anal. Chim. Acta* **372**, 147–160.
9. Oh, J., Pyo, J. H., Jo, E. H., Hwang, S. I., Kang, S. C., Jung, J. H., Park, E. K., Kim, S. Y., Choi, J. Y., and Lim, J. (2004) Establishment of a near-standard two-dimensional human urine proteomic map. *Proteomics* **4**, 3485–3497.
10. Pieper, R., Gatlin, C. L., McGrath, A. M., Makusky, A. J., Mondal, M., Seonarain, M., Field, E., Schatz, C. R., Estock, M. A., Ahmed, N., Anderson, N. G., and Steiner, S. (2004) Characterization of the human urinary proteome: a method for high-resolution display of urinary proteins on two-dimensional electrophoresis gels with a yield of nearly 1400 distinct protein spots. *Proteomics* **4**, 1159–1174.
11. Spahr, C. S., Davis, M. T., McGinley, M. D., Robinson, J. H., Bures, E. J., Beierle, J., Mort, J., Courchesne, P. L., Chen, K., Wahl, R. C., Yu, W., Luethy, R., and Patterson, S. D. (2001) Towards defining the urinary proteome using liquid chromatography-tandem mass spectrometry. I. Profiling an unfractionated tryptic digest. *Proteomics* **1**, 93–107.
12. Davis, M. T., Spahr, C. S., McGinley, M. D., Robinson, J. H., Bures, E. J., Beierle, J., Mort, J., Yu, W., Luethy, R., and Patterson, S. D. (2001) Towards defining the urinary proteome using liquid chromatography-tandem mass spectrometry. II. Limitations of complex mixture analyses. *Proteomics* **1**, 108–117.
13. Kiernan, U. A., Tubbs, K. A., Nedelkov, D., Niederkofler, E. E., McConnell, E., and Nelson, R. W. (2003) Comparative urine protein phenotyping using mass spectrometric immunoassay. *J. Proteome Res.* **2**, 191–197.
14. Wittke, S., Fliser, D., Haubitz, M., Bartel, S., Krebs, R., Hausadel, F., Hillmann, M., Golovko, I., Koester, P., Haller, H., Kaiser, T., Mischak, H., and Weissinger EM. (2003) Determination of peptides and proteins in human urine with capillary electrophoresis-mass spectrometry, a suitable tool for the establishment of new diagnostic markers. *J. Chromatogr. A* **1013**, 173–181.
15. Theodorescu, D., Fliser, D., Wittke, S., Mischak, H., Krebs, R., Walden, M., Ross, M., Eltze, E., Bettendorf, O., Wulfing, C., and Semjonow, A. (2005) Pilot study of capillary electrophoresis coupled to mass spectrometry as a tool to define potential prostate cancer biomarkers in urine. *Electrophoresis* **26**, 2797–2808.
16. Ullsten, S., Danielsson, R., Backstrom, D., Sjoberg, P., and Bergquist, J. (2006) Urine profiling using capillary electrophoresis-mass spectrometry and multivariate data analysis. *J. Chromatogr. A* **1117**, 87–93.
17. Hampel, D. J., Sansome, C., Sha, M., Brodsky, S., Lawson, W. E., and Goligorsky, M. S. (2001) Toward proteomics in uroscopy: urinary protein profiles after radiocontrast medium administration. *J. Am. Soc. Nephrol.* **12**, 1026–1035.
18. Rogers, M. A., Clarke, P., Noble, J., Munro, N. P., Paul, A., Selby, P. J., and Banks, R. E. (2003) Proteomic profiling of urinary proteins in renal cancer by surface enhanced laser desorption ionization and neural-network analysis: identification of key issues affecting potential clinical utility. *Cancer Res.* **63**, 6971–6983.
19. Adachi, J., Kumar, C., Zhang, Y., Olsen, J. V., and Mann, M. (2006) The human urinary

proteome contains more than 1500 proteins, including a large proportion of membrane proteins. *Genome Biol.* 7, R80.

20. Pang, J. X., Ginanni, N., Dongre, A. R., Hefta, S. A., and Opitek, G. J. (2002) Biomarker discovery in urine by proteomics. *J. Proteome Res.* **1**, 161–169.
21. Schaub, S., Rush, D., Wilkins, J., Gibson, I. W., Weiler, T., Sangster, K., Nicolle, L., Karpinski, M., Jeffery, J., and Nickerson, P. (2004) Proteomic-based detection of urine proteins associated with acute renal allograft rejection. *J. Am. Soc. Nephrol.* **15**, 219–227.
22. Thongboonkerd, V., and Malasit, P. (2005) Renal and urinary proteomics: current applications and challenges. *Proteomics* **5**, 1033–1042.
23. Tantipaiboonwong, P., Sinchaikul, S., Sriyam, S., Phutrakul, S., and Chen, S. T. (2005) Different techniques for urinary protein analysis of normal and lung cancer patients. *Proteomics* **5**, 1140–1149.
24. Lokeshwar, V. B., and Selzer, M. G. (2006) Urinary bladder tumor markers. *Urol. Oncol.* **24**, 528–537.
25. Ye, B., Skates, S., Mok, S. C., Horick, N. K., Rosenberg, H. F., Vitonis, A., Edwards, D., Sluss, P., Han, W. K., Berkowitz, R. S., and Cramer, D. W. (2006) Proteomic-based discovery and characterization of glycosylated eosinophil-derived neurotoxin and COOH-terminal osteopontin fragments for ovarian cancer in urine. *Clin. Cancer Res.* **12**, 432–441.
26. Lafitte, D., Dussol, B., Andersen, S., Vazi, A., Dupuy, P., Jensen, O. N., Berland, Y., and Verdier, J. M. (2002) Optimized preparation of urine samples for two-dimensional electrophoresis and initial application to patient samples. *Clin. Biochem.* **35**, 581–589.
27. Thongboonkerd, V., McLeish, K. R., Arthur, J. M., and Klein, J. B. (2002) Proteomic analysis of normal human urinary proteins isolated by acetone precipitation or ultracentrifugation. *Kidney Int.* **62**, 1461–1469.
28. Thongboonkerd, V., Chutipongtanate, S., and Kanlaya, R. (2006) Systematic evolution of sample preparation methods for gel-based human urinary proteomics: quantity, quality, and variability. *J. Proteome Res.* **5**, 183–191.
29. Hiratsuka, N., Shiba, K., Shinomura, K., and Hosaki, S. (1996) Rapid and highly sensitive colloidal silver staining on cellulose acetate membrane for analysis of urinary proteins. *J. Clin. Lab. Anal.* **10**, 403–406.
30. Sharma, K., Lee, S., Han, S., Lee, S., Francos, B., McCue, P., Wassell, R., Shaw, M. A., and RamachandraRao, S. P. (2005) Two-dimensional fluorescence difference gel electrophoresis analysis of the urine proteome in human diabetic nephropathy. *Proteomics* **5**, 2648–2655.
31. Khan, A., and Packer, N. H. (2006) Simple urinary sample preparation for proteomic analysis. *J. Proteome Res.* **5**, 2824–2838.

Chapter 6

Isolation and Purification of Exosomes in Urine

Patricia A. Gonzales, Hua Zhou, Trairak Pisitkun, Nam Sun Wang, Robert A. Star, Mark A. Knepper, and Peter S.T. Yuen

Abstract

Exosomes represent an important and readily isolated subset of the urinary proteome that has the potential to shed much insight on the health status of the kidney. Each segment of the nephron sheds exosomes into the urine. Exosomes are rich in potential biomarkers, especially membrane proteins such as transporters and receptors that may be up- or downregulated during disease states. Two differential centrifugation methods are available for simple purification of exosomes: one uses ultracentrifugation, and the other uses a nanomembrane concentrator. Validation methods include western blots of pan-exosome markers and segment-specific exosome markers, and negative staining electron microscopy.

Key words: Exosomes, Endocytosis, Biomarker, Proteomics, Uromodulin/Tamm-Horsfall protein, Podocyte, Ultracentrifugation

1. Introduction

Urinary exosomes are protein-containing vesicles (<100 nm) that originate from glomerular podocytes or epithelial cells lining the renal tubules (1). After endocytosis from the plasma membrane, endocytic vesicles, which have an inside-out orientation, can fuse with the outer membrane of a multivesicular body (MVB). The outer membrane of MVB can invaginate, generating internal vesicles that have a right side-out orientation. When the outer membrane of a MVB fuses with the plasma membrane of a podocyte or the apical plasma membrane of a renal epithelial cell, the vesicles are released from the cell into the urinary space as exosomes. Proteins, such as ALIX and TSG101, which are involved in the maturation of MVB, have been found through the proteomic analysis of human urinary exosomes (2). These proteins are used

Alex J. Rai (ed.), *The Urinary Proteome: Methods and Protocols*, Methods in Molecular Biology, vol. 641, DOI 10.1007/978-1-60761-711-2_6, © Springer Science+Business Media, LLC 2010

as markers for identifying exosomes. Under normal conditions exosomes originate from podocytes and epithelial cells, where one can detect the presence of proteins that are either segment specific such as aquaporin 2 (AQP2, collecting duct), sodium-proton exchanger 3 (NHE-3, proximal tubule) or podocalyxin (PODXL) found in podocytes.

The isolation of these vesicles provides a simple enrichment strategy of potential disease biomarkers by reducing the complexity of the urinary proteome with minimal loss of yield or purity. In addition, exosomes are a relatively small proportion of whole urine; isolating exosomes can reduce the amount of high-abundance proteins while enriching low-abundance proteins that may have pathophysiological significance.

The collection, storage, and handling of human urine samples have an impact on the isolation of urinary exosomes. Zhou et al. (3) recommend the use of protease inhibitors to preserve exosomes-associated proteins. When shipping and/or storage of urine is necessary, it is recommended to freeze fresh urine samples at –80°C instead of –20°C. After thawing, samples should be vortexed extensively to increase recovery of the exosomes from frozen urine. These recommendations will facilitate the use of urinary exosomes in a clinical setting (3).

The Tamm-Horsfall protein (uromodulin) is the most abundant soluble protein in normal urine, which can interfere with proteomic discovery methods (1). Its polymerization can entrap urinary exosomes and other membrane elements present in the urine. Tamm-Horsfall protein can be depolymerized with reducing agents and heat (Fig. 1), thereby denaturing the zona pellucida (ZP) domains in the Tamm-Horsfall protein (4). For validation studies of urinary exosomal biomarkers, depletion of Tamm-Horsfall protein is typically not necessary.

We describe two alternative methods for isolation of urinary exosomes, as well as validation by negative staining electron microscopy (Fig. 2) or by western blot.

2. Materials

2.1. Urine Collection Solutions (See Note 1)

2.1.1. Home-Made Cocktail (Volume per 50 ml of Urine, See Fig. 3, Note 2)

1. 1.67 ml 100 mM NaN_3.
2. 2.5 ml AEBSF (2.75 mg/ml in ddH_2O, stable at 1 week at 4°C, 6 months at –20°C).

   Or

2. 2.5 ml PMSF (2.5 mg/ml in isopropanol, stable at –20°C).
3. 50 µl Leupeptin (1 mg/ml in dd-H_2O, stable 1 week 4°C, 6 months –20°C).

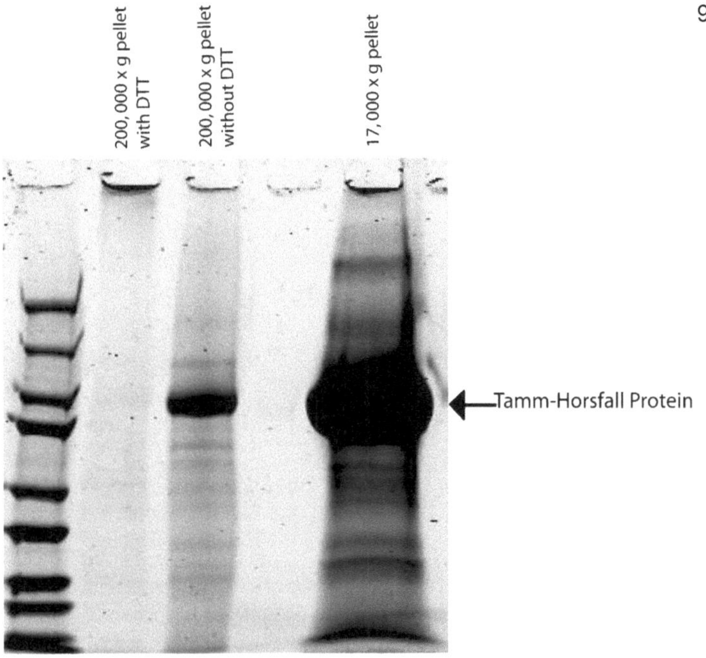

Fig. 1. Depletion of Tamm-Horsfall protein by treating with dithiothreitol (DTT). Fresh urine was processed by differential centrifugation, 50 ml treated with DTT and 50 ml without DTT. *Lane 1*: 200,000 × *g* pellet was obtained from 50 ml of sample and treated with DTT (200 mg/ml), heated at 95°C for 2 min. *Lane 2*: 200,000 × *g* pellet obtained from 50 ml of sample without DTT treatment. *Lane 3*: 17,000 × *g* pellet of 100 ml fresh urine. Gel loading is by equal fraction of volume (5% of total resuspension volume). Tamm-Horsfall protein (92 kDa) shown by *arrow*.

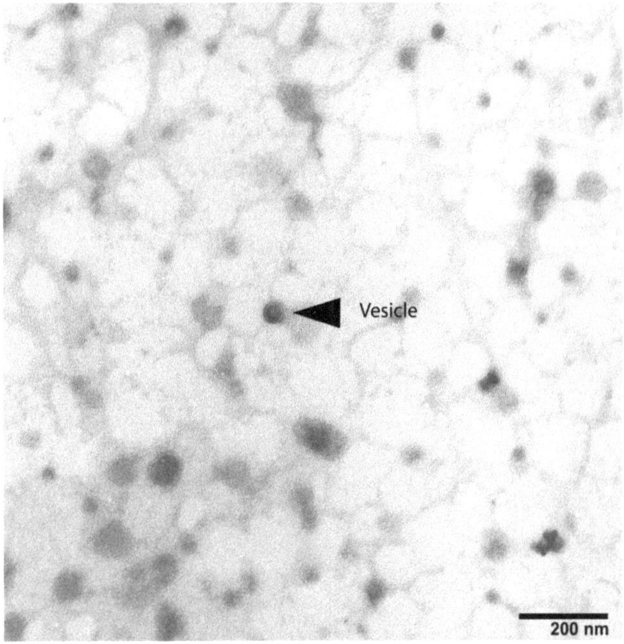

Fig. 2. Electron microscopy of urinary exosomes. Urinary exosomes obtained by differential centrifugation. *Black bar* denotes 200 nm.

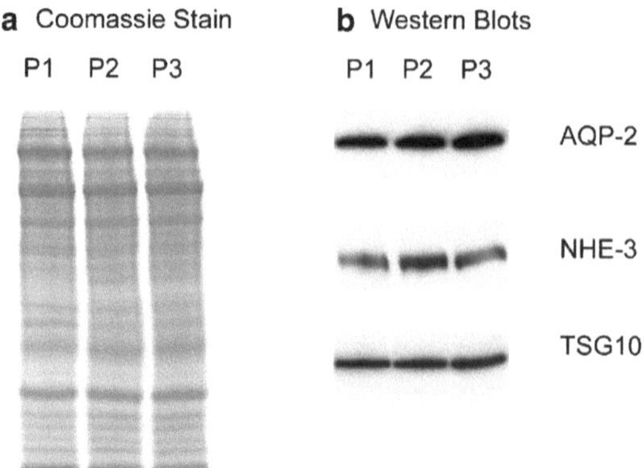

P1- Mixture solution of NaN_3, PMSF and Leupeptin
P2- Sigma Cocktail
P3- Roche Mini-Tablet

Fig. 3. Comparison of different protease inhibitors. (**a**) Coomassie gel of urinary exosomes treated with three protease inhibitor mixtures, P1 (3.25 mM NaN_3, 1 μg/ml leupeptin, 1.5 mM PMSF), P2 (Sigma P2714), and P3 (Roche Complete Mini tablet). (**b**) Western blot of urinary exosomes treated with three protease inhibitor mixtures and probed for AQP-2, NHE-3, and TSG101.

2.1.2. Protease Inhibitor Cocktail (Sigma P2714) (Volume per 50 ml of Urine, See Note 3)

1. Add one bottle of cocktail to 5 ml of dd-H_2O.
2. Add 625 μl to urine sample (12.5 μl/ml of urine).

2.1.3. Complete Mini (Roche 11 853 153 001, See Note 4)

1. Add one tablet per 10 ml of fresh urine.

2.2. Isolation Solution (50 ml)

1. 10 mM Triethanolamine (MW 185.7)	0.093 g
2. 250 mM Sucrose (MW 342.3)	4.28 g
3. dd-H_2O to	45 ml
4. Adjust pH to 7.6 with 1 N NaOH	~220 μl
5. Add dd-H_2O to	50 ml
6. Add protease inhibitors day of isolation	

2.3. 5× SDS-Laemmli (for Western Blot)

1. SDS	3.75 g
2. Glycerol	15 ml
3. 1 M Tris-HCL, pH6.8	2.5 ml
4. Bromophenol Blue	dab
5. Dithiothreitol (DTT)	60 mg/ml
6. dd-H_2O to	50 ml

2.4. Nanomembrane Concentrator, (Sartorius Inc.)

1. Vivaspin 20-PES 100,000 MWCO (VS2041)	20 ml

2.5. Negative Staining for Electron Microscopy

1. 4% Paraformaldehyde in PBS, pH 7.4.
2. 1% Uranyl acetate in dd-H_2O.
3. Parafilm.
4. 200 mesh nickel grid.

2.6. In-Gel Digestion Protocol for Proteomic Analysis

Destaining

1. 25 mM NH_4HCO_3	100 mg NH_4HCO_3, 50 ml dd-H_2O
2. 25 mM NH_4HCO_3 in 50% acetonitrile (ACN)	100 mg NH_4HCO_3, 25 ml ACN, 25 ml dd-H_2O
3. Stock solution: 1% formic acid (FA) in dd-H_2O	990 µl dd-H_2O, 10 µl 100% FA
4. 0.1% FA in dd-H_2O	100 µl Stock solution, 900 µl dd-H_2O
Reduction and Alkylation	
5. 10 mM DTT	1.5 mg DTT per ml 25 mM NH_4HCO_3
6. 55 mM iodoacetamide (maintain in dark)	10 mg iodoacetamide per ml 25 mM NH_4HCO_3
Enzyme digestion	
7. Trypsin (Promega V5113, see Note 12)	12.5 ng trypsin per µl 25 mM NH_4HCO_3
Extraction of peptides	
8. 50% ACN/0.5% FA	500 µl Stock solution, 500 µl ACN
ZipTip® cleanup	
9. Millipore ZipTip® C18 Pipette Tips (ZTC18S096)	
10. Stock solution: 1% FA	
11. Equilibration and washing: 0.1% FA in dd-H_2O	
12. Wetting and elution: 50% ACN/0.1% FA	100 µl Stock solution, 500 µl ACN, 400 µl dd-H_2O

3. Methods

The two methods discussed here use differential centrifugation as a means to isolate urinary exosomes from urine. The first step in both methods is a 17,000 × g centrifugation step that removes whole cells, casts, and debris. The resulting supernatant fraction is subjected to a second centrifugation step (a) by ultracentrifugation at 200,000 × *g* or (b) by filtration with a commercially available nanomembrane concentrator at 3,000 × *g* (5). The ultracentrifugation step is currently the standard method to purify urinary exosomes, but the equipment is not necessarily available, especially in clinical laboratories. Nanomembrane concentrators are commercially available and simplify the isolation procedure, but recovery of exosomal proteins is not uniform, which can pose problem for proteomics discovery experiments, including potentially retaining non-exosomal proteins that are aggregates or high molecular weight complexes (5).

The nanomembrane concentrator contains a polyethersulfonate membrane and is manufactured to have a uniform pore size of 13 mm, as opposed to conventionally manufactured centrifugation-based concentrator membranes. Additionally, the orientation of the membranes within the receptacle minimizes the shearing forces that can introduce artifacts. The weakness of the method is that the recovery of proteins from the concentrator is reproducible for a given protein, but varies from one protein to another. This variation may be due to differential trapping of exosome subpopulations and/or non-specific binding of proteins and exosomes to the membrane (5).

3.1. Purification of Urinary Exosomes

3.1.1. Ultracentrifugation

1. Add protease inhibitors to urine sample.
2. Centrifuge urine sample at 17,000 × *g* for 10 min at 25°C for fresh urine, or at 4°C for previously frozen urine (see Note 5).
3. Transfer 17,000 × *g* supernatant to one or more high speed tubes, and mark each tube so that the expected location of the pellet on the tube can be found after the spin.
4. Ultracentrifuge the supernatant at 200,000 × *g* for 1 h at 25°C, or at 4°C for previously frozen urine (see Notes 6, 7).
5. Discard supernatant and resuspend the first tube's pellet (typically light yellow, but not always visible for lower volumes) with isolation solution. Vortex for 30 s, and transfer the suspension (as much as possible) to the second tube, resuspend pellet and repeat for every tube (see Notes 8, 9, and 10).

3.1.2. Nanomembrane Concentrator

1. Remove glycerol and other preservatives by washing with one volume of PBS, centrifuge at 3,000 × *g* in a swinging bucket rotor at 25°C (see Note 11).
2. Centrifuge the urine samples at 17,000 × *g* for 10 min and at 25°C for fresh urine, and at 4°C for previously frozen urine.
3. Add 20 ml of urine supernatant to nanomembrane concentrator. Centrifuge at 3,000 × *g* at 25°C for 30 min.

4. Add an equal volume of 2× solubilizing buffer to retentate while still in concentrator. Shake at room temperature for 30 min.

3.2. Preparation of Exosomes for Western Blot

1. Prepare 5× SDS-Laemmli with dithiothreitol (DTT) (60 mg/ml).
2. Add 5× SDS-Laemmli-DTT to suspension (1:4), vortex thoroughly.
3. Heat samples on heat block at 60°C for 10 min.
4. Store at –80°C.

3.3. Preparation of Sample for Electron Microscopy-Negative Staining (See Fig. 2)

1. Perform all procedures in a covered box to prevent contamination.
2. Mix sample and 4% paraformaldehyde (in PBS pH 7.4) in 1:1 ratio.
3. Spot 10 μl mixture onto parafilm.
4. Float 200 mesh nickel grid on a droplet of the sample for 10 min.
5. Float grid on PBS droplet (20 μl) for 5 min. Repeat.
6. Float grid on water (dd-H_2O) droplet (20 μl) for 5 min. Repeat.
7. Negative stain by floating grid on droplet (20 μl) of 1% uranyl acetate (in dd-H_2O) for 1 min.
8. Use a torn edge of filter paper to absorb excess fluid from grid.
9. Let grid dry on filter paper (coated side up) in a covered dish 15 min.
10. Ready for EM.

3.4. In-Gel Digestion of Proteins After SDS PAGE and Coomassie Blue Staining

1. Cut each gel slice into small cube pieces (1–2 mm^2) and place into a 1.5-ml autoclaved presiliconized centrifuge tubes (see Notes 12–14).
2. Destain the Coomassie stain from gel:
   (a) Add 100 μl of 25 mM NH_4HCO_3/50% ACN, vortex, spin, and let sample stand for 10 min.
   (b) Using pipette, extract the supernatant and discard (see Note 15).
   (c) Repeat step 2(a)–2(b) twice. At this point, the gel pieces shrink and become white. This visual criterion should be used to determine whether or not additional washes should be performed.
   (d) Vacuum dry the gel pieces to complete dryness (~20 min).
3. Reduction and alkylation:
   (a) Prepare fresh solutions.
   (b) Add 50 μl (or enough volume to cover gel pieces) of 10 mM DTT in 25 mM NH_4HCO_3 to dried gels. Vortex

and spin tubes. Allow reaction to proceed at 56°C for 1 h. Cool to room temperature and briefly spin.

(c) Remove supernatant, immediately add 50 μl of 55 mM iodoacetamide to gel pieces (turn light off). Vortex and spin tubes. Allow reaction to proceed for 45 min in the dark at room temperature.

(d) Remove supernatant. Wash gels with 100 μl of 25 mM NH_4HCO_3, vortex, spin, and let samples stand for 10 min.

(e) Remove supernatant. Dehydrate gels with 100 μl (or enough volume to cover gel pieces) of 25 mM NH_4HCO_3 in 50% ACN, vortex, spin, and let samples stand for 10 min. Repeat once.

(f) Vacuum dry gel pieces to complete dryness (approx. 20 min).

4. Enzyme digestion:

   (a) Add 3× volume of gel piece of enzyme solution to gel pieces. Reconstitute gel pieces on ice or at 4°C for 30 min. Remove the remaining trypsin solution containing excess trypsin. Wash briefly with 50 μl of 25 mM NH_4HCO_3 (just add and remove) to remove the trypsin on the outside of the gel pieces. Add 50 μl of 25 mM NH_4HCO_3 to cover gel pieces (see Note 16).

   (b) Spin briefly and incubate at 37°C overnight (see Note 17).

5. Extraction of peptides.

   (a) Transfer the digest solution (aqueous extraction) into a clean 1.5 ml tube.

   (b) To the gel pieces, add 30 μl (or enough volume to cover gel pieces) of 50% ACN/0.5% FA, vortex and stand for 30 min sonicate for 5 min in a water bath, and spin. Pipette the supernatant and pool to tube from 5.1. Repeat extraction once.

   (c) Vortex the extracted digests, spin, and vacuum dry to reduce volume to about 5–10 μl (~30 min, each sample may not equally dry). Add 0.1% FA 15 μl, vortex, and spin.

   (d) Proceed with ZipTip® (Millipore) cleanup.

6. ZipTip® Protocol for Peptide and Protein Analysis (see Note 18).

   (a) Aliquot washing solution 50 μl to 96-well plate, one well per sample.

   (b) Aliquot elution solution 20 μl to new clean tube, one tube per sample.

   (c) Wetting ZipTip®:

      1. Aspirate 10 μl wetting solution into tip.

      2. Dispense to waste. Repeat.
   (d) Equilibrate ZipTip®:
      1. Aspirate equilibration solution.
      2. Dispense to waste. Repeat.
   (e) Binding:
      1. Aspirate and dispense sample in tube for 10 cycles.
   (f) Washing:
      1. Aspirate washing solution.
      2. Dispense to waste.
      3. Repeat steps 1 and 2 twice.
7. Elution:
   (a) Aspirate and dispense elution solution in new clean tube (from step 6(b)) ten times without introducing air.
8. For LC MS/MS, vacuum dry to reduce volume to 5 μl. Add 15 μl of 0.1% FA vortex, spin, and transfer sample to LC MS/MS tube.

4. Notes

General

1. Updated protocols can be found at the UroProt website: http://intramural.niddk.nih.gov/research/uroprot/.

Protease Inhibitors

2. As an alternative to PMSF, AEBSF offers lower toxicity, improved solubility in water, and improved stability in aqueous solutions.
3. The Sigma P2714 protease inhibitor cocktail should be stored in small aliquots at −20°C.
4. The Roche Complete Mini tablets need to be stored at 4°C. When dissolving the tablet, vortex the sample thoroughly.
5. For fresh urine, maintain the sample at room temperature to avoid Tamm-Horsfall protein precipitation at cold temperatures (4°C or on ice). Tamm-Horsfall protein aggregates can trap exosomes and decrease recovery. For previously frozen samples, yield can be drastically reduced unless thawed samples are vigorously mixed.

Ultracentrifugation

6. The second step removes whole cells, large membrane fragments, and other debris. We use the Sorvall RC2-B refrigerated centrifuge and a SS-34 rotor and tubes.

7. Each step should be kept consistent by keeping the number of tubes used per sample the same.
8. After the ultracentrifugation step: To increase the yield of exosomes, discard supernatant, transfer another 8 ml of 17,000 × *g* supernatant to the same high speed tube, vortex, and then repeat ultracentrifuge at above setting. We use the Beckman L8-70M ultracentrifuge with the 70.1 Ti rotor. The high speed tube we use is the thick-walled, polycarbonate tube #355630 from Beckman.
9. The volume of isolation solution used is 50 μl for 100 ml of urine and 100 μl for 200 ml of urine. To remove Tamm-Horsfall protein (THP) from the exosomes, discard supernatant, to each tube add 50 μl isolation solution, vortex 30 s, and pool all samples in an eppendorf tube. Add DTT to the sample to a final concentration of 200 mg/ml, and incubate at 95°C for 2 min. Transfer the suspension to a high speed tube, fill up isolation solution to 8 ml, and repeat ultracentrifugation. Discard supernatant; resuspend the pellet in 50 μl isolation solution (see Fig. 1).
10. The least number of tubes used in the ultracentrifugation step increases exosome recovery. Otherwise, resuspension is harder when more tubes are used. However, there is a trade-off between using fewer tubes and more spins. For example, one spin with five tubes would be faster, but lower yield. On the other extreme, five consecutive spins with one tube would maximize yield, but would be very time-consuming. Two spins in three tubes would be a reasonable compromise.

Nanomembrane

11. Prewashing with 1× PBS is essential for the preparation of the nanomembrane concentrator. Without a prewash step, there would be a large variability of urinary exosome yield (4).

In-Gel Digestion

12. Wearing non-latex gloves, wipe down ALL areas with 70% ethanol, including the outside of all your tubes, the outside and inside of the Vacuum Dryer, centrifuge, tube racks, bottles, etc. Wipe razor blades with ethanol.
13. 20 μg trypsin (1 vial) in 1.6 ml 25 mM NH_4HCO_3. Dissolve trypsin in ice-cold buffer (to reduce auto-proteolysis). Diluted enzyme solutions should be freshly made or aliquot (50 μl/tube) and stored at –80°C for later uses. Do not freeze-thaw more than once.
14. In step 1, a small gel particle size facilitates the removal of SDS (and coomassie) during the washes, and improves enzyme access to the gel.
15. In step 2(b), beware not to suck up the gel pieces.

16. In step 4(a), this volume will vary from sample to sample, but on average approx. 5–25 µl is sufficient.
17. In step 4(b), the time to digest is anywhere between 4 h and overnight.
18. In steps 6(a)–6(f), the resin bed of the ZipTip® provides back pressure, so set pipette to 10 µl, depress plunger to a dead stop and slowly release or dispense plunger throughout the entire ZipTip® procedure.

References

1. Pisitkun, T., Johnstone, R., and Knepper, M. A. (2006) Discovery of urinary biomarkers. *Mol. Cell. Proteomics* **5** 1760-1771
2. Pisitkun, T., Shen, R. F., and Knepper, M. A. (2004) Identification and proteomic profiling of exosomes in human urine. *Proc. Natl. Acad. Sci. U S A* **101** 13368-13373
3. Zhou, H., Yuen, P. S. T., Pisitkun, T., Gonzales, P. A., Yasuda, H., Dear, J. W., Gross, P., Knepper, M. A., and Star, R. A. (2006) Collection, storage, preservation, and normalization of human urinary exosomes for biomarker discovery. *Kidney Int.* **69** 1471-1476
4. Fernández-Llama, P., Khositseth, S., Gonzales, P. A., Star, R. A., Pisitkun, T., and Knepper, M. A. Tamm-Horsfall protein and urinary exosome isolation. Kidney Int. 2010 Feb 3. [Epub ahead of print] PMID: 20130532 [PubMed - as supplied by publisher]
5. Cheruvanky, A., Zhou, H., Pisitkun, T., Kopp, J. B., Knepper, M. A., Yuen, P. S. T., and Star, R. A. (2007) Rapid isolation of urinary exosomal biomarkers using a nanomembrane ultrafiltration concentrator. *Am. J. Physiol. Renal Physiol.* **292** F1657-F1661

Chapter 7

Bioinformatics of the Urinary Proteome

Laurence D. Parnell and Christine M.E. Schueller

Abstract

Proteomics-based biomarker discovery studies usually entail the isolation of peptide fragments from candidate biomarkers of interest. Detection of such peptides from biological or clinical samples and identification of the corresponding full-length protein and the gene encoding that protein provide the means to gather a wealth of information. This information, termed annotation because it is attached to the gene or protein sequence under study, describes relationships to human disease, cytogenetic map position, protein domains, protein–protein and small molecule interactions, tissues or cell types in which the gene is expressed, as well as several other aspects of gene and protein function. Bioinformatics tools are employed and genome databases are mined to retrieve this information. Coupled with extensive gene and protein annotation, detected peptides are better placed in a biological context with respect to the health status of the subject. Examples of the status include cancers (bladder, kidney), metabolic disorders (diabetes and kidney function), and the nutritional state of the subject.

Key words: Bioinformatics, Genome annotation, Protein annotation, Protein sequence analysis, Polymorphism, Interactome, Database mining

1. Introduction

The objective of this chapter is to give an overview of bioinformatics analyses and genome databases that the proteomics analyst can mine in order to gain a more complete understanding of function and role in disease of the full-length protein containing an identified peptide fragment and of the gene encoding that protein. Such investigations follow the natural course of progression for protein characterization in biomarker discovery experiments after protein identity has been delineated. The sequence of the peptide fragment provides the identification of the full-length protein. In turn, the protein leads to the gene and with the gene identifier one can obtain annotation from a variety of sources.

Alex J. Rai (ed.), *The Urinary Proteome: Methods and Protocols*, Methods in Molecular Biology, vol. 641, DOI 10.1007/978-1-60761-711-2_7, © Springer Science+Business Media, LLC 2010

2. HUGO Gene Symbol and Official Name

The Human Genome Organisation (HUGO, http://www.hugo-international.org) is a global organization with a keen interest in genomes and genome variation. Of its several committees, the HUGO Gene Nomenclature Committee (HGNC) seeks standardization in gene symbols and official full names of genes (see http://www.genenames.org). These symbols and names are those that should be used in scientific writing, because use of a standardized term eliminates confusion as to which gene one is referring and allows for more accurate and more complete database querying and text mining. Additionally, the official gene symbol and name are the proper queries by which one can obtain more information on that gene and its encoded protein(s). The National Center for Biotechnology Information's (NCBI) EntrezGene page is a source for these official terms and serves as a gateway to an abundance of practical information regarding the gene encoding the protein of interest, that protein, and its known and proposed biological and physiological functions. See http://www.ncbi.nlm.nih.gov/sites/entrez?db=gene. Additionally, the official gene symbol is listed under the *HGNC field* in SwissProt entries (http://ca.expasy.org/sprot/).

2.1. Method 1

At times, the proteomics data analysis software yields results that are of a textual nature, meaning a gene name or symbol is given. Depending on when that software incorporated those gene data, the current, official name, or symbol may not be given in the results. Therefore, in order to obtain official HUGO gene symbols and names of genes encoding proteins whose peptide fragments were identified, one can do the following:

1. At http://www.ncbi.nlm.nih.org, with the *Search* window set to *Gene*, enter a term that pertains to the identified peptide, its gene or gene family, *for example*, "*DBCCR1*", "perforin," or "bladder cancer." The server will return a list of links to those EntrezGene entries relevant to the query term. Use the "Limits" tab to narrow the returned results to a specific set of organisms or types of record.
2. Select the entry to be viewed in detail, being mindful of the organism. For example, human gene *DBCCR1* is an alternate name for gene *DBC1*, *deleted in bladder cancer 1*. *DBC1* is the proper symbol for this gene and DBC1 is the correct term to refer to the encoded protein. *DBC1/Dbc1* is also known in other vertebrates: chimpanzee, bovine, dog, mouse, and chicken.

3. The EntrezGene page that is returned will contain information pertaining to that particular gene. The official gene symbol and gene name are listed at the top of this page.

2.2. Method 2

At times, proteomics data analysis software returns results that consist solely of peptide sequence data without any gene identifier. In this instance, one must use that peptide sequence as a query in a sequence similarity search to detect the full-length protein. A convenient tool for this is BLAST, specifically BLASTP for a search with a peptide query against a database of protein sequences. Other useful sequence search algorithms are FASTA and Smith–Waterman. To perform a BLASTP search, do the following:

1. At http://www.ncbi.nlm.nih.gov/BLAST, scan the list of options for searching a sequence database. Generally, the two most useful searches will be either a search of the human genome ("*Human*" link under "*BLAST Assembled Genomes*") or a search of a general protein database ("*protein blast*").
2. In the search window at the top of the next page, enter or paste the sequence of the peptide fragment.
3. Set the other options for the search. If searching the human genome set the *Database* to "RefSeq protein," set the *Program* to "BLASTP: Compare protein sequences," and under *Advanced options* enter the text "-W 2" in order to set the word size to 2. Actually, it is important to perform a BLASTP search with a smaller window size of 2 when the query is a short sequence. All other options can be maintained at the default settings. Alternatively, if performing a "protein blast," set the *Database* to "Nonredundant protein sequences (nr)," set the *Algorithm* to "BLASTP," and under *Algorithm parameters* set *Word size* to 2. Use "composition-based statistics," the default setting under *Compositional adjustments*. Do not check off any box under *Filter*. Keep the *Expect threshold* value set at 10. Do change the "Word Size" to a value of 2. All other options can be maintained at the default settings. With a "protein blast" search, the user has the option of limiting the returned sequence matches to an organism or taxon or interest: examples include *Homo sapiens*, mammals, *Mammalia*, or placental mammals. Under *Choose Search Set*, the appropriate term can be entered into the *Organism* box.
4. Click the "Begin Search" button and await the return of the results.
5. On the results page, you will see displayed a graphical distribution of the BLAST hits along with the query sequence followed by a textual listing of sequences producing significant alignments.

These are then followed by alignments of the query to each of those sequences. It is beyond the scope of this chapter to discuss in detail all aspects of sequence similarity searching. Suffice to state that the important measures of quality of the BLAST search are the score (bit score, a measure of information content of the aligned sequences) and the *E* value (expect value, essentially a measure of the likelihood that the observed alignment could occur by chance given the size and composition of the database that was searched). Thus, higher bit scores and lower *E* values represent alignments of greater statistical significance.

6. Scroll through the results page. Note any entry with an accession number of NP_xxxxxx or NP_xxxxxxxxx that signifies the reference sequence (known as RefSeq) of that particular protein. The reference sequences are derived from genome sequencing and for human these sequence data represent data of the highest quality. One can easily follow hyperlinks to arrive at the gene encoding that protein and its associated annotation.

2.3. Method 3

One can utilize resources of the International Protein Index (IPI) available at http://www.ebi.ac.uk/IPI/IPIhelp.html, which describes the proteomes of higher eukaryotic organisms (1). Many publications report IPI accessions of identified peptides.

3. Human Disease

Information relating to human disease can be found using gene name or symbol as a query term at other NCBI databases such as Online Mendelian Inheritance in Man (OMIM), Online Mendelian Inheritance in Animals (OMIA), and PubMed (abstracts of scientific literature). A search of OMIM, a database of human genes and genetic disorders, with a gene term can return medical and physiological information on genetic disorders related to the gene encoding the detected peptide. Alternatively, a disease term can serve as query. Searching OMIA returns information on genes, inherited disorders and traits in animal species. This can be useful in that there is often information or data on a particular gene or disorder from an animal species that adds insight to the human gene or disorder. Lastly, a search of NCBI's PubMed is the portal to scientific literature pertaining to the gene under consideration.

Method:

1. Go to http://www.ncbi.nlm.nih.gov and set the *Search* window to *OMIM*, *OMIA*, or *PubMed*.

2. Enter the gene symbol (preferred) or name retrieved from EntrezGene and click "Go."
3. Follow the links within the returned results that are of interest.

4. Cytogenetic Map

Querying the appropriate databases, one can learn to which chromosome and which location on that chromosome a gene maps. With the completion of sequencing of the euchromatic regions of the human genome, nearly all human genes can be placed on a cytogenetic map relative to genetic markers and banding patterns that serve to subdivide the chromosomes.

Knowledge of a gene's cytogenetic map position can be important when relating a gene of interest to other data derived by genetic means. That is, the gene encoding the peptide detected in the proteomics experiment can be placed on a genetic map relative to genetic markers. Genetic markers are used to define the location of genetic elements responsible for disease phenotypes or genetic disorders. Many OMIM entries contain data on the cytogenetic map position of the described disorder, a phenotype or a quantitative trait locus (QTL) affecting a phenotype. A QTL is defined as a region of the genome associated with a certain measurable, phenotypic trait. This region typically is described in terms of genetic markers. For example, the trait *essential hypertension* maps to several different regions of the human genome because several different genetic factors are responsible for this phenotype. One such region is the QTL *HYT3*, mapping to 2p25-p24, between markers D2S2278 and D2S168 (2). Human QTLs are often defined rather broadly, encompassing many genes, but at times are more finely mapped to regions covering fewer genes. In the case of *HYT3*, the corresponding segment of the genome encompasses 7.5 cM, or ~124 genes, from *FAM110C* to *ATAD2B* along the short arm of chromosome 2, meaning that within this chromosomal region there is a genetic element at least partially responsible for observed differences in blood pressure.

Additionally, the ability to access and manipulate the cytogenetic map of a relevant region of the human genome allows for the viewing of neighboring genes and other genetic elements in the vicinity of the query gene. This can also be important in locating the region of synteny between, for example, the genome of the mouse where a particular QTL may be more finely mapped and the human genome. Synteny is simply defined as conserved gene order along two related genomes, for example, human and mouse.

Genome-wide association studies have the ability to interrogate from 100,000 to 1,000,000 genetic variants across the human genome in order to associate a given variant with a (disease) phenotype. One such study suggests that a genetic locus mapping to chromosome 15q26 associates with hypertension in a population from the UK (3). It would be noteworthy if peptide fragments encoded by the genes (such as *LOC100132798* and *LOC100130976*) near the variants which associate with hypertension were identified by proteomics. Such data can be queried at http://www.genome.gov/26525384.

Method:

1. Go to the EntrezGene page at NCBI for the gene encoding the detected peptide.
2. Under the heading *Genomic context*, click on "See *GENE* in MapViewer" in order to display a genetic and physical map of the gene.
3. The MapViewer images are highly customizable; display options are accessed by the button in the left panel – "Maps & Options." At the top of the "MapViewer options window" that opens are listed the species, chromosome, and region displayed. The displayed region is usually defined by base pair positions on the chromosome. To zoom in or out, change the values in the "Region Shown" windows. Click the "Apply" button.
4. One can also add and remove various vertical tracks in the display. Select "Ideogram," for example, and that track is added in parallel to the list of displayed maps. An ideogram is a schematic drawing of the chromosome bands. Many other options are possible and help pages are also available.

5. Alternative mRNA Splicing

A single human gene can give rise to one, two, or several different transcription products. When more than one messenger RNA or mRNA is encoded by a gene, that gene's primary transcript is said to undergo alternative splicing, yielding different mRNA isoforms. This process, which adds a layer of control to gene expression as well as diversity in the absence of gene duplication, can be specific to a certain tissue or developmental stage. On average, a human gene is spliced into 3.7 different transcripts, which in turn increase the diversity of expressed proteins. It has recently been proposed that as many as 73% of genes exhibit tissue-specific alternative splicing (4).

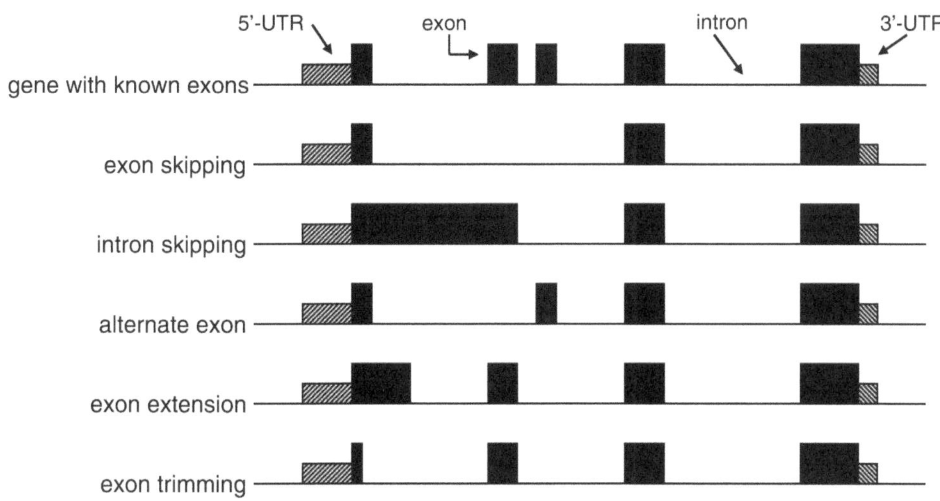

Fig. 1. Schematic showing different types of alternative mRNA splicing.

One potential consequence of alternate splicing is different mRNA isoforms produce different translated protein sequences. This can occur under several different scenarios: exon skipping, intron skipping, use of an alternate exon, exon extension, or exon trimming (Fig. 1). The effect of any of these changes to the mRNA structure and resulting protein is different peptides may result from the proteomics analysis. Alternate mRNA isoforms correspond to different protein sequences, each yielding differences in the pattern of digested peptides. Essentially, although many exons are shared between two mRNA isoforms, one or more exons are unique to one isoform. Thus, the encoded protein sequence is different in some places and so the peptide fragments from these regions are also different. An exception to this is where alternate splicing is confined solely to UTRs (untranslated regions of an mRNA). Additionally, alternate mRNA splicing can result in differences in sites of posttranslational modification to the mature protein with the result that a given peptide exhibits a shifted migration through the separation medium, for example, slower migration through an electrophoretic gel. In summary, alternate mRNA isoforms may yield different proteins and those proteins can undergo different posttranslational modifications, altering peptide mobility and retention times.

Of several proteins expressed differentially in murine kidney inner medullary collecting duct 3 (IMCD3) cells grown under isotonic or chronic hypertonic conditions, one was Mllt4 (myeloid/lymphoid or mixed lineage-leukemia translocation to 4 homolog (Drosophila)), also known as afadin or AF6 (5). In humans, the *MLLT4* gene is transcribed into at least three different transcripts, each encoding unique protein isoforms. For example, isoforms 1 and 2 retain exon 9 while isoform 3 skips this exon.

Table 1
Differences in peptides arising from alternative mRNA splicing of the human *MLLT4* gene

Peptide seq.	Residue(s)	Isoform	Exon source
ADGSGYGSTLPPEK	369–382	1	8
	370–383	2	8
	369–382	3	8
LPYLVELSPGR	383–393	1	8,9
	384–394	2	8,9
LPYLVELSPDGSDSR	383–397	3	8,10
R	394	1	9
	395	2	9
NHFAYYNYHTYEDGSDSR	395–412	1	9,10
	396–413	2	9,10

A simulated trypsin digest of the portion of human MLLT4 protein encoded by exons 8, 9, and 10 reveals differences that can be used as a diagnostic of which isoforms are present in the sample (see Table 1). For example, the occurrence of peptide LPYLVELSPDGSDSR is indicative of isoform 3 while peptide NHFAYYNYHTYEDGSDSR could only arise from isoforms 1 and 2.

It is fundamental to draw a connection from a peptide detected in the proteomics experiment to the full-length protein to the gene encoding such. At the same time, it is also important to specify that the detected peptide is encoded by one, some, or all of the mRNA isoforms originating from the gene. For example, the *CFH* gene, *complement factor H*, encodes at least two alternatively spliced mRNAs. CFH protein has been detected in the urine of type 2 diabetics with macroalbuminuria (6). The accession numbers for the mRNA isoforms are NM_000186.3 (isoform alpha, 19 exons) and NM_001014975.2 (isoform beta, 9 exons), encoding proteins NP_000177.2 and NP_001014975.1, respectively. Exons 9–19 of isoform alpha are not found in isoform beta. Conversely, exon 9 of isoform beta sits in intron 8 of isoform alpha. Therein are encoded the diagnostic peptides that allow one to determine which, or if both, isoforms are present in a sample.

Method:

1. To simply learn if a gene encodes alternatively spliced mRNA products, visit the EntrezGene page and scroll down to the

"Genomic regions, transcripts, and products" section. The Ensembl database (see below) is also an excellent source of gene annotation and different mRNA isoforms.

2. The line drawing(s) with bars representing exons (red for protein-coding, blue for UTRs) each represent different mRNA isoforms. These line drawings are flanked by an NM_xxxxxx or NM_xxxxxxxxx mRNA accession number on the left and the corresponding NP_xxxxxx or NP_xxxxxxxxx protein accession number on the right.
3. To align two sequences to see where the differences and similarities reside – use a BLAST tool called Bl2seq. The url for Bl2seq is http://www.ncbi.nlm.nih.gov/blast/bl2seq/wblast2.cgi. One can align two mRNA or two protein sequences with this tool.
4. Some genes are transcribed into three or more mRNA isoforms. In order to compare three or more mRNAs or proteins in a multiple sequence alignment, use the tool CLUSTALW (http://www.ebi.ac.uk/clustalw/) or MUSCLE (http://www.drive5.com/muscle/).
5. Alternatively, one can query the Ensembl database (see below) to find gene entries with multiple mRNA isoforms.

6. Other Sources of Annotation

There are several good databases of the human genome and numerous sources of very specific information pertinent to a particular gene family or characteristic of a protein of interest. Space does not permit a comprehensive review of all sources for gene and protein annotation. The following table is a short list of some useful resources.

6.1. Multiple Species Genome Annotation

1. UCSC – University of California at Santa Cruz genome browser – http://genome.ucsc.edu.
2. Ensembl – European Bioinformatics Institute and the Wellcome Trust Sanger Institute – http://www.ensembl.org.
3. KEGG – Kyoto Encyclopedia of Genes and Genomes, pathways – http://www.genome.jp/kegg/.
4. Reactome – Pathways of key biological processes – http://www.reactome.org/.

6.2. Human Genome Annotation

1. GeneCards® http://www.genecards.org.
2. CGAP – Cancer Genome Anatomy Project http://cgap.nci.nih.gov/.

6.3. Biomarker Database

1. Osteoarthritis Initiative http://www.oai.ucsf.edu./datarelease/.

6.4. Gene Expression Data

1. SymAtlas – http://symatlas.gnf.org/SymAtlas/.
2. GEO – Gene Expression Omnibus http://www.ncbi.nlm.nih.gov/geo/.

6.5. Specialized Databases

1. Druggable genome – likely targets for drugs and drug design – http://users.ox.ac.uk/~magd1983/Supplementary%20materials.html.

7. Protein–Protein, Protein–DNA, and Protein–Small Molecule Interactions

Proteins do not exist nor function in a vacuum, but instead form associations and interactions with a range of molecules from water and ions to drugs and food components to macromolecules like protein, RNA, and DNA. Yet another way to annotate the detected peptide fragment and its full-length protein in order to understand its function is to examine its interactome. An interactome is a network of interacting biological entities with proteins (or mRNA or DNA) at the nodes and interactions depicted as edges connecting those nodes.

Chronic allograft nephropathy (CAN) is a primary cause of kidney graft failure following transplantation. Levels of connective tissue growth factor (CTGF) in urine of transplant recipients have been shown to correlate with the histological presence of CAN (7). While urinary levels of CTGF have been proposed as a biomarker of CAN, it is also helpful to consider the network of proteins that interact with CTGF (Fig. 2). This CTGF-centric interactome contains eight proteins and indicates that CTGF and transforming growth factor, beta 1 (TGFB1) form an interaction (8). Furthermore, urinary TGFB1 levels are also important markers of various renal diseases, including glomerulonephritis and diabetic nephropathy (9). Thus, the interactome indicates a functional

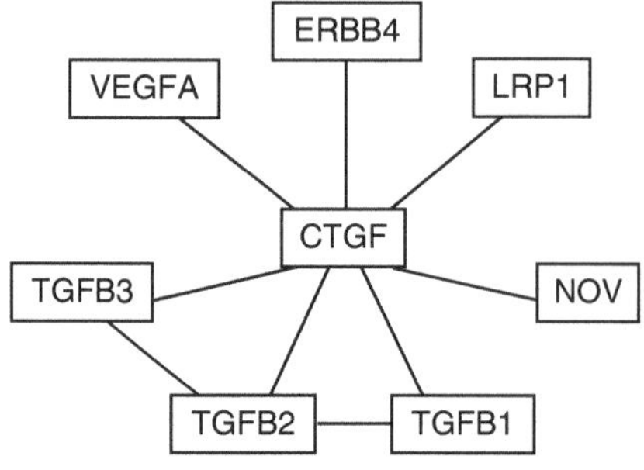

Fig. 2. Protein–protein interaction network for CTGF, connective tissue growth factor.

relationship between CTGF and TGFB1, a relationship that is also extended to kidney dysfunction and urinary proteomics.

Method:

1. Protein–protein interaction data, as well as some data on protein–DNA and protein–small molecule interactions, are available at the EntrezGene page at NCBI. At the EntrezGene page for a gene of interest, scroll down to the heading *Interactions* and view the table of interacting proteins under the "Interactant" column.
2. Under the "Product" column are listed those elements derived from the gene whose EntrezGene page is displayed. Typically these elements are the encoded protein (often beginning with an accession number NP_xxxxxx indicating the reference or standard protein sequence), but which may also include a genomic DNA segment or an mRNA, thereby indicating a protein–DNA or protein–mRNA interaction.
3. Alternatively, these data are also available at other Web-accessible databases. BioGRID is described as a general repository for interaction datasets and is available at http://www.thebiogrid.org/. These data also appear at NCBI but it is unclear how often NCBI mirrors the BioGRID data.
4. The Database of Interacting Proteins (DIP, http://dip.doe-mbi.ucla.edu/) is another good source for experimentally determined protein–protein interaction data.
5. The Small Molecule Interaction Database is a subscription-based small molecule interaction database where a query can be either a protein or a small molecule. This type of query is increasingly being made available at other sites such as PubChem (http://pubchem.ncbi.nlm.nih.gov/) and ChemBank (http://chembank.broad.harvard.edu/).

8. Gene Expression

Knowing the full-length sequence of the mRNA or encoded protein in addition to mining the appropriate databases can yield gene expression data in the absence of performing a Northern blot. Although such data are probable and not absolutely definitive, this approach can reveal that a gene is very likely expressed in certain tissues at certain developmental times. For example, if one learns that libraries of expressed genes (cDNA or EST library; cDNA is a DNA copy of an mRNA, EST indicates "expressed sequence tag," both originate from mRNA) constructed from liver and pancreatic β-islet cells yields some clones whose sequence represents a gene of interest, it is quite reasonable to state with confidence that the gene is expressed in liver and β-islet cells of the pancreas. This information can also distinguish between fetal and adult expression. Thus, the objective is to take sequences of the detected peptides or the corresponding full-length protein and use them to determine in which cell types or tissues the analogous gene is expressed, whether it is highly, moderately, or lowly expressed, and when in a developmental timeframe. Two caveats about EST libraries that should be noted are (1) size information for full-length transcripts is generally not known and (2) only partial sequence information is available. Hence, the presence of a particular sequence in the database does not guarantee the identity of the corresponding full-length transcript, and additional experimentation such as RNAseq may be required.

Method:

1. First, the gene must be properly identified in order to query the databases and to use the tools described below. This is described in Subheading 2.
2. Databases such as Gene Cards, NCBI's GEO and the Novartis® expression dataset can be queried in order to learn in which tissue(s) a gene is expressed. The Gene Cards site (http://www.genecards.org/) gives some basic information regarding tissue expression of genes. A more sophisticated approach is to query an expression database such as NCBI's GEO (Gene Expression Omnibus, http://www.ncbi.nlm.nih.gov/geo/) or the Novartis® gene expression atlas (http://symatlas.gnf.org/SymAtlas/). Such a query can describe a gene whose expression in a tissue distant from the kidney and bladder may indicate an adverse health condition, for example, hemoglobin is expressed in blood cells while its presence in urine is a cause for concern. There is also some very notable and useful information available at CGAP (Cancer Genome Anatomy Project, http://cgap.nci.nih.gov/) in terms of expression in some 50 tumor cell lines. Cornell's Institute for Computational BioMedicine also has some interesting databases and Web-based analysis programs (http://icb.med.cornell.edu/).
3. An alternative method is to use the sequence similarity search tool BLAST to identify human ESTs corresponding to the gene encoding the identified peptide. Launching BLAST requires a query sequence and the preferred query for this analysis is either a protein or an mRNA sequence without the polyA-tail. Inclusion of the poly(A)n in the query overwhelms the BLAST process; avoid this situation. There are hyperlinks on the EntrezGene page for the reference sequences (RefSeq) of the protein and mRNA encoded by the gene of interest. A RefSeq mRNA has an accession number in the form NM_

xxxxxx or NM_xxxxxxxxx and a RefSeq protein accession number has the format NP_xxxxxx or NP_xxxxxxxxx.

4. At the BLAST page at NCBI, the most useful search will be to compare to the human genome ("*Human*" link under "*BLAST Assembled Genomes*"). This will then deliver the BLAST search window.
5. In the search window, enter or paste in either a single query sequence (protein or mRNA) or the accession number of that sequence. Be careful – when BLAST retrieves an mRNA accession to use as a query, that sequence often contains a poly(A) segment.
6. Set the other options for the search. When comparing to the human genome with a nucleotide query, set the *Database* to "ESTs," and set the *Program* to either "BLASTN: Compare nucleotide sequences" or "megaBLAST: Compare highly related nucleotide sequences." When searching the human genome with a protein query, set the *Database* to "ESTs," and set the *Program* to "TBLASTN: Compare a protein sequence against a nucleotide database."
7. Click the "Begin Search" button and await return of the results.
8. As of 29 January 2010, there were 64,727,557 different human EST entries deposited in NCBI and so a similarity search may return numerous hits. When the search is complete, scroll through the results and note perfect or near perfect matches to the query. These are the relevant hits. Click on the accession number of the EST shown in the BLAST alignment to be taken to the NCBI entry for that EST, where more information about that EST is available.
9. The source library for an EST sequence indicates from which tissue the mRNAs that were used to build that EST library were derived. A gene is expressed in a cell line or tissue and mRNA from those cells is made into an EST library and, therefore, detecting that EST by similarity searching indicates with high likelihood that gene is expressed in that cell line or tissue. For example, a reading of the NCBI entry for EST accession AV660423, which originated from the *GCKR* gene, indicates expression in adult liver. Pooled libraries represent a mixture of mRNA from different cell types or tissues and so are not useful in determining where and when a gene is likely expressed.
10. It is best not to try to ascertain whether a gene represented by a set of EST clones identified from a similarity search such as BLAST is expressed to a certain degree, but rather to simply indicate that a gene is expressed in the cell line or tissue from which the EST library was made. Expression levels of a gene of interest are best deduced with data from the GEO or Symatlas sites.

9. Phenotypes of Gene Knockout and Overexpression

The ability to manipulate the genomes of model or laboratory organisms offers the possibility of studying the effects of complete or partial removal or overexpression of a targeted gene. With respect to a better understanding of human disease and the roles played by certain genes, rodent examples of targeted gene knockout and overexpression are invaluable. Although these types of experiments can be conducted in human cells in culture, that is often an unsatisfactory substitute for performing the same research in a living organism with functioning organs and physiological processes and events that can be measured and monitored. This is especially so when done in conjunction with variations in a second gene, for example, a double knockout. In addition, it is now a rather standard practice to produce mice where the targeted knockout or overexpression is induced only under certain conditions, be they developmental, physiological, or anatomical. Hence, these mice and rats, whose similarities to humans make them useful models and which carry a targeted gene knockout or overexpression system, can provide valuable insight into the function of the homologous gene in human and its encoded protein(s).

Method:

1. At http://www.ncbi.nlm.org, with the *Search* window set to *HomoloGene*, enter the symbol of the human gene encoding the detected peptide fragment. This will return a set of homologous, or highly similar, genes. Distinctions are made, when possible, between paralogs and orthologs. Paralogs are highly similar genes that arose by duplication within a single genome and typically have evolved new functions. Orthologs, on the other hand, are genes in different organisms that arose from a single common ancestral gene via speciation. Frequently, orthologs in different species have retained an identical function and hence, their identification is important in translating data from a model organism to human.
2. Follow the link corresponding to the gene entry from an organism of interest, typically mouse.
3. Using the HomoloGene database at NCBI is a quick and easy way to view the entry for the same gene in mouse or rat. In order to find phenotype data for gene knockouts or overexpression in mouse, find the *MGI* (Mouse Genome Informatics, http://www.informatics.jax.org) hyperlink under the *Links* column at the far right of the EntrezGene entry. This will link directly to the MGI page for this gene. The equivalent link for rat genes is that leading to the *RGD* (Rat Genome Database, http://rgd.mcw.edu).

4. For a gene entry at MGI, the row in the "Gene Detail" table entitled "Phenotypes" provides links to data on knockouts and overexpression. Similarly, data at the RGD can be queried for phenotype data of rat genes. Overall, these two genome databases are quite extensive, offering much more information on genes and gene annotation, all of which can be helpful in ascertaining the function of a detected peptide.
5. Phenotypes assigned to human genes can be obtained from the EntrezGene page at NCBI. At the EntrezGene page for a gene of interest, scroll to the heading *General gene information* and view the list of assigned phenotypes under the "Phenotypes" subheading.

10. Protein Domain Searching

Identification of the protein encoding the detected peptides offers the possibility to search for conserved motifs. Because peptide sequence dictates structure and structure dictates function, a search for conserved motifs or protein domains is useful in gaining an approximation of the secondary and tertiary structure of the protein.

Important protein domain-searching tools are Pfam, available at http://pfam.janelia.org/, where one can search over 9,300 protein family signatures; PROSITE, at http://www.expasy.ch/prosite/, generally used to locate shorter sequence patterns; and ELM, at http://elm.eu.org/links.html, which offers a suite of analysis tools.

11. Protein Structure

The string of amino acids comprising a protein sequence is known as the primary structure. That sequence determines the three-dimensional structure of a protein and structure dictates function. Secondary structure is defined as typical, basic conformations adopted by a polypeptide chain. These include alpha helix, beta sheet, and random coil. Tertiary structure is the overall three-dimensional shape of the entire protein. Prediction of protein secondary and tertiary structure has advanced notably in recent years. Useful to a number of bioinformatics endeavors, structure prediction is most practical to the proteomics analyst for the identification of regions of the protein that exist on the surface. Because amino acid residues oriented toward the surface are more likely to be recognized by antibodies, one can exploit this information to generate the appropriate peptide and raise antibodies against it for use as a molecular probe.

11.1. Method (to Query for a Known Structure)

1. Databases at the Protein Data Bank (PDB, http://www.pdb.org), NCBI (http://www.ncbi.nlm.nih.gov/sites/entrez?db=Structure) and SwissProt either contain outright or have links to data for solved structures of proteins. These solved structures often are complexed with a ligand that is relevant to the activity of that protein.
2. Although some structures can be found at NCBI and the PDB site supports similarity searching in order to identify a structure, one easy way to access a solved structure is to follow links within a SwissProt protein entry. Under the section entitled *Cross-references*, look for links under the *3D structure databases* subsection for links to solved structures for that protein or for similar proteins. This approach can provide a direct link to PDB without the need to query PDB.

11.2. Method (to Predict the Structure)

1. An excellent, easy-to-use tool to predict protein secondary structure is JPRED3 (http://www.compbio.dundee.ac.uk/~www-jpred/). A protein sequence serves as query. In addition to predicting the likelihood that an amino acid residue comprises helix, sheet, or coil, JPRED also predicts solvent accessibility and burial of residues. Residues accessible to the solvent are located on the surface of the protein molecule while buried residues are located within the bulk of the three-dimensional structure.

12. Antigenicity Index

For the fully folded protein, the amino acids on the surface are the ones that interact with the environment. The protein surface may also come in contact with immune system molecules like immunoglobulins. A determination of the antigenicity index of a polypeptide chain combines measures of surface probability, flexibility in the polypeptide chain, and hydrophilicity. The less flexible and hydrophobic regions of a peptide tend to be buried within the protein structure. Generally, the few algorithms available require downloading the program. One online tool is available at http://bioinformatics.org/JaMBW/3/1/7/index.html.

13. Subcellular Localization

It should be no surprise that a determination of which proteins are present in a given assayable bodily fluid, such as urine, tears, sweat, or saliva, would turn up a plethora of secreted proteins. Alternatively, detection of proteins described as nuclear, cytoplasmic, or mitochondrial, for example, could be indicative of cellular damage and apoptotic events. At the same time, localization information can be used to assist in determination of protein structure. Cysteine residues, for example, can form a covalent bond with one another, thus bringing together two different sections of the polypeptide chain. However, cysteine–cysteine bond formation typically occurs only in the oxidizing environment outside the cell; the reducing environment of the cytosol destabilizes such bonds.

Method:

1. As before, the gene must be properly identified in order to use the tools described below. This is described in Subheading 2.
2. Although there are numerous tools for assigning a putative subcellular locale for a detected protein, two Web sites of collected tools are worth noting: the proteomics tools at the ExPASy Web site of SwissProt (http://ca.expasy.org/tools/#ptm) and the prediction servers at the Center of Biological Sequence Analysis (CBS) from the Technical University of Denmark (http://www.cbs.dtu.dk/services/).
3. Select an appropriate tool, for example, *TargetP* from the CBS server (http://www.cbs.dtu.dk/services/TargetP/) and enter the full-length protein sequence corresponding to the detected peptide. TargetP assigns a predicted subcellular location for eukaryotic proteins.
4. Submit and await the return of the results. Help pages are extensive and will guide the user in the interpretation of results.

14. Nonhuman Peptide Sequences Identified in the Sample

Urinary tract infections (UTI) are a common human affliction. Reports have estimated that between 1988 and 1994, 34% of adults in the United States aged 20 or older (approximately 62 million people) self-reported having had at least one occurrence of a UTI or cystitis (10, 11). About 80% of all cases are in females. The vast majority of UTIs are infections of the gram-negative bacteria *Escherichia coli*. Thus, the presence of peptide sequence

Table 2
Comparison of two bacterial allantoinases

Species	Protein	Accession	Sequence	Length/predicted MW/predicted pI
E. coli	allB	NP_415045.1	GGITTMIEMPLNQLPATVDR	20/2157.53/4.37
K. pneumoniae	PyrC	NP_943454.1	GTIATESAAAVMGGITSFME-MPNVTPPTTTR	31/3140.59/4.53

Both of the displayed peptides are taken from functionally similar regions of the allantoinase protein

data from *E. coli* can indicate either sloppy technique in preparing the sample for analysis or the likelihood of a UTI.

The product of the *allB* gene of *E. coli* encodes an allantoinase, which catalyzes one step in the conversion of urate to urea. An example of a peptide fragment, generated by trypsin digest, from the allB protein is GGITTMIEMPLNQLPATVDR. This peptide is 20 amino acids in length, has a theoretical pI of 4.37 and a theoretical molecular weight (MW) of 2157.53. These values were calculated with the Compute pI/MW tool at http://ca.expasy.org/tools/ (12). The genome of the bacterium *Klebsiella pneumoniae* also encodes a similar allantoinase gene, *PyrC*. Because the allB and PyrC sequences differ, trypsin digests of these proteins generate fragments of different sizes. The *E. coli* and *K. pneumoniae* fragments are compared with respect to differences in sequence length and composition in Table 2. In summary, peptide sequences from bacterial species can be used to identify species present in the sample preparation.

Alternatively, a second likely source of nonhuman peptide sequence data in the sample could arise from a viral infection. Consider human papillomavirus (HPV) which has been linked to cervical cancer and which may be found in some clinical preparations. HPV is known to exist in numerous distinct genotypes with genotypes 16 and 18 often associated with cervical cancer. Other genotypes are less well characterized. Thus, the peptide data could be used to rapidly assess the HPV genotype of an infected individual. For example, three different trypsin fragments of the L1 capsid protein from HPV genotypes 16 and 6 demonstrate how this process works (Fig. 3). These fragments are displayed in order of increasing sequence divergence such that the first set of peptides (Fig. 3a) offers no distinguishing sequence features while the third peptide set exhibits significant sequence divergence (Fig. 3c).

The fragments aligned in Fig. 3c, therefore, are distinct and more likely to serve as genotype-specific signatures. Lastly, the sensitivity of the proteomics assays means that other

```
a   HPV-16   AQGHNNGICWGNQLFVTVVDTTR
    HPV-6    AQGHNNGICWGNQLFVTVVDTTR

b   HPV-16   QTQLCLIGCKPPIGEHWGK
    HPV-6    QTQLCMVGCAPPLGEHWGK

c   HPV-16   FGFPDTSFYNPDTQR
    HPV-6    FALPDSSLFDPTTQR
```

Fig. 3. Comparison of three different peptide fragments of the L1 capsid protein from HPV genotypes 16 and 6. Residues different between the two peptides are *highlighted in bold.*

nonhuman peptides, such as from house pets, or the patient's surroundings or diet may also be detected. Fortunately, the ever-growing genome databases generally allow at least a partial, if not complete, identification of most of these peptides. In summary, detection of nonhuman peptides in a preparation is entirely plausible and may be indicative of a bacterial or viral infection.

15. Variation in the Human Genome

Although we are a single species, the genomes of any two *H. sapiens* are remarkably different from one another. There is approximately 1.0% difference between the genomic sequences of two unrelated individuals. Regions of sequence difference range in size from a single base pair (single nucleotide polymorphism or SNP) to insertions or deletions of a few base pairs to over a million base pairs (copy number variants or CNV). In fact, there are such differences between the genetic contributions of the mother and father whereby a single individual carries two "half genomes" that can be remarkably different from one another. These genetic differences, whether large or small in size, can also be observed in proteomics data. Differences in the sequence of a gene or in gene copy number can translate into different peptides produced by gene copies of different parental origin or different levels of protein expressed from genes of copy number other than two.

The *A1BG* gene, encoding alpha-1-B glycoprotein, contains several polymorphisms. One of these is SNP rs893184, also known as NP_570602.2p.Arg52His, which changes amino acid 52 from arginine to histidine. This SNP has two alleles: the genome contains either a G residue at the particular position within the *A1BG* gene or an A residue. G is the common or major allele and the minor or less common allele is A. The SNP is classified as nonsynonymous or missense, because the alleles of the SNP change the protein sequence. A useful term is the minor allele frequency (MAF) that gives a measure of the frequency of occurrence of the minor allele in a given population. In Caucasian populations, the MAF for SNP rs893184 is ~2.5% and in a Yoruba (Africa) population the MAF is ~22%. With a G, codon 52 is CGC coding for arginine and an A makes the codon CAC for histidine. Individuals who are homozygous for G at this genetic locus inherited a G residue from each parent and express A1BG protein with only arginine at position 52. On the other hand, individuals who are heterozygous at this locus possess a G from one parent and an A from the other and likely express a mixture of two isoforms of the A1BG protein, which are either arginine or histidine at position 52.

The alpha-1B-glycoprotein has been identified in human urine samples (13). The Arg52His SNP may be important in this respect as an arginine at position 52 in the protein allows cleavage by trypsin while histidine does not. Thus, different peptides are produced by digestion with this enzyme (Fig. 4). The His52 or "H" peptide gives one fragment from protein residues 1–60 and the Arg52 or "R" peptide yields two fragments.

Curiously, concerning the above peptides of A1BG, the SwissProt accession P04217 codes for arginine at residue 52 and the NCBI RefSeq accession NP_570602.2 codes for histidine. Both are correct because position 52 is a point of variation naturally found in human populations. In a clinical setting, variants in specific proteins can dictate the treatment regimen. Examples include hemoglobin, KRAS-activating missense mutations at codons 12 and 13 for colorectal cancer, codon 315 of BCR-ABL predicts resistance to Gleevac®, variations in the *epidermal growth factor receptor* (*EGFR*) gene can distinguish between responders and nonresponders in the treatment of nonsmall cell lung cancer and include EGFR T790M and EGFR L858R, and cytochrome P450 genes, encoding enzymes that metabolize drugs, such as *CYP2D6* (~20% of all drugs) and *CYP2C9* (warfarin).

Method:

1. To find all known nonsynonymous SNPs for a gene of interest, go to http://www.ncbi.nlm.nih.org, with the *Search* window set to *Gene*, and enter a gene symbol or name of interest.

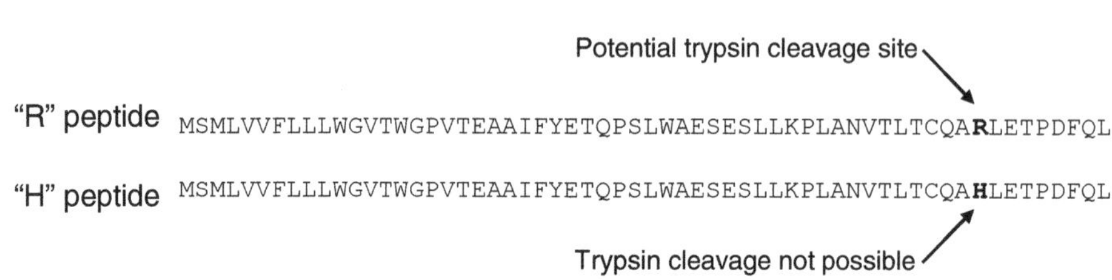

Fig. 4. Consequence of *A1BG* SNP rs893184 on the nature of detected peptides.

2. Click on the hyperlinked gene symbol to display the EntrezGene page.
3. At the right side of this page is a list of *Links*. Select *SNP:GeneView* to see those nonsynonymous SNPs for the gene under study.

Differences in gene copy number between individuals may have little impact on proteomics data. However, it is quite likely that copy number variation (CNV) results in a corresponding change in protein content. Three copies of a gene could produce higher levels of the encoded protein while a genome with a single copy of that gene would be expected to produce an amount lesser than the normal diploid genome. Thus, methods that measure peptide levels, such as an antibody array, may present data indicating increased or decreased expression of a given protein – relative to an internal standard – and it is possible that the differential expression arose from a variation in the genome, for example, a SNP or CNV. A good source to query for CNVs is the database of genomic variants at http://projects.tcag.ca/variation/. The SNP database at NCBI is at http://www.ncbi.nlm.nih.gov/SNP/index.html.

References

1. Kersey, P. J., Duarte, J., Williams, A., Karavidopoulou, Y., Birney, E., and Apweiler, R. (2004) The International Protein Index: An integrated database for proteomics experiments. *Proteomics.* **4**, 1985–1988
2. Angius, A., Petretto, E., Maestrale, G. B., Forabosco, P., Casu, G., Piras, D., Fanciulli, M., Falchi, M., Melis, P. M., Palermo, M., and Pirastu, M. (2002) A new essential hypertension susceptibility locus on chromosome 2p24-p25, detected by genomewide search. *Am. J. Hum. Genet.* **71**, 893–905
3. Wellcome Trust Case Control Consortium. (2007) Genome-wide association study of 14,000 cases of seven common diseases and 3,000 shared controls. *Nature.* **447,** 661–678
4. Clark, T. A., Schweitzer, A. C., Chen, T. X., Staples, M. K., Lu, G., Wang, H., Williams, A., and Blume, J. E. (2007) Discovery of tissue-specific exons using comprehensive human exon microarrays. *Genome Biol.* **8**, R64
5. Lanaspa, M. A., Almeida, N. E., Andres-Hernando, A., Rivard, C. J., Capasso, J. M., and Berl, T. (2007) The tight junction protein, MUPP1, is up-regulated by hypertonicity and is important in the osmotic stress response in kidney cells. *Proc. Natl. Acad. Sci. U S A.* **104**, 13672–13677
6. Rao, P. V., Lu, X., Standley, M., Pattee, P., Neelima, G., Girisesh, G., Dakshinamurthy, K. V., Roberts Jr., C. T., and Nagalla, S. R. (2007) Proteomic identification of urinary biomarkers of diabetic nephropathy. *Diabetes Care.* **30**, 629–637
7. Cheng, O., Thuillier, R., Sampson, E., Schultz, G., Ruiz, P., Zhang, X., Yuen, P. S., and Mannon, R. B. (2006) Connective tissue growth factor is a biomarker and mediator of kidney allograft fibrosis. *Am. J. Transplant.* **6**, 2292–2306
8. Mori, T., Kawara, S., Shinozaki, M., Hayashi, N., Kakinuma, T., Igarashi, A., Takigawa, M., Nakanishi, T., and Takehara, K. (1999) Role and interaction of connective tissue growth factor with transforming growth factor-beta in persistent fibrosis: a mouse fibrosis model. *J. Cell. Physiol.* **181**, 153–159
9. Tsakas, S., and Goumenos, D. S. (2006) Accurate measurement and clinical significance of urinary transforming growth factor-beta 1. *Am. J. Nephrol.* **26**, 186–193
10. Griebling, T. L. (2004) Urinary tract infection in women, in *Urologic Diseases in America* (Litwin, M. S., and Saigal, C. S., eds.) NIH

publication 04-5512, DHHS, PHS, NIH, NIDDK, Washington, DC, pp. 153–183

11. Griebling, T. L. (2004) Urinary tract infection in men, in *Urologic Diseases in America* (Litwin, M. S., and Saigal, C. S., eds.) NIH publication 04-5512, DHHS, PHS, NIH, NIDDK, Washington, DC, pp. 187–209

12. Gasteiger, E., Hoogland, C., Gattiker, A., Duvaud, S., Wilkins, M. R., Appel, R. D., and Bairoch A. (2005) Protein identification and analysis tools on the ExPASy server, in *The Proteomics Protocols Handbook* (Walker, J. M., ed.), Humana, Totowa, MJ, pp. 571–608

13. Kreunin, P., Zhao, J., Rosser, C., Urquidi, V., Lubman, D. M., and Goodison, S. (2007) Bladder cancer associated glycoprotein signatures revealed by urinary proteomic profiling. *J Proteome Res.* **6**, 2631–2639

Chapter 8

PROTIS: Use of Combined Biomarkers for Providing Diagnostic Information on Disease States

Walter Hofmann, Cornelia Sedlmeir-Hofmann, Miroslav Ivandić, Dagmar Ruth, and Peter Luppa

Abstract

We describe herein PROTIS, a software package that allows for the combination of urine biomarkers (proteins, small molecules, cells) to provide accurate diagnostic information on renal disease states. Such an approach exemplifies the advantage conferred by using multiple markers, and can be generalized for identification of biomarker signatures to separate distinct sample groups.

Key words: PROTIS, Multiple markers, Bioinformatics, Renal disease, Biomarker signatures

1. Introduction

Proteinuria and hematuria are the most often found symptoms of renal disease. Recently we have learned that urine protein patterns mirror the sites and mechanisms of the events. Renal dysfunctions are excluded with high certainty when Cystatin C in serum, total protein, albumin and α_1-microglobulin, hemoglobin, and leukocyte esterase excretion are normal (1–3). Figure 1 shows the strategy as workflow.

By quantifying single proteins with different molecular weight (IgG, albumin, α_1-microglobulin, α_2-macroglobulin) in urine, prerenal forms can be separated from glomerular and tubular-interstitial and postrenal ones. Figure 2 shows a complete report of urine protein differentiation produced by the result interpretation software PROTIS (4, 5).

The kidney assessment package consists of different components: the kidney workflows (Dilution Workflow, Screening for Kidney Disease and Exclusion of Kidney Disease) and the kidney assessment.

Alex J. Rai (ed.), *The Urinary Proteome: Methods and Protocols*, Methods in Molecular Biology, vol. 641, DOI 10.1007/978-1-60761-711-2_8, © Springer Science+Business Media, LLC 2010

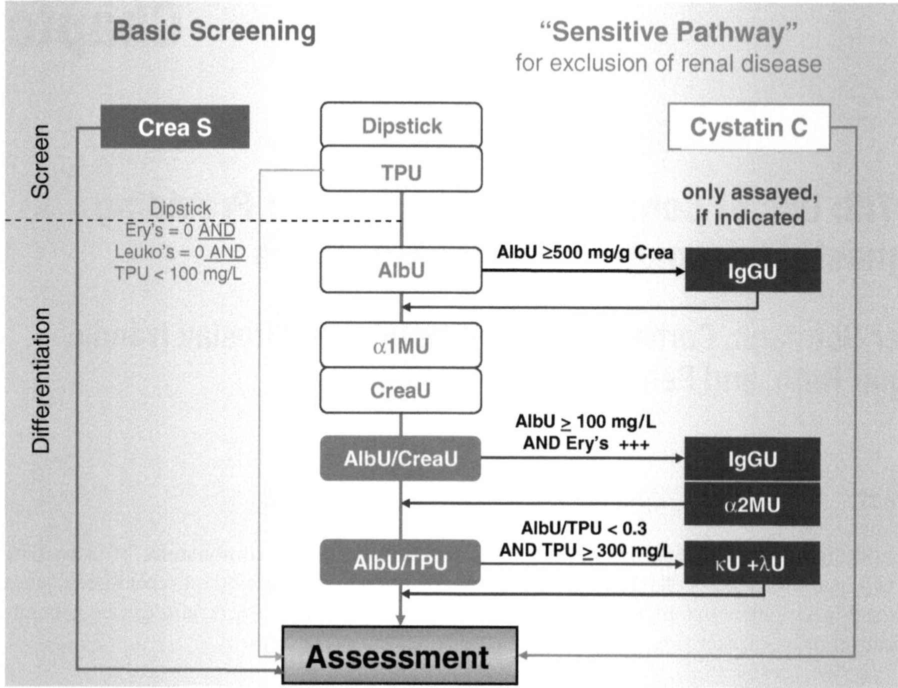

Fig. 1. Proposed combination for serum and urine screening tests for the exclusion and differentiation of renal diseases. TPU: Total Protein Urine; AlbU: Albumin Urine; α_1-MU: α_1-Microglobulin Urine; α_2-MU: α_2-Macroglobulin Urine; CreaU: Creatinine Urine; Ery's: Erythrocytes; Leuko's: Leukocytes.

The workflows can be used to order appropriate dilutions for marker proteins (Dilution Workflow) or initiate kidney assessments with varying reliability (Screening for Kidney Disease and Exclusion of Kidney Disease).

The kidney assessment evaluates the results by using rules and generates the text (2, 4). See Fig. 2 as an example.

In this manuscript, we discuss individually each of the multiple parameters that are used for the assessment of renal dysfunction (these are listed below). Several case reports generated using the PROTIS software are included to demonstrate these differences in levels of various combinatorial parameters.

2. Parameters for Interpretation of Urine Protein Profiles

- Albumin (Alb)
- α_1-Microglobulin (α_1-m)
- α_2-Macroglobulin (α_2-m)
- Immunoglobulin G (IgG)
- Total Protein (TP)

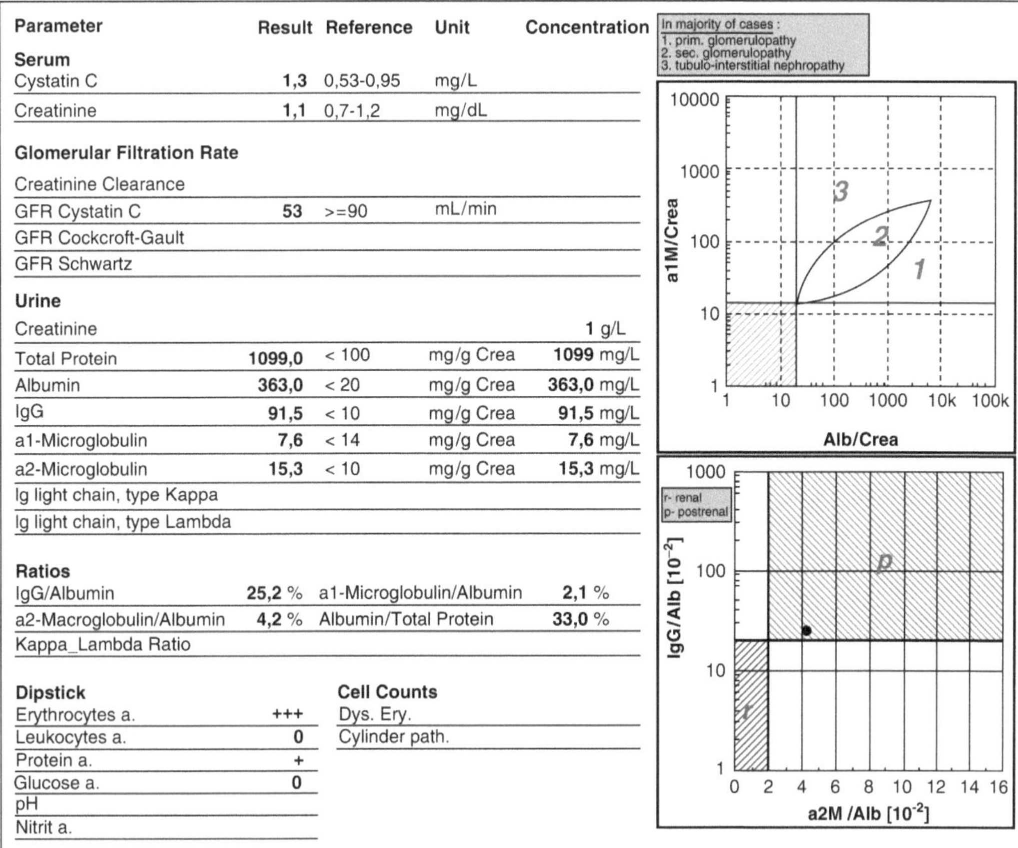

Parameter	Result	Reference	Unit	Concentration
Serum				
Cystatin C	**1,3**	0,53-0,95	mg/L	
Creatinine	**1,1**	0,7-1,2	mg/dL	
Glomerular Filtration Rate				
Creatinine Clearance				
GFR Cystatin C	**53**	>=90	mL/min	
GFR Cockcroft-Gault				
GFR Schwartz				
Urine				
Creatinine				**1** g/L
Total Protein	**1099,0**	< 100	mg/g Crea	**1099** mg/L
Albumin	**363,0**	< 20	mg/g Crea	**363,0** mg/L
IgG	**91,5**	< 10	mg/g Crea	**91,5** mg/L
a1-Microglobulin	**7,6**	< 14	mg/g Crea	**7,6** mg/L
a2-Microglobulin	**15,3**	< 10	mg/g Crea	**15,3** mg/L
Ig light chain, type Kappa				
Ig light chain, type Lambda				

Ratios			
IgG/Albumin	**25,2** %	a1-Microglobulin/Albumin	**2,1** %
a2-Macroglobulin/Albumin	**4,2** %	Albumin/Total Protein	**33,0** %
Kappa_Lambda Ratio			

Dipstick		**Cell Counts**
Erythrocytes a.	+++	Dys. Ery.
Leukocytes a.	0	Cylinder path.
Protein a.	+	
Glucose a.	0	
pH		
Nitrit a.		

Interpretation

GFR estimation indicates a moderate decrease in glomerular filtration, which clearly indicates a pathologic level of kidney function loss. (Based on the Cystatin C formula)

Despite an inconspicuous serum-creatinine concentration, a possible GFR reduction cannot be ruled out.

Albuminuria is present. This is classified as significant (presence of macroalbuminuria). No additional, tubular proteinuria exists.

The presence of postrenal hematuria is highly probable; however, an additional renal protein excretion cannot be excluded. Control is recommended following remission of hematuria. Tubulo-interstitial dysfunction can be ruled out with high probability, whereas a glomerular permeability disorder cannot.

Fig. 2. Typical assessment report created by the result interpretation software PROTIS.

- Creatinine (Crea)
- Cystatin C (Cys C)
- Use of urine creatinine to correct for urine concentration
- Exclusion of an active renal parenchymal disorder
- Differentiation of a proteinuria (prerenal/renal proteinuria, selectivity, protein gap)
- Interpretation of a leukocyturia
- Interpretation of a hematuria (prerenal/renal/postrenal hematuria)

3. Albumin

The quantitatively most important plasma protein, albumin (molecular weight 69 kD, molecular radius 36×10^{-10} m) has been introduced as a sensitive indicator of a dysfunction of glomerular filtration. It dominates the excretion pattern of primary or secondary glomerulopathies. An albuminuria between 20 and 200 mg/l is termed "microalbuminuria" and corresponds to a negative result in a normal dipstick. As albumin is also physiologically filtered to a small extent, its excretion is increased when tubular re-absorption is disturbed, as observed in tubulo-interstitial nephropathies. Therefore, a glomerulopathy and a tubulo-interstitial nephropathy can only be distinguished by means of urinalysis when a tubular marker, e.g., α_1-microglobulin, is considered in addition to albumin (see interpretation of a proteinuria); see graph in Fig. 3.

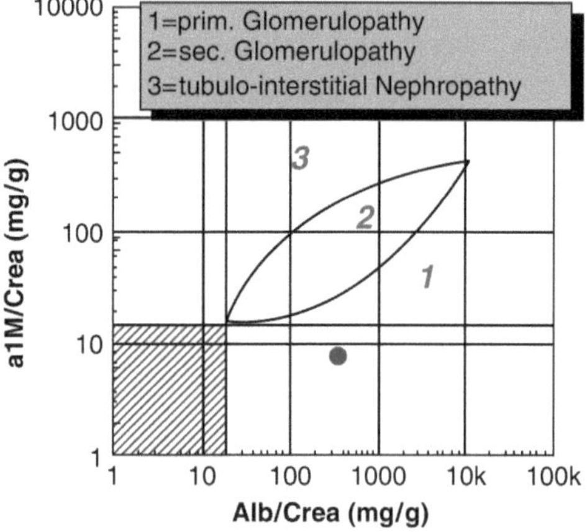

Fig. 3. Urinary α_1-microglobulin excretion rates in comparison to albuminuria in primary glomerulopathies (1), secondary glomerulopathies (2) and tubulo-interstitial diseases (3) separated by two hyperbolic curves. The two lines characterize the upper reference limits for albumin (20 mg/g creatinine) and α_1-microglobulin (14 mg/g creatinine).

4. α_1-Microglobulin (Protein HC)

α_1-Microglobulin was isolated for the first time in 1975 and has been recommended as a marker of tubular reabsorption. The unbound share of the glycoprotein is freely filtered because of its small size (molecular weight approximately 33 kD). More than 99% is reabsorbed in the proximal tubules. Whenever this energy-dependent process of protein reabsorption is impaired, α_1-microglobulin can be found elevated in urine. Therefore, the excretion of α_1-microglobulin is increased in various kinds of tubulo-interstitial nephropathies. In glomerulopathies, high urine concentrations of α_1-microglobulin indicate an interstitial involvement. In patients with nephrotic glomerular proteinuria (albuminuria > 3,000 mg/g creatinine), the share of this tubular proteinuria must be calculated as follows:

$$\alpha_1\text{-microglobulin (tubulo-interstitial)} = \alpha_1\text{-microglobulin (measured)} - 4.7 \times e^{(0.00022 \times \text{albumin [mg/g creatinine]})}$$

As α_1-microglobulin can be elevated both in tubulo-interstitial and in glomerular disorders, it should always be interpreted with respect to a glomerular marker, e.g., (albumin, see interpretation of a proteinuria).

PROTIS

Five examples for protein description show how PROTIS, the result interpretation software works:

If the condition in the case of a patient is correct, one of the following text-sequences is printed on the report (see Table 1).

Table 1
Examples for protein description

Condition	Text
Selectivity of glomerular proteinuria (Alb > 500 mg/g)	
AlbU > 500 mg/g Crea and IgG/Alb > 0.03 and glomerular indication	Nonselective glomerular proteinuria is present
Classification of albuminuria or glomerular proteinuria	
AlbU > 20 mg/g Crea and AlbU ≤ 30 mg/g Crea	This is classified as borderline
Classification of additional tubular indication	
AlbU > 20 mg/g Crea and α1mU ≤ 14 mg/g Crea	No additional, tubular proteinuria exists
Classification of stand-alone tubular indication	
AlbU ≤ 20 mg/g Crea and α1mU > 14 mg/g Crea and α1mU ≤ 20 mg/g Crea	Borderline tubular proteinuria exists
Isolated IgG-Uria	
α1mU ≤ 14 mg/g Crea AlbU ≤ 20 mg/g Crea IgG > 10 mg/g Crea No hematuria and no inflammation	As far as can be assessed, examination of the urine measurement variables reveals no indication of glomerular protein filtration or tubular reabsorption disorder. The isolated increase of IgG in urine may occur in increased serum IgG concentrations

5. α_2-Macroglobulin

As only very small amounts of α_2-macroglobulin are filtered under physiological conditions because of its size (molecular weight 720 kD), elevated concentrations in urine most often indicate a postrenal proteinuria. Therefore, it is used in combination with IgG to differentiate renal and postrenal hematuria. In cases of massive malfunction of the glomerular filter causing nephrotic proteinuria (e.g., due to a rapid progressive glomerulonephritis), α_2-macroglobulin can also be found elevated in urine. See graph in Fig. 4.

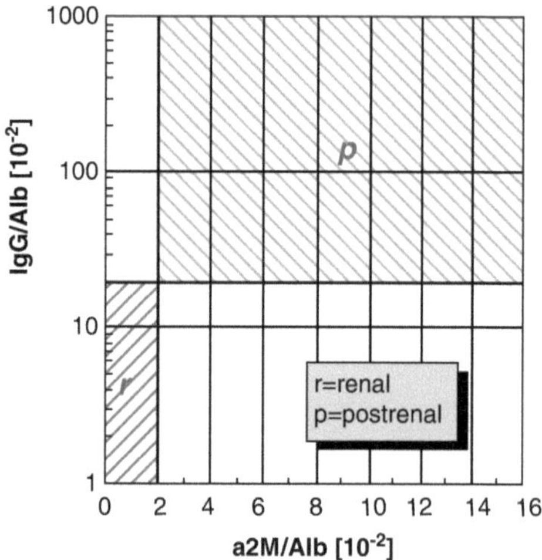

Fig. 4. Comparison of IgG/albumin and α_2-macroglobulin/albumin ratios in hematuric urines from patients with renal and postrenal diseases. The two lines separate renal from postrenal bleedings.

6. Immunoglobulin G

Immunoglobulin G (molecular weight 150 kD, molecular radius 55×10^{-10} m) can be helpful to evaluate proteinuria by determination of the selectivity index. Besides, it is used in combination with α_2-macroglobulin to differentiate hematuria and together with α_1-microglobulin, albumin, and total protein for the classification of leukocyturia.

7. Total Protein

The measurement of total protein in urine can be used as a cost-efficient initial screening in patients without clinical signs of renal disease or risk factors to develop kidney disease (e.g., diabetes mellitus, hypertension). The various proteins excreted contribute to a different extent to the total protein concentration, depending on the method used for total protein measurement. The analyte total protein can be used as a plausibility control when it is compared to the sum of single proteins. A so-called "protein gap" can be detected using this ratio.

8. Creatinine

8.1. Urine

Quantitative urinary analytes can be related to creatinine to correct for urine dilution. As a product of the muscle metabolism, the urine concentration of creatinine is mainly dependent on muscle mass. Very high and very low concentrations can be found in urines of bodybuilders and children, respectively. In these cases, analytes such as albumin, α_1-microglobulin, IgG, or total protein might be underestimated or overestimated when related to creatinine.

8.2. Serum/Plasma

Creatinine can be used as an endogenous marker of renal clearance function. Numerous influence factors limit the diagnostic value of this analyte. Various formulas have been suggested to improve the assessment of the glomerular filtration rate based on creatinine measurements.

9. Cystatin C (Gamma-Trace, Post Gamma Globulin) Serum/Plasma

Cystatin C (molecular weight 13.3 kD, 120 aminoacids), is a member of the cystatin family of cystein protease inhibitors. It seems to be a sensitive endogenous marker to adequately assess the glomerular filtration rate with some advantages in comparison with other low-molecular-weight serum markers. It is expressed in most nucleated cells at a relatively steady rate ("housekeeping gene"), that seems to stay constant also in inflammatory and other pathological processes. However, elevated concentrations have been described in patients with lung cancer and immunological diseases.

9.1. Assessment of the Glomerular Filtration Rate

The glomerular filtration rate (GFR) can be estimated by clearance methods. The endogenous creatinine clearance can be used for this purpose if the GFR is over 20%. The creatinine clearance is calculated from creatinine concentrations in a timed urine and serum sample considering urine volume and collection time:

$$\text{creatinine clearance [ml/min]} = (\text{creatinine}_{\text{urine}}[\text{mg/dl}] \times \text{urine volume [ml]})/(\text{creatinine}_{\text{serum}}[\text{mg/dl}]\times\text{collection time [min]})$$

The creatinine clearance can be related to the body surface of a patient:

$$\text{creatinine clearance [ml/min/1.73 m}^2] = \text{creatinine clearance [ml/min]}\times 1.73\ /\ \text{body surface [m}^2]$$

The body surface can be estimated from a patient's weight and height using a nomogram.

The concentration of creatinine in serum is influenced by numerous physiological, metabolic, and analytical factors. Moreover, creatinine is not found elevated before GFR is already diminished by approximately 50%. To improve the diagnostic power of a single serum creatinine measurement, various formulas have been suggested that have been evaluated with different groups of patients:

PROTIS

Different formulas to calculate GFR show how PROTIS works. They are shown below in Table 2.

9.2. Interpretation of Elevated Creatinine and Cystatin C Results

If the GFR is impaired according to serum analytes, but urine protein differentiation is normal, these findings might be due to a loss of functioning nephrons that is completely compensated by an increased protein reabsorption of the remaining nephrons. This constellation can be found e.g., after a "burnt-out" glomerulonephritis that has left scarred glomeruli. However, a presently active renal parenchymal disease can be excluded with high certainty in such a case.

10. Use of Urine Creatinine to Correct for Urine Concentration

As the individual renal excretion of creatinine is quite constant, the concentration of urine creatinine can be used to estimate the concentration of the sample. By referencing all quantitative urinary analytes to creatinine, their variability caused by changes in diuresis can be diminished. Extreme creatinine concentrations imply a possible under- and overestimation when referring urinary analytes to grams creatinine. If the concentration of urine creatinine is very high or very low, this procedure might cause an

Table 2
Different formulas to calculate GFR

Value	Formula
Creatinine clearance (collected urine)	Correct: $$\text{GFR} = \frac{\text{Creatinine U} \backslash \text{times Sample} \backslash \text{times Volume} \backslash \text{times } 1.73}{\text{Creatinine S} \backslash \text{times Sample Collection Duration} \backslash \text{times Patient Surface}} [\text{ml / min / } 1.73\text{m}^2$$ No age and gender specific factors to be considered [Sample: Collection duration] in minutes
Creatinine clearance estimation (Cockcroft-Gault)	Estimation by Cockcroft-Gault formula: $$\text{male : GFR} = \frac{(140 - \text{Patient} \times \text{Age}[\Upsilon]) \times \text{Patient} \times \text{Weight[kg]}}{72 \times \text{Creatinine S[mg / dl]}} [\text{ml / min}]$$ $$\text{female : GFR} = 0.85 \times \frac{(140 - \text{Patient} \times \text{Age}[\Upsilon]) \times \text{Patient} \times \text{Weight[kg]}}{72 \times \text{Creatinine S[mg / dl]}} [\text{ml / min}]$$ Valid for patients older than 18 years Note: GFR results above 60 ml/min are reported as >60 ml/min
Creatinine clearance estimation (Schwartz)	Estimation by Schwartz formula: $$\text{GFR} = k \times \frac{\text{Patient} \times \text{Height[cm]}}{\text{Creatinine S[mg / dl]}} [\text{ml / min / } 1.73\text{m}^2]$$ For infants <2 years: $k = 0.45$ For males/females between 2 and 13 years and Females between 13 and 18 years: $k = 0.55$ For males between 13 and 18 years: $k = 0.70$ Note: GFR results above 60 ml/min/1.73 m^2 are reported as >60 ml/min/1.73 m^2
Creatinine clearance estimation (MDRD)	For males: GFR = 186 (CreaS[mg/dl]) − 1.154 (Age) − 0.203 [ml/min/1.73 m^2] For females: GFR = 186 (CreaS[mg/dl]) − 1.154 (Age) − 0.203 × 0.742 [ml/min/1.73 m^2] Racefactor: GFR = GFR 1.21 [ml/min/1.73 m^2] (for Africans) Valid for patients older than 18 years
Creatinine clearance estimation (MDRD IDMS)	For males: GFR = 175 (CreaS[mg/dl]) − 1.154 (Age) − 0.203 [ml/min/1.73 m^2] For females: GFR = 175 (CreaS[mg/dl]) − 1.154 (Age) − 0.203 × 0.742 [ml/min/1.73 m^2] Racefactor: GFR = GFR 1.21 [ml/min/1.73 m^2] (for Africans) Valid for patients older than 18 years Note: Results above 60 ml/min/1.73 m^2 are reported as >60 ml/min/1.73 m^2
Cystatin C clearance estimation (Cystatin C)	Formula 1: $$\text{GFR} = \frac{74.835}{\text{CysC (Serum)[mg / l]}^{1/0.75}} [\text{ml / min}]$$ No age and gender specific factors to be considered Formula 2 (Larsson formula): GFR = 77.24 CysC(Serum)[mg/l] $^{1.2623}$ [ml/min] No age and gender specific factors to be considered Note: GFR results above 90 ml/min are reported as >90 ml/min

overestimation or underestimation of analytes. This fact must be considered especially when dealing with urines from individuals with high muscle mass and patients with low muscle metabolism (newborn, young children, elderly people). To prevent misinterpretations of concentrations of urine proteins, Protis lists urinary analytes in mg/l, when urine creatinine concentration is too low (<0.2 g/l).

11. Exclusion of an Active Renal Parenchymal Disorder

If there is neither hematuria nor leukocyturia and urine concentration of total protein, albumin, and α_1-microglobulin are within their reference intervals, then a relevant renal parenchymal disease can be excluded with very high certainty. See Fig. 5 as an example.

The measurement of total protein in urine can be regarded as a more cost-efficient alternative for initial screening for proteinuria, but only in patients without clinical signs of a renal disease or risk factors to develop kidney disease (e.g., diabetes mellitus, hypertension). However, slight tubular proteinuria or albuminuria might not be detected with this approach.

12. Differentiation of a Proteinuria

A prerenal cause can be distinguished from a renal (glomerular and tubular) cause of proteinuria by measuring total protein, albumin, and α_1-microglobulin in urine. Moreover, a renal proteinuria can be assigned to a tubulo-interstitial nephropathy, a primary glomerulopathy (glomerulonephritis), or a secondary glomerulopathy (e.g., diabetic or hypertensive nephropathy).

13. Proteinuria

13.1. Pre-renal Proteinuria

A prerenal proteinuria should be suspected, when a "protein gap" is detected using the ratio of total protein to the sum of albumin and α_1-microglobulin: If the sum of the single proteins accounts for less than 30% of total protein, a prerenal cause for this protein gap should be considered. Using the dipstick result for hemoglobin/myoglobin and an immunofixation test in such a case, a Bence Jones proteinuria of a patient with a monoclonal gammopathy (e.g., multiple myeloma, M. Waldenström, primary amyloidosis)

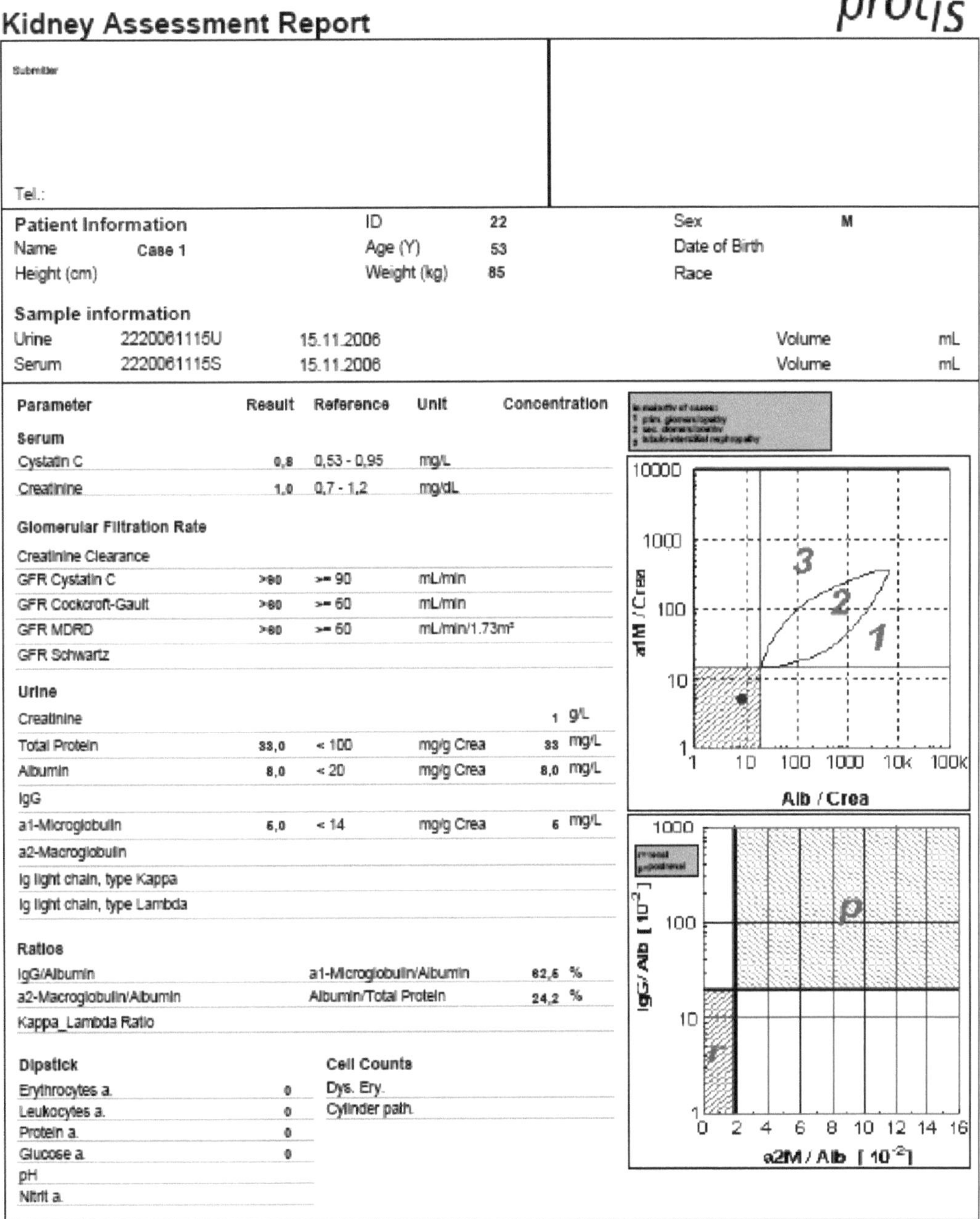

Kidney Assessment Report

protis

Submitter

Tel.:

Patient Information

Name	Case 1	ID	22	Sex	M
		Age (Y)	53	Date of Birth	
Height (cm)		Weight (kg)	85	Race	

Sample information

Urine	2220061115U	15.11.2006	Volume	mL
Serum	2220061115S	15.11.2006	Volume	mL

Parameter	Result	Reference	Unit	Concentration	
Serum					
Cystatin C	0,8	0,53 - 0,95	mg/L		
Creatinine	1,0	0,7 - 1,2	mg/dL		
Glomerular Filtration Rate					
Creatinine Clearance					
GFR Cystatin C	>90	>= 90	mL/min		
GFR Cockcroft-Gault	>60	>= 60	mL/min		
GFR MDRD	>60	>= 60	mL/min/1.73m²		
GFR Schwartz					
Urine					
Creatinine				1	g/L
Total Protein	33,0	< 100	mg/g Crea	33	mg/L
Albumin	8,0	< 20	mg/g Crea	8,0	mg/L
IgG					
a1-Microglobulin	5,0	< 14	mg/g Crea	5	mg/L
a2-Macroglobulin					
Ig light chain, type Kappa					
Ig light chain, type Lambda					

Ratios

IgG/Albumin	a1-Microglobulin/Albumin	62,5	%
a2-Macroglobulin/Albumin	Albumin/Total Protein	24,2	%
Kappa_Lambda Ratio			

Dipstick		Cell Counts
Erythrocytes a.	0	Dys. Ery.
Leukocytes a.	0	Cylinder path.
Protein a.	0	
Glucose a.	0	
pH		
Nitrit a.		

Interpretation

GFR estimation indicates no impairment of glomerular filtration. (Based on the Cystatin C formula)

Analyses of the marker proteins in urine do not indicate any dysfunction of glomerular protein filtration and tubular reabsorption. There is no indication of hemorrhage and/or granulocytic inflammation reaction in the descending urinary tract.

Fig. 5. Case 1 without pathological finding.

Table 3
Different conditions for Bence Jones evaluation

Condition	Text
TP > 300 mg/g Crea, no hematuria, Alb/TP < 0.3 and IgG is not defined	The discrepancy between the albumin concentration and the concentration of total protein (albumin/total protein < 0.3) indicates pre-renal proteinuria. For confirmation, additional tests (Kappa/Lambda free light chains, immunofixation) are recommended
TP > 300 mg/g Crea and Alb > 14 mg/g Crea and α1m/Alb > 0.3	This constellation indicates restricted tubulo-interstital reabsorption with high probability

can be distinguished from a myoglobinuria (rhabdomyolysis, trauma) or a hemoglobinuria after intravasal hemolysis. A protein gap in combination with highly elevated excretion of tubular markers can be found in urines of children with an inborn tubulopathy (e.g., DeToni-Debré-Fanconi-syndrome).

PROTIS

Different conditions for Bence Jones evaluation show how PROTIS works. See Table 3 for two different conditions.

13.2. Renal Proteinuria

Using albumin and α_1-microglobulin as marker proteins, a renal proteinuria can be described quantitatively and qualitatively according to its glomerular and/or tubular shares and classified as tubulo-interstitial nephropathy, or primary or secondary glomerulopathy.

Glomerular proteinuria results from a dysfunction of the glomerular filter. Its permeability for macromolecules is dependent on their size, charge, and steric configuration. Large and negatively charged molecules are held back to a larger extent than small and cationic or neutral molecules. As a qualitative measure of glomerular dysfunction, the degree of "selectivity" of a proteinuria has been defined based on the excretion rate of immunoglobulin G: the IgG-clearance as a marker for size selectivity is divided by the albumin clearance as a marker for charge selectivity. An isolated albuminuria as a "selective" proteinuria (IgG/albumin < 0.03) is an early sign of an impaired charge selectivity of the glomerular filter which can be observed in a minimal-change glomerulonephritis or a beginning perimembranous glomerulonephritis, for instance. A "nonselective" proteinuria with a high share of IgG (IgG/albumin > 0.03) is found when there is already a pronounced structural damage of the filter pores (see Fig. 6).

Apart from describing site and extent of renal damage, a proteinuria pattern can be assigned to different renal diseases.

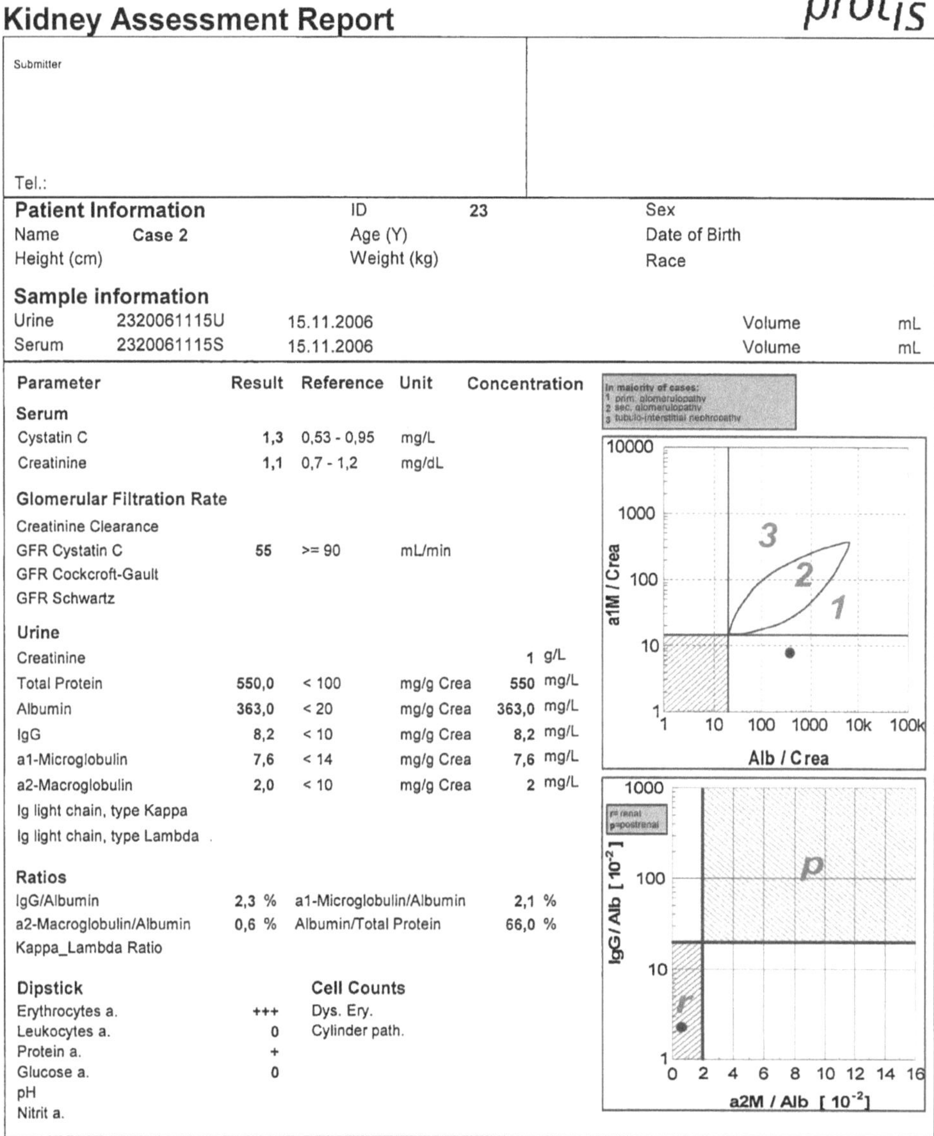

Kidney Assessment Report

protis

Submitter

Tel.:

Patient Information

Name	Case 2	ID	23	Sex	
		Age (Y)		Date of Birth	
Height (cm)		Weight (kg)		Race	

Sample information

Urine	2320061115U	15.11.2006	Volume	mL
Serum	2320061115S	15.11.2006	Volume	mL

Parameter	Result	Reference	Unit	Concentration	
Serum					
Cystatin C	1,3	0,53 - 0,95	mg/L		
Creatinine	1,1	0,7 - 1,2	mg/dL		
Glomerular Filtration Rate					
Creatinine Clearance					
GFR Cystatin C	55	>= 90	mL/min		
GFR Cockcroft-Gault					
GFR Schwartz					
Urine					
Creatinine				1	g/L
Total Protein	550,0	< 100	mg/g Crea	550	mg/L
Albumin	363,0	< 20	mg/g Crea	363,0	mg/L
IgG	8,2	< 10	mg/g Crea	8,2	mg/L
a1-Microglobulin	7,6	< 14	mg/g Crea	7,6	mg/L
a2-Macroglobulin	2,0	< 10	mg/g Crea	2	mg/L
Ig light chain, type Kappa					
Ig light chain, type Lambda					

Ratios

IgG/Albumin	2,3 %	a1-Microglobulin/Albumin	2,1 %
a2-Macroglobulin/Albumin	0,6 %	Albumin/Total Protein	66,0 %
Kappa_Lambda Ratio			

Dipstick		Cell Counts
Erythrocytes a.	+++	Dys. Ery.
Leukocytes a.	0	Cylinder path.
Protein a.	+	
Glucose a.	0	
pH		
Nitrit a.		

Interpretation

GFR estimation indicates a moderate decrease in glomerular filtration, which clearly indicates a pathologic level of kidney function loss. (Based on the Cystatin C formula)

Despite an inconspicuous serum-creatinine concentration, a possible GFR reduction cannot be ruled out.

Glomerular proteinuria is present. This is classified as significant (presence of macroalbuminuria). No additional, tubular proteinuria exists.

This constellation is compatible with primary or secondary glomerulopathy. (e.g. glomerulonephritis, diabetic or hypertensive nephropathy).
The presence of renal hematuria is highly likely; additional erythrocytes from postrenal sources cannot be ruled out.

23.01.2007 Released by:
Release date:

Signature

Fig. 6. Case 2 with a glomerular proteinuria.

According to the functional parts of a nephron, tubulo-interstitial nephropathies can be distinguished from primary glomerulopathies (glomerulonephritis) and secondary glomerulopathies (e.g., diabetic or hypertensive nephropathy). Based on the excretion pattern of the marker proteins albumin and α_1-microglobulin, conclusions can be drawn about the involvement of glomeruli and tubulo-interstitial space in the pathological process. With albumin and α_1-microglobulin making up the *x*- and *y*-axes, respectively, clusters of disease groups can be defined and separated in logarithmic coordinates.

Given the same albuminuria, excretion of α_1-microglobulin is, on an average, higher in patients with tubulo-interstitial nephropathies than with glomerulopathies because of impaired tubular protein re-absorption. The higher urine concentrations of this tubular marker protein in patients with secondary glomerulopathies in comparison with primary glomerulopathies indicate a stronger and different involvement of the interstitial space in the underlying glomerular disease (e.g., fibrosis of tubules in a diabetic glomerulosclerosis), which is consistent with the histopathological observations. See Fig. 7 as an example.

In glomerulopathy with nephrotic albuminuria (>3 g/g creatinine), it should be taken into consideration that an increase of α_1-microglobulin in urine is caused at least partly by an "overload" of the tubule system due to the excessive glomerular proteinuria. Therefore, there is always an elevated concentration of the tubular marker when a nephrotic albuminuria is found. A significant tubular proteinuria, on the other hand, can also be interpreted as caused by an involvement of the interstitial space in the original glomerular disease, for instance, if there is an interstitial fibrosis in a case of glomerulonephritis. In order to decide how much of the excretion of α_1-microglobulin is due to a structural tubular damage, the tubular marker can be "corrected" using an exponential formula. This formula was derived from excretion patterns of selected patients with glomerulonephritis whose renal interstitial space was devoid of major histopathological findings. It describes approximately the lower margin of the cluster of primary glomerulopathies with a nephrotic albuminuria (see Fig. 8).

The following tables lists various types of proteinuria and the possible disease causes:

13.3. Selective Glomerular Proteinuria

1. Minimal-change glomerulopathy
2. Membranous glomerulonephritis, stage I
3. Focal segmental glomerulonephritis, stage I
4. IgA-Nephritis
5. Early stage of a diabetic nephropathy

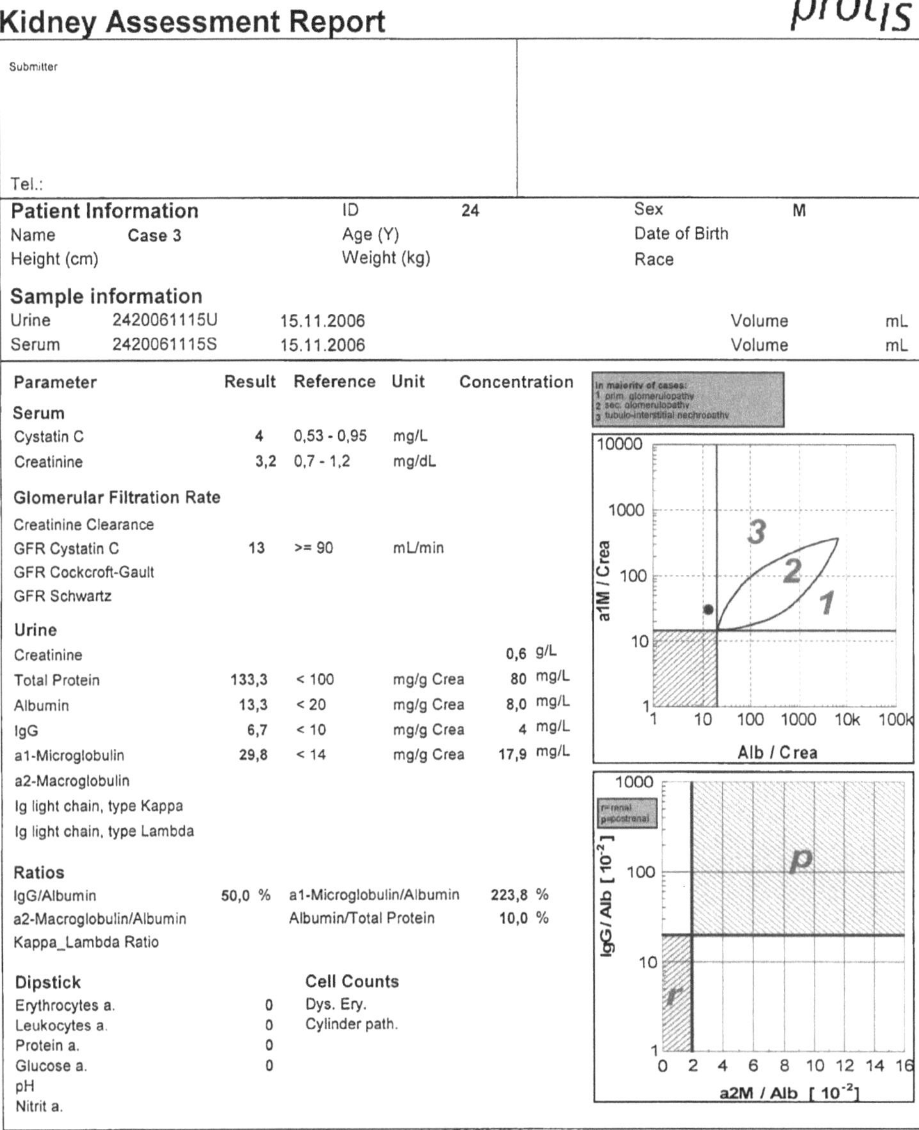

Kidney Assessment Report

protis

Submitter

Tel.:

Patient Information

Name	Case 3	ID	24	Sex	M
Height (cm)		Age (Y)		Date of Birth	
		Weight (kg)		Race	

Sample information

Urine	2420061115U	15.11.2006	Volume	mL
Serum	2420061115S	15.11.2006	Volume	mL

Parameter	Result	Reference	Unit	Concentration	
Serum					
Cystatin C	4	0,53 - 0,95	mg/L		
Creatinine	3,2	0,7 - 1,2	mg/dL		
Glomerular Filtration Rate					
Creatinine Clearance					
GFR Cystatin C	13	>= 90	mL/min		
GFR Cockcroft-Gault					
GFR Schwartz					
Urine					
Creatinine				0,6	g/L
Total Protein	133,3	< 100	mg/g Crea	80	mg/L
Albumin	13,3	< 20	mg/g Crea	8,0	mg/L
IgG	6,7	< 10	mg/g Crea	4	mg/L
a1-Microglobulin	29,8	< 14	mg/g Crea	17,9	mg/L
a2-Macroglobulin					
Ig light chain, type Kappa					
Ig light chain, type Lambda					

Ratios

IgG/Albumin	50,0 %	a1-Microglobulin/Albumin	223,8 %
a2-Macroglobulin/Albumin		Albumin/Total Protein	10,0 %
Kappa_Lambda Ratio			

Dipstick		Cell Counts
Erythrocytes a.	0	Dys. Ery.
Leukocytes a.	0	Cylinder path.
Protein a.	0	
Glucose a.	0	
pH		
Nitrit a.		

Interpretation

GFR estimation indicates a drastic decrease in glomerular filtration, which is classified as renal failure. (Based on the Cystatin C formula)

Slight tubular proteinuria exists.

This constellation is compatible with tubulo-interstitial dysfunction. The slight proteinuria is attributable to remaining nephron overload (overflow proteinuria) with existing renal insufficiency.

24.01.2007

Signature

Released by:
Release date:

Fig. 7. Case 3 with a tubular proteinuria.

Kidney Assessment Report

protis

Submitter

Tel.:

Patient Information

Name	Case 4	ID	25	Sex	M
		Age (Y)	66	Date of Birth	
Height (cm)		Weight (kg)		Race	

Sample information

Urine	2520061115U	15.11.2006	Volume	mL
Serum	2520061115S	15.11.2006	Volume	mL

Parameter	Result	Reference	Unit	Concentration	
Serum					
Cystatin C	0,6	0,53 - 0,95	mg/L		
Creatinine	0,7	0,7 - 1,4	mg/dL		
Glomerular Filtration Rate					
Creatinine Clearance					
GFR Cystatin C	>90	>= 90	mL/min		
GFR Cockcroft-Gault					
GFR MDRD	>60	>= 60	mL/min/1.73m²		
GFR Schwartz					
Urine					
Creatinine				0,6	g/L
Total Protein	4936,7	< 100	mg/g Crea	2962	mg/L
Albumin	4105,0	< 20	mg/g Crea	2463,0	mg/L
IgG	226,0	< 10	mg/g Crea	135,6	mg/L
a1-Microglobulin	300,0	< 14	mg/g Crea	180	mg/L
a2-Macroglobulin					
Ig light chain, type Kappa					
Ig light chain, type Lambda					

Ratios

IgG/Albumin	5,5 %	a1-Microglobulin/Albumin	7,3 %
a2-Macroglobulin/Albumin		Albumin/Total Protein	83,2 %
Kappa_Lambda Ratio			

Dipstick		**Cell Counts**
Erythrocytes a.	+++	Dys. Ery.
Leukocytes a.	0	Cylinder path.
Protein a.	+++	
Glucose a.	0	
pH		
Nitrit a.		

In majority of cases:
1 prim. glomerulopathy
2 sec. glomerulopathy
3 tubulo-interstitial nephropathy

a1M / Crea
Alb / Crea

r=renal
p=postrenal

IgG / Alb [10^{-2}]
a2M / Alb [10^{-2}]

Interpretation

GFR estimation indicates no impairment of glomerular filtration. (Based on the Cystatin C formula)

Nonselective glomerular proteinuria is present. This is classified as nephrotic. Additional, substantial tubular proteinuria exists.

This constellation is compatible with primary or secondary glomerulopathy. (e.g. glomerulonephritis, diabetic or hypertensive nephropathy). Substantial restriction of tubulo-interstital reabsorption is simultaneously found. Differentiation of hematuria is only possible if Alpha2-macroglobulin and Albumin have been measured or by means of phase contrast microscopy!

Fig. 8. Case 4 a patient with a nephrotic syndrome.

13.4. Nonselective Glomerular Proteinuria

1. Rapid progressive glomerulonephritis
2. Proliferative glomerulonephritis (vasculitis)
3. Membranoproliferative glomerulonephritis
4. Epimembranous glomerulonephritis, stage II and III
5. Focal segmental glomerulonephritis, stage II and III
6. Stage III and IV of diabetic nephropathy
7. Arterial hypertension, benign nephrosclerosis
8. EPH-gestosis

13.5. Nonselective Glomerular Plus Tubular Proteinuria

1. Renal amyloidosis
2. Gold-nephropathy, D-penicillamin-glomerulonephritis
3. Diabetic nephropathy (stage IV and V)
4. Membranoproliferative glomerulonephritis
5. Systemic vasculitis with renal involvement
6. Acute rejection of renal transplants

13.6. Tubular Proteinuria

1. "Pyelonephritis," interstitial nephritis
2. Analgetics nephropathy
3. Tubulotoxic nephropathy (aminoglycosides, cisplatin, cadmium, mercury, lead, lithium)
4. Fanconi syndrome, renal tubular acidosis (type II)
5. Myeloma kidney
6. Chromoprotein kidney (malaria tropica, rhabdomyolysis)

14. Interpretation of a Leukocyturia

Using total protein as well as urine concentrations of the marker proteins albumin, α_1-microglobulin, and immunoglobulin G, a renal cause of leukocyturia can be distinguished from a postrenal source of a leukocyturia. A contamination is probable if a normal urine protein pattern is combined with leukocyturia. Whereas in case of a postrenal leukocyturia the excretion of albumin and IgG might be slightly elevated; a renal involvement should be considered when a significant mixed proteinuria is found (see Fig. 9).

15. Interpretation of a Hematuria

A differentiation of a prerenal, renal and postrenal cause of hematuria is possible by means of urine protein differentiation, if there is a significant albumin excretion. At lower albumin concentrations,

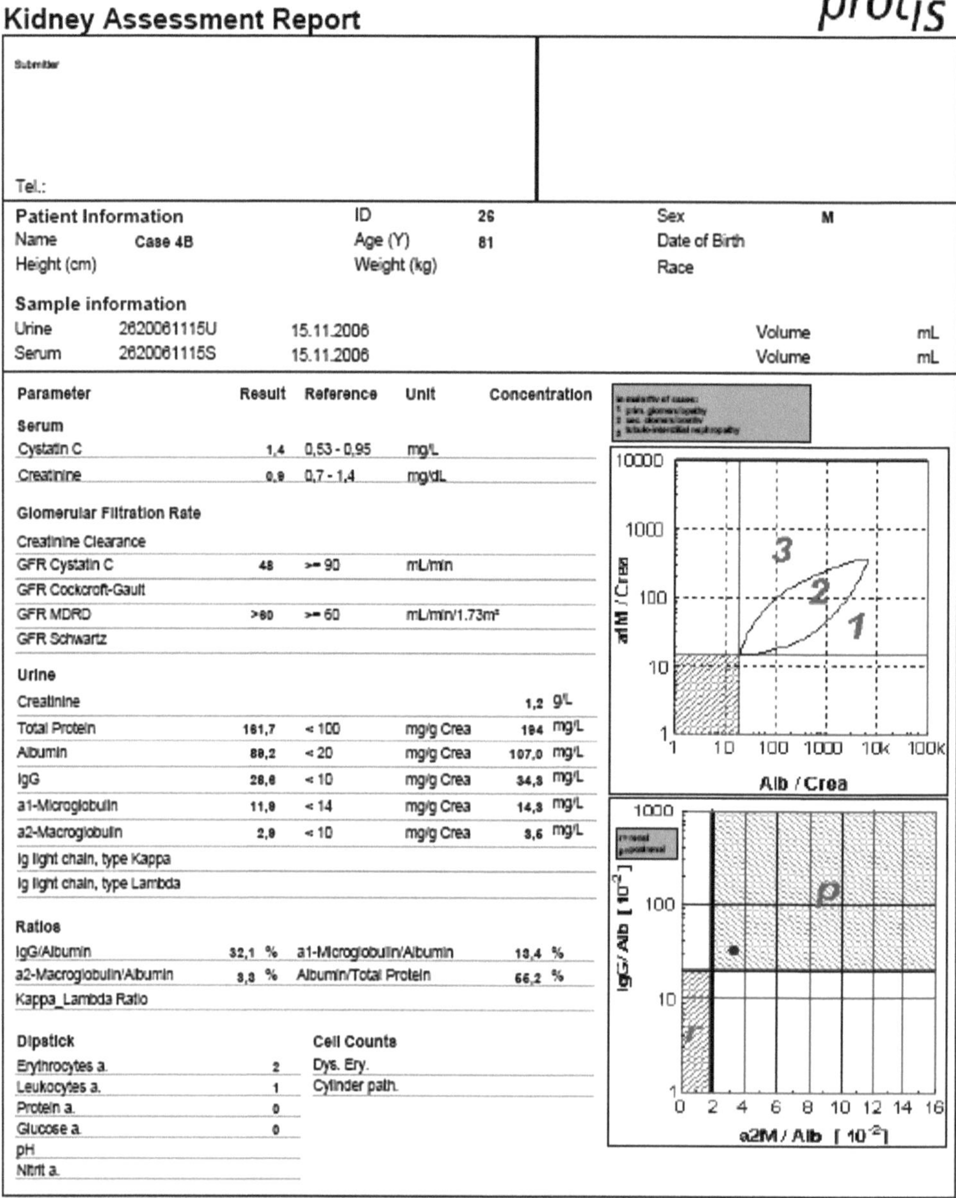

Kidney Assessment Report

protis

Submitter

Tel.:

Patient Information

Name	Case 4B	ID	26	Sex	M
Height (cm)		Age (Y)	81	Date of Birth	
		Weight (kg)		Race	

Sample information

Urine	2620061115U	15.11.2006	Volume	mL
Serum	2620061115S	15.11.2006	Volume	mL

Parameter	Result	Reference	Unit	Concentration	
Serum					
Cystatin C	1,4	0,53 - 0,95	mg/L		
Creatinine	0,9	0,7 - 1,4	mg/dL		
Glomerular Filtration Rate					
Creatinine Clearance					
GFR Cystatin C	48	>= 90	mL/min		
GFR Cockcroft-Gault					
GFR MDRD	>60	>= 60	mL/min/1.73m²		
GFR Schwartz					
Urine					
Creatinine				1,2	g/L
Total Protein	161,7	< 100	mg/g Crea	194	mg/L
Albumin	89,2	< 20	mg/g Crea	107,0	mg/L
IgG	28,6	< 10	mg/g Crea	34,3	mg/L
a1-Microglobulin	11,9	< 14	mg/g Crea	14,3	mg/L
a2-Macroglobulin	2,9	< 10	mg/g Crea	3,6	mg/L
Ig light chain, type Kappa					
Ig light chain, type Lambda					

Ratios

IgG/Albumin	32,1 %	a1-Microglobulin/Albumin	13,4 %
a2-Macroglobulin/Albumin	3,3 %	Albumin/Total Protein	66,2 %
Kappa_Lambda Ratio			

Dipstick

Erythrocytes a.	2
Leukocytes a.	1
Protein a.	0
Glucose a.	0
pH	
Nitrit a.	

Cell Counts

Dys. Ery.	
Cylinder path.	

Interpretation

GFR estimation indicates a moderate decrease in glomerular filtration, which clearly indicates a pathologic level of kidney function loss. (Based on the Cystatin C formula)

Despite an inconspicuous serum-creatinine concentration, a possible GFR reduction cannot be ruled out.

Albuminuria is present. This is classified as slight (presence of microalbuminuria). IgG excretion is additionally increased. No additional, tubular proteinuria exists.

Leucocyturia and, with high probability, postrenal hematuria, are present; however, an additional renal protein excretion cannot be excluded. Control is recommended following remission of hematuria and leucocyturia. Tubulo-interstitial dysfunction can be ruled out with high probability, whereas a glomerular permeability disorder cannot.
Urine protein differentiation should be repeated following the remission of leukocyturia, as inflammations in the descending urinary tracts may also lead to a slight increase in Albumin and/or IgG.
The urine measurement variables make an infection of the urinary tract probable. A urine culture (mid-stream) is recommended.

Fig. 9. Case 5 urinary tract infection.

Kidney Assessment Report

protis

Submitter

Tel.:

Patient Information

Name	Case 6	ID	27	Sex	M
Height (cm)		Age (Y)	69	Date of Birth	
		Weight (kg)		Race	

Sample information

Urine	2720061115U	15.11.2006	Volume	mL
Serum	2720061115S	15.11.2006	Volume	mL

Parameter	Result	Reference	Unit	Concentration	
Serum					
Cystatin C	4,3	0,53 - 0,95	mg/L		
Creatinine	5,6	0,7 - 1,4	mg/dL		
Glomerular Filtration Rate					
Creatinine Clearance					
GFR Cystatin C	12	>= 90	mL/min		
GFR Cockcroft-Gault					
GFR MDRD	11	>= 60	mL/min/1.73m²		
GFR Schwartz					
Urine					
Creatinine				0,5	g/L
Total Protein	744,0	< 100	mg/g Crea	372	mg/L
Albumin	460,0	< 20	mg/g Crea	230,0	mg/L
IgG	31,6	< 10	mg/g Crea	15,8	mg/L
a1-Microglobulin	70,8	< 14	mg/g Crea	35,4	mg/L
a2-Macroglobulin	4,0	< 10	mg/g Crea	2	mg/L
Ig light chain, type Kappa					
Ig light chain, type Lambda					

Ratios

IgG/Albumin	6,9 %	a1-Microglobulin/Albumin	15,4 %
a2-Macroglobulin/Albumin	0,9 %	Albumin/Total Protein	61,8 %
Kappa_Lambda Ratio			

Dipstick		Cell Counts
Erythrocytes a.	++	Dys. Ery.
Leukocytes a.	0	Cylinder path.
Protein a.	++	
Glucose a.	0	
pH		
Nitrit a.		

In majority of cases:
1 prim. glomerulopathy
2 sec. glomerulopathy
3 tubulo-interstitial nephropathy

a1M / Crea
Alb / Crea
10000
1000
100
10
1
1
10
100
1000
10k
100k
3
2
1

r=renal
p=postrenal
IgG / Alb [10^{-2}]
a2M / Alb [10^{-2}]
1000
100
10
1
0 2 4 6 8 10 12 14 16
p
r

Interpretation

GFR estimation indicates a drastic decrease in glomerular filtration, which is classified as renal failure. (Based on the Cystatin C formula)

Glomerular proteinuria is present. This is classified as significant (presence of macroalbuminuria). Additional, significant tubular proteinuria exists.

This constellation is compatible with glomerulopathy with restricted tubulo-interstitial reabsorption.
The presence of renal hematuria is highly likely; additional erythrocytes from postrenal sources cannot be ruled out. The postrenal hematuria associated with acute prostatitis may lack the expected elevation of a2-macorglobulin concentration; phase contrast microscopy (fresh, first morning urine) should be performed for differential diagnosis of hematuria.

Fig. 10. Case 6 hematuria with mixed proteinuria.

phase-contrast microscopy is recommended to look for signs of a renal (glomerular) hematuria such as dysmorphic erythrocytes, especially acanthocytes and erythrocyte casts.

A prerenal cause of a positive dipstick for blood should be considered if a "protein gap" is found. Additional tests can exclude or confirm a hemoglobinuria or myoglobinuria as possible causes. A hemoglobinuria occurs after intravascular hemolysis when the binding capacity of haptoglobin is exhausted. Myoglobinuria is a sign of lysis of muscular cells caused, e.g., by rhabdomyolysis after muscle trauma or by infarction.

Differentiation of renal and postrenal hematuria is based on the observation that urine concentrations of α_2-macroglobulin, immunoglobulin G, and α_1-microglobulin allow a classification of the source of hematuria, if they are referred to the excreted amount of albumin. Under physiological conditions, α_2-macroglobulin (molecular weight 720 kD) and IgG (molecular weight 150 kD) pass the glomerular filter only in traces because of their molecular size. Therefore, renal bleeding is characterized by a low albumin-ratio of these proteins. In postrenal hematuria, the ratios of IgG and α_2-macroglobulin to albumin in urine are similar to those in plasma (α_2-macroglobulin/albumin > 0.02 and IgG/albumin > 0.2). As a postrenal bleeding might mask an additional renal proteinuria, urine protein differentiation should be repeated after postrenal hematuria has ceased in order to exclude active renal disease.

If the hematuria cannot be clearly classified using marker protein ratios, phase-contrast microscopy should be used to clarify the source of bleeding (see Fig. 10).

References

1. Hofmann W, Guder W (1989) A diagnostic programme for quantitative analysis of proteinuria. J Clin Chem Clin Biochem **27**, 589–600
2. Ivandic M, Hofmann W, Guder WG (1996) Development and evaluation of a urine protein expert system. Clin Chem **42**, 1214–1222
3. Kouri T, Fogazzi G, Gant V, Hallander H, Hofmann W, Guder WG (2000) European urinalysis guidelines. Scand J Clin Lab Invest **60(Supp 231)** 1–96
4. Hofmann W, Ruth D, Guder W (2003) Urine protein differentiation and Protis, a new expert system for its interpretation. Riv Med Lab-JLM **4**, 67–68
5. Dati F, Metzmann E (2005) Proteins. Laboratory testing and clinical use. Publisher: Diasys 279–295

Chapter 9

Statistical Contributions to Proteomic Research

Jeffrey S. Morris, Keith A. Baggerly, Howard B. Gutstein, and Kevin R. Coombes

Abstract

Proteomic profiling has the potential to impact the diagnosis, prognosis, and treatment of various diseases. A number of different proteomic technologies are available that allow us to look at many proteins at once, and all of them yield complex data that raise significant quantitative challenges. Inadequate attention to these quantitative issues can prevent these studies from achieving their desired goals, and can even lead to invalid results. In this chapter, we describe various ways the involvement of statisticians or other quantitative scientists in the study team can contribute to the success of proteomic research, and we outline some of the key statistical principles that should guide the experimental design and analysis of such studies.

Key words: Blocking, Data preprocessing, Experimental design, False discovery rate, Image processing, Mass spectrometry, Peak detection, Randomization, Spot detection, 2D gel electrophoresis, Validation

1. Introduction

The field of proteomics has been rapidly advancing in recent years, and shows tremendous promise for research in various diseases, including cancer. Proteomic studies have been conducted to search for proteins and combinations of proteins that can be used as biomarkers to diagnose cancer, to provide more detailed prognostic information for individual patients, and even to identify which patients will respond to which treatments, with the ultimate hope of developing molecular-based personalized therapies that are more effective than the treatment strategies currently employed. Some arguments for the use of protein-based methods over mRNA or DNA-based methods to accomplish these goals are that proteins are more relevant to the biological functioning

Alex J. Rai (ed.), *The Urinary Proteome: Methods and Protocols*, Methods in Molecular Biology, vol. 641,
DOI 10.1007/978-1-60761-711-2_9,

of the cell, being typically downstream from the DNA and mRNA, and are subject to posttranslational modifications, plus proteomic assays can be applied to readily available biological samples like serum and urine, while mRNA-based methods in general cannot be applied in these settings, because these fluids have no mRNA source and thus should not contain nucleic acids.

Various proteomic instruments have been developed that are able to survey a biological sample and reveal a slice of their proteomic profile. One of the oldest technologies in this field is 2d gel electrophoresis (2DE), which was developed in the 1970s and has been extensively used since (1). While this technology has been around for a while, computational limitations and other factors have prevented it from reaching its potential impact in biomedical research. The advanced computing tools now available and the significantly improved preprocessing algorithms being developed for this technology may make it possible for 2DE to overcome some of its perceived shortcomings and make a much stronger contribution to biomedical research in the near future. In recent years, mass spectrometry-based techniques have gained popularity in proteomic profiling studies as an alternative to 2DE. Methods used include matrix-assisted laser desorption and ionization mass spectrometry (MALDI-MS) and the closely related surface-enhanced laser desorption and ionization mass spectrometry (SELDI-MS) commercially developed by Ciphergen, Inc. (Fremont, CA). The interest in these technologies was fueled by some early studies reporting incredible results, for example 100% sensitivity and 94% specificity for diagnosing ovarian cancer based on serum samples (2). Unfortunately, many of these results have not held up to scrutiny, with experimental design flaws in these studies drawing their results into question (3–8). Subsequent well-designed studies have been performed that, while falling short of the incredible results of the initial studies, have identified potential biomarkers that may be useful for improving the early diagnosis of certain cancers (9). In recent years, liquid chromatography coupled to mass spectrometry (LC-MS) has been increasingly used for proteomic profiling. This technology has some advantages over MALDI-MS, with its 2-dimensional nature allowing one to resolve more proteins and simultaneously discern their identities, but at the price of reduced throughput capabilities.

While quite different in many ways, all of these technologies share some common characteristics. First, all of these instruments are very sensitive. This sensitivity is good when it comes to detecting subtle proteomic changes in a sample, but unfortunately this sensitivity also extends to small changes in sample handling or processing. Because of this, experimental design considerations are crucial in order to obtain valid results from proteomic studies. Second, all of these technologies yield complex, high dimensional data, either noisy spectra (MALDI-MS, SELDI-MS) or images (2DE, LC-MS), which must be preprocessed. This preprocessing is a challenge, and if not done properly, can prevent one from identifying the significant proteomic patterns in a given study. Third, these technologies all produce simultaneous measurements of hundreds or thousands of peptides or proteins in a sample. Given these large numbers, multiple testing and validation need to be done properly in order to ensure that reported results are likely to be from real changes in the proteome, and not false positive results due to random chance alone.

All of these characteristics raise quantitative issues that must be adequately addressed in order to perform successful proteomic studies. The goal of this chapter is to demonstrate the benefit of involving statisticians or other quantitative scientists in all areas of proteomic research, and to elucidate some of the statistical principles and methods we have found to be essential for this field. We will illustrate these principles using a series of case studies, some from studies performed at The University of Texas M.D. Anderson Cancer Center, and some performed at other institutions.

There are four key areas in which a statistician can make a major impact on proteomic research: experimental design, data visualization, preprocessing, and biomarker identification. We will organize this chapter according to these four areas, and finish with some general conclusions.

2. Experimental Design

In this section, we will describe three case studies, point out important experimental design principles highlighted by them, and then provide some general experimental design guidelines for proteomic studies.

2.1. Case Study 1: Brain Cancer MALDI Study

A group of researchers conducted an experiment at M.D. Anderson Cancer Center on tissue samples from 50 patients with brain cancer, which were believed to include two subtypes of the disease (10). The disease subtype information was “stripped out” and the resultant blinded dataset consisting of MALDI spectra from these samples was brought to our group for analysis. The aim of the analysis was to perform unsupervised clustering of the data to see if the two subtypes could be identified correctly and blindly. After preprocessing the data using in-house routines of simultaneous peak detection and baseline correction (SPDBC, see (11)), we performed hierarchical clustering on the set of peaks consistently detected across spectra. This clustering produced two distinct groups, which excited us (See Fig. 1).

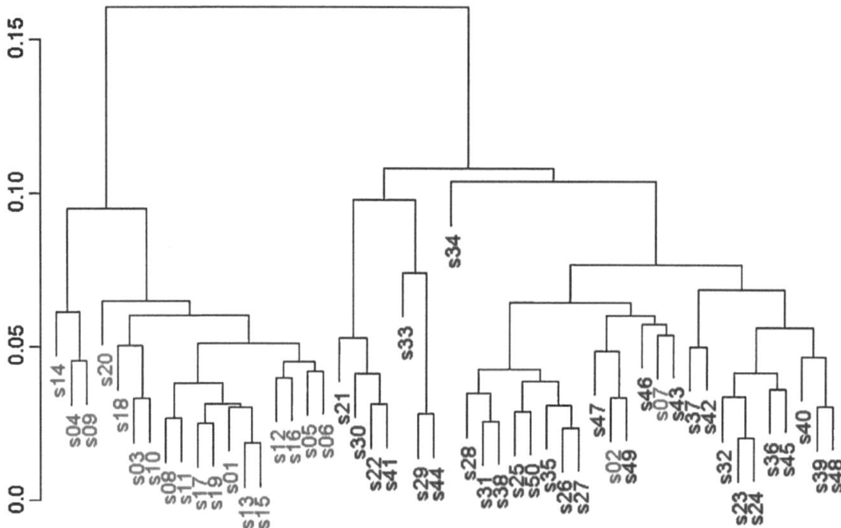

Fig. 1. Discovery of clusters in data from bsa70 fraction of brain tumor samples.

Upon being unblinded, however, we found that these groups did not cluster according to disease subtypes, but according to sample number, with samples 1–20 roughly clustering in one group and 21–50 clustering into the other. We found out that the sample collection protocol had been changed after the first 20 patients had been collected. This change in protocol resulted in systematic and reproducible changes in the serum proteome that were stronger than the changes between the subtypes of brain cancer present in the study. While mass spectrometry can be exquisitely sensitive to changes in the part of the proteome it can measure accurately, it can be even more sensitive to changes in sample collections and handling protocols. Thus, it is crucial with these sensitive technologies to ensure consistent protocols throughout a given study, and especially take care not to differentially handle cases and controls – since this can induce large scale differences between them that are not driven by the biology of interest, but rather the nuisance factor of handling. Sample handling and processing is not the only nuisance factor that can impact a study in this way.

2.2. Case Study 2: Leukemia SELDI Study

In another study performed at MD Anderson Cancer Center, SELDI spectra were obtained from blood serum of 122 Leukemia patients, some with chronic myelogenous leukemia (CML), and others with acute lymphoblastic leukemia (ALL). Since with leukemia, the "tumors" are in the blood, we expected that it should not be difficult to distinguish these two subtypes using these blood serum spectra. Again, in a blinded fashion as the previous

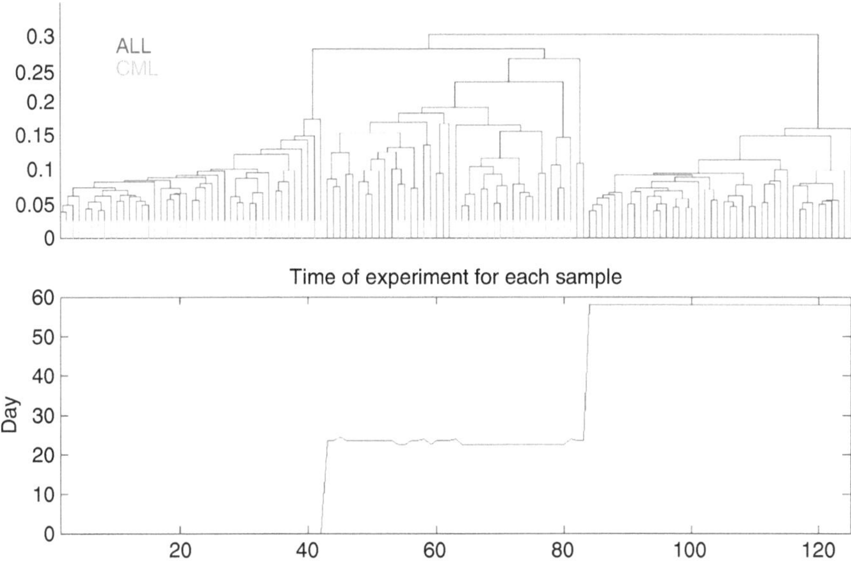

Fig. 2. Unsupervised clustering of Leukemia spectra reveals clustering driven by run date, not type of leukemia.

example, we preprocessed the data and performed hierarchical clustering. Rather than two clusters, we found three very strong clusters (see Fig. 2, top panel).

Upon unblinding, we were encouraged that cluster 1 was 100% CML, cluster 3 was 100% AML, showing good discrimination, and cluster 2 was split between the two. Did cluster 2 represent a subtype of CML and ALL that were proteomically similar, or was something else going on here? We plotted the sample number versus run date (Fig. 2, bottom panel), and found that the spectra were obtained in three distinct blocks over a period of 2 months. Block 1 contained only CML samples, block 3 contained only ALL samples, and block 2 contained a mix of CML and AML samples. These three processing blocks corresponded to the three clusters our unsupervised analysis discovered. This demonstrates that the SELDI instrument can vary systematically over time, especially over the scope of weeks and months. This systematic difference was stronger than the proteomic difference between ALL and CML in blood serum. Time effects are not just limited to SELDI-MS instruments, nor are they limited to long-term effects over weeks and months – sometimes these instruments can give very different results from day-to-day.

2.3. Case Study 3: Ovarian Cancer High Resolution MALDI Study

A study was performed by the NCI-FDA group (12) using high resolution Qstar profiling of serum for ovarian cancer detection, with 100% sensitivity and 100% specificity reported. In Fig. 6a of their paper, they plotted a QA/QC measure for each spectrum against sample number, with different symbols indicating the 3 days on which the samples were run. As the authors noted, this figure clearly showed that something was going wrong on the third day. Figure 7 of their paper showed the record numbers of the spectra deemed of high enough quality for data analysis, with controls on the left and cancers on the right. Superimposing these figures (see Fig. 3, adapted from (13)), we see that they coincide perfectly.

This indicates that almost all of the controls were run on day 1, with a small number on day 2, and that cancers were run on days 2 and 3. This is a serious problem, since any systematic changes in the machine from day-to-day can systematically bias the results by distorting one group more than the other. The authors noted this type of trend, since the reduced quality signals on the third day they mentioned would affect only cancer spectra. Thus, it is possible that the observed 100% sensitivity and specificity was driven by the machine effect, and not the case/control status.

The previous two case studies highlight the danger of *confounding* in experimental design – in which by design (usually inadvertently) a nuisance factor is inseparably intermingled with a

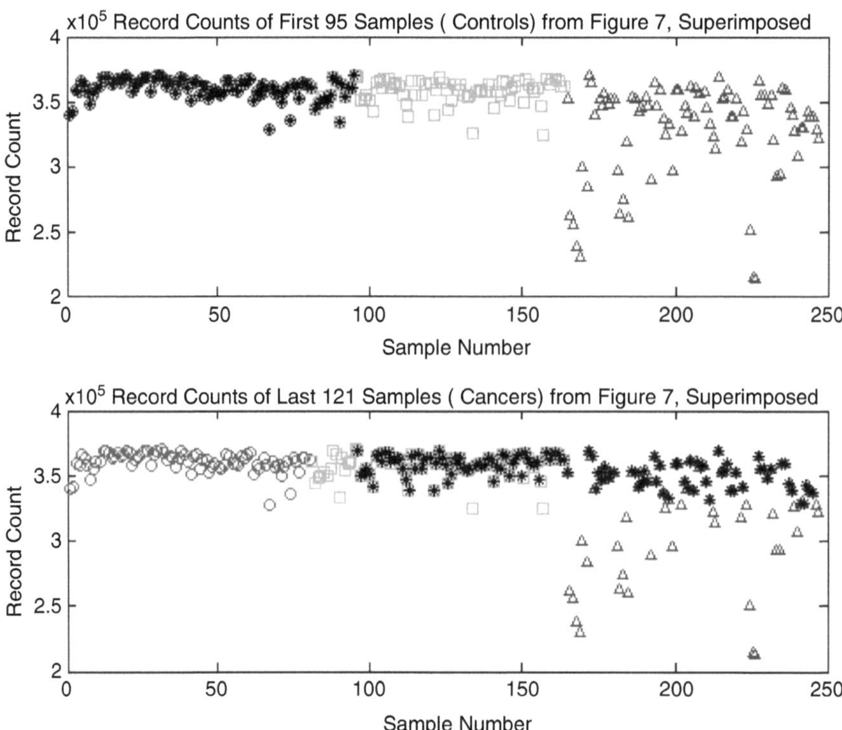

Fig. 3. Plot reveals confounding between cancer status and run date in the study design, which is especially problematic given the quality control problems evident on day 3. Adapted from (13).

factor of interest in the study. In case study 2, block was confounded with CML/ALL type and in case study 3, run day was confounded with case/control status. This confounding can also occur at the sample collection and handling level as well – if all cases come from one center, and all controls from another, then there is no way to tell whether differences between the groups are due to case/control status or to factors specific to the centers. This is an especially sinister problem because it can make a study appear to have strikingly strong effects for the factor of interest, when in fact these effects may be driven by the confounding nuisance factor, not the factor of interest. This can be ultimately revealed when later studies are not able to replicate the earlier findings. For example, in case study 3, it is impossible to tell whether the observed remarkable separation was due to the day effect or the case versus control effect. When the two effects are completely confounded, no statistical modeling can fix this problem – the data set is essentially worthless unless one is willing to arbitrarily assume that the confounded nuisance factor is not significant, which would clearly be a wrong assumption in case studies 2 and 3.

How can one prevent confounding from ruining their proteomics studies? Confounding can be prevented by following certain experimental design principles during the planning phase of the study. There has been extensive statistical work in the area of experimental design – some principles can be found in standard textbooks in the area (14). Two key principles are *blocking* and *randomization*. Blocking is necessary when for whatever reason it is not possible to perform the entire experiment at one time. Each *block* is then a portion of the experiment, for example, a certain point in time where a batch of gels or spectra are run, a given set of samples collected from a given site at a given time, or a given SELDI chip is used. In practice, the blocks are typically constructed merely by convenience, not based on any sense of balance between study groups nor using any statistical principles. As seen in the case studies above, this practice can lead to confounding of study factor with the blocking factor. Instead of convenience, blocks should be thoughtfully constructed trying to balance the samples within each block with respect to the study factors of interest. For example, one design is to have balanced numbers of cases and controls within each block. In this case, strong block effects, if they exist, will only add random *noise* to the data, not systematic *bias*, since the block effect will impact both cases and controls equally. For a given proteomic technology, one should identify which factors and steps of the process introduce the most variability, and then block on these factors. At the point where further blocking is not possible, then randomization should be used to determine run order, sample position, etc. This randomization protects against bias caused by some unknown or unexpected source. The randomization can be done using a random number generator, which is an easy exercise for any quantitative scientist. Note that picking what seems to be a "random-like" sequence is not the same as randomization. Studies have clearly shown that individuals do a poor job acting as random number generators; so true randomization should be done. If the studies in case 2 and case 3 had randomized the run order of their experiments, or used block designs, they would have prevented the confounding that for all practical purposes invalidated these data sets.

In summary, the key design issues to remember include:

- Sample collection and handling must be carefully controlled.
- Block on factors that are likely to impact on the data, including sample collection and handling.
- At whatever level you stop blocking, randomize.

In many areas of science, these principles are routinely incorporated into study conduct. In our experience, we have found this not to be the case in biological laboratory science, including proteomics. Even in the best-run laboratory, given the extreme sensitivity of proteomic instruments, it is crucial to take these factors into consideration. This point cannot be made too strongly: to ignore them is to risk obtaining data that are completely worthless for detecting group differences, i.e., achieving the goals of the study.

3. Data Visualization

One of the most important statistical principles in analyzing proteomic data is also the simplest – look at the data! Frequently, simple plots of the data can reveal experimental design problems like those discussed in the previous section or other issues that need to be addressed before the data can be properly analyzed. In this section, we will look at three case studies that illustrate the benefit of data visualization techniques, and then we will discuss some tools we have found useful for this purpose.

3.1. Case Study 4: Duke University Lung Cancer MALDI Data

These data are from the First Annual Proteomics Data Mining Competition at Duke University, and consist of 41 MALDI spectra obtained from blood serum from 24 lung cancer patients and 17 normal controls (15). The goal is to identify proteomic peaks from serum that can distinguish individuals with and without lung cancer.

We plotted a *heat map* of the entire set of 41 spectra. A heat map is an image plot of the data, with rows corresponding to

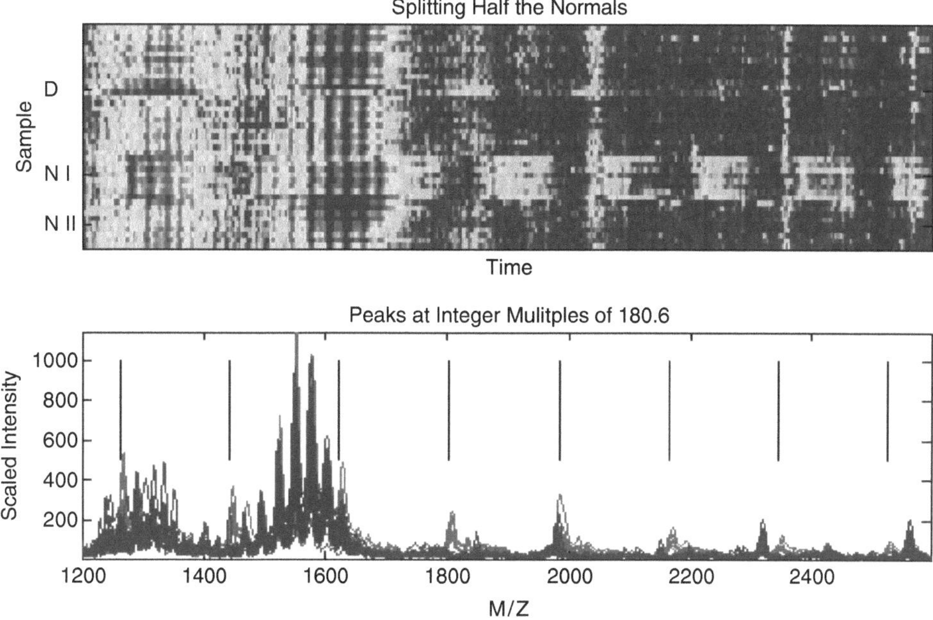

Fig. 4. A simple heat map revealed peaks at integer multiples of 180.6, which is unlikely to be driven by biology, so we removed these peaks from consideration when discriminating between cancers and normals.

individual spectra and columns corresponding to individual clock ticks within the spectra. Upon looking at the heat map, an anomaly jumped out at us – we noticed a distinguishing pattern in the low mass region of the first 8 spectra from normal controls. Specifically, we saw that these samples contained recurring peaks at precise integer multiples of 180.6 Da, suggesting the presence of some unknown substance with identical subunits (see Fig. 4). This is unlikely to be a protein, since no combination of amino acids sums to this mass, so what we are seeing here may be a matrix or detergent effect that affected a single batch (or block) of samples. Thus, we removed this region of the spectra (1,200–2,500 Da) from consideration in our search for differentially expressed peaks. Had we not looked at the heat map, we likely would have missed this pattern, and would have mistakenly reported these peaks as potential lung cancer biomarkers, as done by many other groups analyzing this data set as part of the competition.

3.2. Case Study 5: NCI-FDA Ovarian Cancer SELDI Study

These data are from a seminal paper appearing in February, 2002, in *The Lancet* (2) that reported finding patterns in SELDI proteomic spectra that can distinguish between serum samples from healthy women and those from women with ovarian cancer. The data consisted of 100 cancer spectra, 100 normal spectra, and 16 "benign disease" spectra. The cancer and normal spectra were randomly split, with 50 cancer and normal spectra used to train a classification algorithm. The resulting algorithm was used to classify the remaining spectra. It correctly classified 50/50 of the cancers, 47/50 of the normals, and called 16/16 of the benign disease spectra "other" than normal or cancer. There have been a large number of criticisms and issues raised with this study (3–8, 16), but here we will focus on the remarkable result that they were able to recognize all 16 of the benign disease spectra as being something other than cancer or normal.

After publication, the authors made their data publicly available on their website. We downloaded these data, and one of our first steps was to plot a heat map of the data, given by the top panel of Fig. 5 (adapted from (4)), with the samples in order from the cancers to the normals to the benign disease spectra (other).

From this heat map, it is not surprising that their classification algorithm was able to distinguish the benign disease spectra from cancer and normal. What is surprising is that the cancer and normal spectra look quite alike to the naked eye, while the benign disease spectra looked completely different from the others. Why would spectra from normals be more similar to those from cancers than those from individuals with benign disease? These spectra, the ones reported in the *Lancet* paper, were obtained using

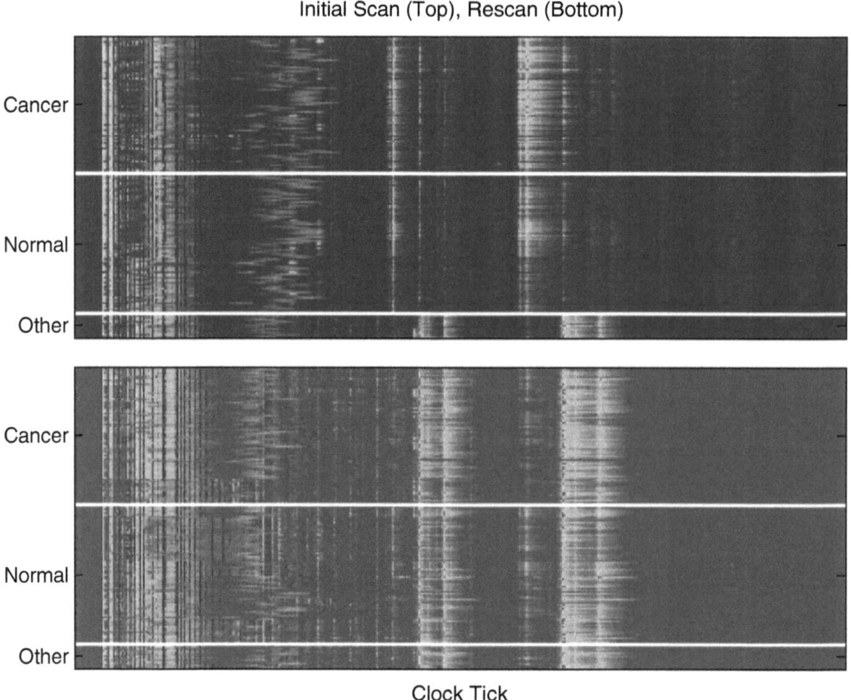

Fig. 5. Heat map of all 216 samples run on the H4 chip (*top*), and on the WCX2 chip (*bottom*). The extreme difference in the "benign disease" group in the spectra on the top, along with the similarity of these profiles to the WCX2 spectra, suggest a change in protocol occurred in the middle of the first experiment. Adapted from (4).

Ciphergen's H4 SELDI chip. The authors also reran the same 216 biological samples on the WCX2 SELDI chip, and posted these data on their website. The bottom figure contains a heatmap of these 216 spectra. Plotting these figures adjacent to one another, we see that the benign spectra reportedly run on the H4 chip are very similar to all of the spectra from the WCX2 chip, suggesting a change in protocol occurred in the middle of the first experiment. This is clearly revealed by a simple heatmap of the raw data.

3.3. Case Study 6: M.D. Anderson Urine SELDI Data

This study was performed at M.D. Anderson Cancer Center, and involved SELDI spectra obtained from urine samples from five groups of individuals: (1) disease-free individuals, (2) patients presenting with low-grade bladder tumors, (3) patients presenting with high-grade bladder tumors, (4) patients with history of low-grade bladder tumors, and (5) patients with history of high-grade bladder tumors. One goal of the study was to find proteins distinguishing the groups. After preprocessing and analysis, we found several peaks that could separate controls from cancers successfully, including the set of three peaks shown in the first three rows of Fig. 6 (adapted from (10)).

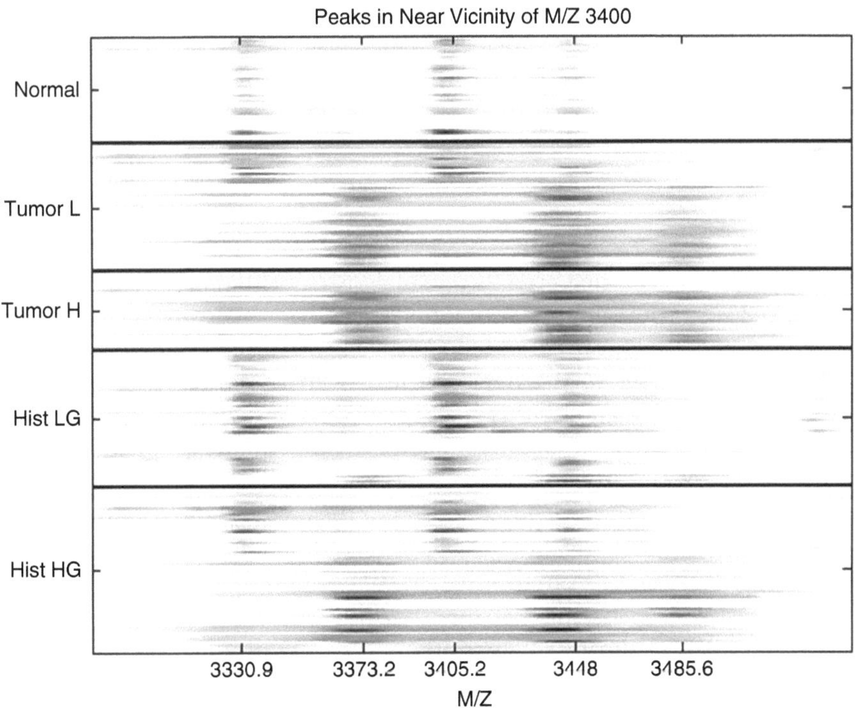

Fig. 6. Comparisons of urine spectra in the vicinity of M/Z 3,400, revealing calibration issues between the spectra in different groups. Adapted from (10).

Upon plotting a heat map of the data, however, we discovered that the spectra from the disease-free individuals contained the same three peaks, but that these peaks were shifted to the left relative to many of the peaks from those with cancer. This revealed that the data had been inadequately calibrated prior to analysis. After correcting the calibration, there was no difference in protein expression in this region of the spectra.

These three case studies all emphasize the importance of plotting a heat map of the data before conducting any analyses. There are other graphical procedures that can also be routinely used for diagnostic and confirmatory purposes. Dendograms produced by hierarchical clustering of the raw and preprocessed spectra, as demonstrated in the previous section, are useful for diagnosing systematic trends in the data, those we want (between treatment groups) and those we don't (between nuisance factors). In the next section, we will discuss the issue of preprocessing. It is always a good idea to view plots of individual spectra before and after preprocessing, to ensure that the preprocessing is sufficient, and to ensure that it did not induce any unwanted artifacts. Also, once identifying peaks in mass spectrometry experiments or spots in 2DE experiments that are differentially expressed, it is also advisable to look at these regions in the preprocessed as well as raw data to ensure that you believe those results are real. Statistical summaries and tests of statistical significance are useful, but it is still important to look at the actual data to confirm that the results look believable and practically significant. Don't just trust the numbers – look at the pictures!

4. Preprocessing

The first step in the analysis of proteomics data is to extract the meaningful proteomic information from the raw spectra or images. We use the term *preprocessing* as a collective term referring to the data and image analysis steps that are necessary to accomplish this goal. Given a set of N spectra or images containing information from a total of p proteins or peptides, the goal of preprocessing is to obtain an $N \times p$ matrix, whose rows correspond to the individual spectra or images, and the columns contain some quantification of a proteomic unit such as a spectral peak or 2DE spot. From this matrix, we can perform a variety of different types of statistical analyses to find out which proteins are associated with outcomes of interest, as discussed in the next section. The different steps in preprocessing include calibration or alignment, baseline or background correction, normalization, denoising, and feature (peak or spot) detection. It is important to find effective methods for performing these steps, since subsequent analyses condition on these determinations. While other methods are available in existing literature and new methods are constantly being developed, here we briefly describe methods we have developed for one-dimensional mass spectrometry methods such as MALDI-MS and SELDI-MS, and for two-dimensional methods such as 2DE. We have found our methods to be very effective and believe they outperform other existing alternatives.

Preprocessing MALDI-MS and SELDI-MS spectra involves a number of steps that interact in complex ways. Some of these steps are as follows:

- Alignment involves adjusting the time of flight axes of the observed spectra so that the features of the spectra are suitably aligned with one another.
- Calibration involves mapping the observed time-of-flight to the inferred m/z ratio.
- Denoising removes random noise, likely caused by electrical noise from the detector.
- Baseline subtraction removes a systematic artifact typically observed for these data, usually attributed to an abundance of ionized particles striking the detector during early portions of the experiment.
- Normalization corrects for systematic differences in the total amount of protein desorbed and ionized from the sample plate.
- Peak detection involves the identification of locations on the time or m/z scale that correspond to specific proteins or peptides striking the detector. Note that this differs from the process of peak identification, which is the process of determining the species molecule that causes a peak to be manifested in the spectra.
- Peak quantification involves obtaining some quantification for each detected peak for all spectra in a sample, and may involve computing heights or areas.
- Peak matching across samples may be necessary if peak detection is applied to the individual spectra, in order to match and align the peaks detected for different spectra.

We view preprocessing as the decomposition of the observed raw spectra into three components: true signal, baseline, and noise. The basic idea is to estimate and remove the baseline and noise, and then decompose the signal into a list of peaks, whose intensities are then recorded. Our procedure is described below and follows two basic principles: (1) Keep it simple; we wish to keep processing to a minimum in order to avoid introducing additional bias or variance into the measurements, and (2) Borrow strength; we wish to borrow information across spectra wherever possible in order to make more accurate determinations. This procedure has evolved, with the principles discussed in published literature (15, 17–20). Software implementing this process is freely available in both standalone form (PrepMS, (21)), which implements method described in (20) and as an R package (msProcess, available at http://cran.r-project.org/web/packages/msProcess/index.html).

1. Align the spectra on the time scale by choosing a linear change of variables for each spectrum in order to maximize the correlation between pairs. We have found that alignment can be done much more simply and efficiently on the time scale rather than the clock tick scale.
2. Compute the mean of the aligned raw spectra.
3. Denoise the mean spectrum using the undecimated discrete wavelet transform (UDWT).
4. Locate intervals containing peaks by finding local maxima and minima in the denoised mean spectrum.
5. Quantify peaks in individual raw spectra by recording the maximum height and minimum height in each interval, which should contain a peak, then taking their difference. This quantification method implicitly removes the baseline artifact.
6. Calibrate all spectra using the mean of the full set of calibration experiments.

Assuming that a total of p peaks are detected on the average gel, this leaves us with an $N \times p$ matrix of peak intensities for the N spectra in the study that can be surveyed for potential biomarkers. Figure 7 contains an illustration of the preprocessing and peak detection process.

A key component of our approach is that we perform peak detection on the average spectrum, rather than on individual spectra, which has a number of advantages (20). First, it avoids the difficult and error-prone peak-matching step that is necessary when using individual spectra, improving the accuracy of the quantifications and eliminating the problem of missing data. Second, it tends to result in greater sensitivity and specificity for peak detection, since averaging across N spectra reinforces the true signal while weakening the noise by a factor of $\sqrt{N}$. This enables us to do a better job of detecting real peaks down near the noise region of the spectra, and decreases the chance of flagging spurious peaks that are in fact just noise. These points are demonstrated by the simulation study presented in (20), which shows that using the average spectrum results in improved peak detection, with the largest improvement being for low abundance peaks of high prevalence. Third, it speeds preprocessing time con-

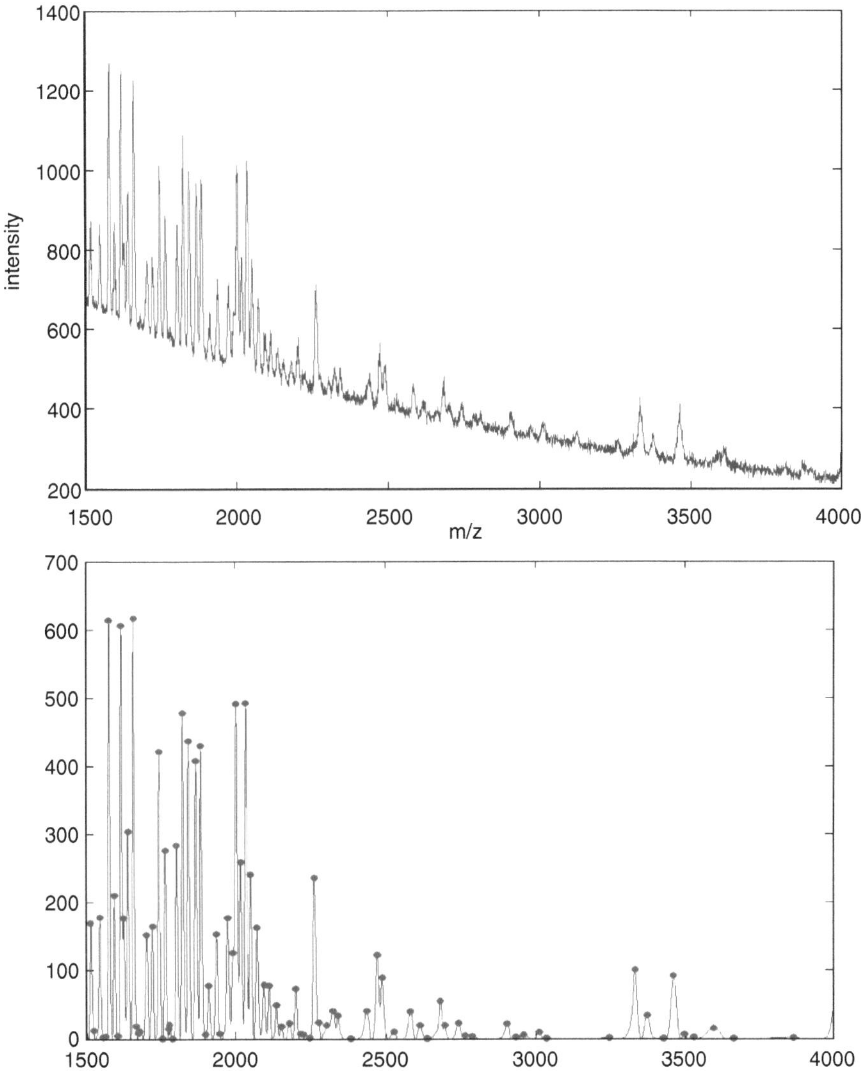

Fig. 7. Portion of a raw MALDI spectrum (*top*), along with the preprocessed version after denoising, baseline correction and normalization (*bottom*) using the method described in (20), with detected peaks indicated by the dots.

siderably, since peak matching is by far the most time-consuming preprocessing step.

Following similar principles, we have also proposed an approach for preprocessing 2DE data that we refer to as *Pinnacle* (22). The name comes from the fact that, unlike most existing methods, this method performs spot quantification using pixel intensities at pinnacles (peaks in two-dimensions) rather than spot volumes. Again, the fundamental principle is to keep the preprocessing as simple as possible, to prevent the extra bias and variance that can result from the propagation of errors, especially when complex methods are used. Following are the steps of the *Pinnacle* method:

1. Align the 2DE gel images.
2. Compute the average gel, averaging the staining intensities across gels for each pixel in the image.
3. Denoise the average gel using the 2-dimensional undecimated discrete wavelet transform (UDWT).
4. Detect pinnacles on the denoised average gel, which are pixel locations that are local maxima in both the horizontal and vertical directions with intensity above some minimum threshold (e.g., the 75th percentile for the average gel).
5. Perform spot quantification on the individual gels by taking the maximum intensity within a stated tolerance of each pinnacle location, and subtracting a local minimum to remove spatially varying background.

Assuming we detect p pinnacles on the average gel, this leaves us with an $N \times p$ matrix of spot intensities that can be surveyed for potential biomarkers. Figure 8 contains the heat map of an average gel, with detected spots marked.

This method is significantly quicker and simpler than the most commonly used algorithms for spot detection in 2DE, which involve performing spot detection on individual gels using complex spot definitions, quantifying using spot volumes, then matching spots across gels. As for mass spectrometry, use of the average gel for spot detection leads to a number of advantages over

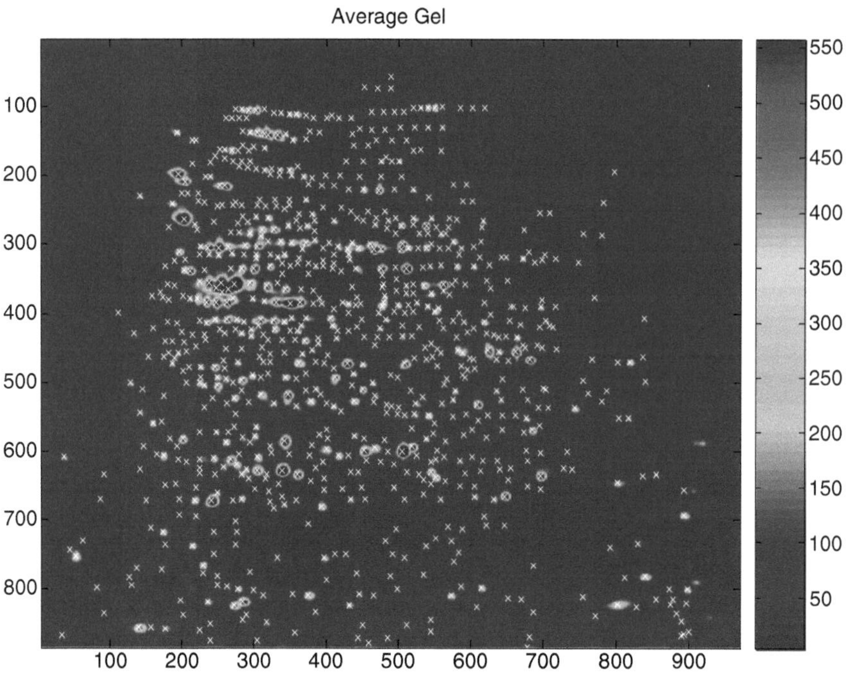

Fig. 8. Average gel from a 2d gel experiment, with detected spots using the Pinnacle method indicated the symbol “x”.

methods performing detection on individual gels than matching spots across gels. First, the use of the average gel leads to greater sensitivity and specificity for spot detection, and leads to robust results, since it can automatically discount artifacts unique to individual gels. Second, the use of the average gel allows one to avoid the difficult and error-prone step of matching spots across gels. As a result, with our approach the accuracy of spot detection actually increases with larger studies, in contrast with other approaches that match across gels, whose accuracy decreases markedly with larger numbers of gels because of the propagation of spot detection and matching errors. Third, use of the average gel eliminates the missing data problem, since quantifications are well defined for every spot on every gel in the study. Fourth, the method is quick and automatic, not requiring any of the subjective, time-consuming hand editing that is necessary with other approaches to fix their errors in spot detection, matching, and boundary estimation.

Two independent dilution series experiments (22, 23) were used to demonstrate that Pinnacle yielded spot quantifications with greater reliability and validity than three popular commercial packages, PDQuest (Bio-Rad, Hercules, CA), Progenesis (Nonlinear Dynamics, Newcastle upon Tyne, UK), and SameSpots (Nonlinear Dynamics, Newcastle upon Tyne, UK), and demonstrated increased power for detecting differentially expressed spots.

Another important feature of Pinnacle is that pinnacle intensities are used in lieu of spot volumes for quantification. We believe that this property explains the dramatically increased reliability as demonstrated by lower CVs in the dilution series experiments. Although the physical properties of the technology suggest that spot volumes are a natural choice for protein quantification, they are also problematic since they require estimation of the boundaries of each individual spot, which is a difficult and error-prone exercise, especially when many spots overlap. The spot boundary estimation variability is a major component in the CVs for individual spots across gels. It can be easily shown that pinnacle intensities are highly correlated with spot volumes, yet can be computed without estimating spot boundaries, which leads to more precise quantifications and lower CVs. Pinnacle is currently being developed into a standalone executable.

5. Biomarker Discovery

The final step of analysis is to discern which protein peaks or spots are associated with the factors of interest, e.g., differentially expressed, and thus may indicate potential biomarkers. This is the analysis step frequently receiving the most attention, but in this chapter, we give it the least. This is because the probability of success in this step depends strongly on the adequacy of the previous steps discussed in lengths in this chapter. If the experimental design and/or the preprocessing have been done poorly, then there is little hope of finding meaningful biomarkers regardless of what methods are used in the biomarker discovery phase. Poor design or preprocessing can mask significant results, making them more difficult to discover, or more ominously can raise many false positive results that can mislead one into thinking they have discovered a potential biomarker, only to be disappointed when subsequent studies fail to replicate or validate the results. That said, we still discuss some statistical principles for guiding the biomarker discovery process.

Given an $N \times p$ matrix of features (peak/spot quantifications) from each spectrum/gel, there are a large number of possible methods that can be used to identify which features are significantly associated with the factor of interest, and thus may indicate potential biomarkers. These methods can be divided into two types. The first is *univariate* methods that look at the features one-at-a-time. One example is to apply a simple statistical test (e.g., a t-test or simple linear regression) to each feature, and then compose a ranked list of the features based on the p-values and/or effect sizes. The second is *multi-factor* methods that seek to build models to predict an outcome of interest using multiple features. In either case, the key statistical challenge is dealing with the multiplicity problem, which results from surveying a large number of features. In the univariate context, we expect a certain number of features to have small p-values even if no proteins are truly related to the factor of interest. In the multivariate context, given the large number of potential features that could be used in the model, we expect it is easy to find a model that is nearly perfect in predicting the desired outcome by random chance alone, even when in fact none of the features are truly predictive of the outcome. In their survey of published microarray studies, Dupuy and Simon (24) found that a large number of them did not adequately control for multiple testing. Methods that adjust for these multiplicities ensure that the reported outcomes of the study are likely to be real, and not simply due to random chance.

In the univariate case, adjusting for multiplicities means finding an appropriate p-value cutoff that has desirable statistical properties. The usual cutoff of 0.05 is typically not appropriate, since we expect about 5% of the features to have p-values of less than 0.05 even if none are in fact truly associated with the feature of interest. A classical approach for dealing with multiplicities is Bonferroni, which would use a cutoff of $0.05/p$ for determining significance, where p is the number of features surveyed. This cutoff would control the *experiment-wise error rate* at 0.05, meaning the total expected number of false discoveries in the study is 0.05. This criterion strongly controls the false positive rate, but results in a large number of false negatives, so is widely considered too conservative for exploratory analyses like these. Other methods control the *false discovery rate*, or FDR (25). Controlling the false discovery rate at 0.05 means that of the features declared significant, we expect 5% or fewer of them to be false positives. There are a number of procedures for controlling FDR (25–36). Many of these methods take advantage of the property that the *p*-values corresponding to features that are not associated with the factor of interest should follow a uniform distribution, while the distribution of *p*-values for predictive features should be characterized by an overabundance of small *p*-values. In most of these methods, one inputs the list of *p*-values and desired FDR, and receives as output a *p*-value cutoff that preserves the desired FDR. Typically, this cutoff is smaller than 0.05, but quite a bit larger than the corresponding Bonferroni bound $0.05/p$.

When building multi-factor models, it is important to validate the model in some way to assess how well the model can predict future outcomes. The key principle is that the same data should not be used to both build the model (called *training*) and assess its predictive accuracy (called *testing*). Testing a model with the same data used to build it results in an overly optimistic sense of the model's predictive ability, with the problem much worse in settings like proteomics in which we have such a large number of potential models. We have performed simulation studies (not published) that demonstrate that building a model with just 14 features from a set of 1,000 features to distinguish 30 cases from 30 controls, we can obtain 100% predictive accuracy on the training set nearly every time even when none of the features are truly predictive, i.e., all of the data are just random noise. This highlights the notion that over-fitting training data when fitting multi-factor models with large numbers of potential features is very easy. It is crucial to perform some type of independent validation when performing multivariate modeling. As discussed by Dupuy and Simon (24), improper validation is a frequent flaw in published reports from many microarray studies, even those published in very high impact journals.

One commonly used validation approach is to split the data into two sets, using the first in building the model, and then using the second to assess its predictive accuracy. While straightforward, this method has its drawbacks. It is inefficient, since it uses only half of the data to build the model, and also it may be difficult to convince skeptical reviewers that you didn't "cheat," i.e., look at various training/test splits and only report the one that yielded good results for both! One way to combat this is to repeat the analysis over for many different random splits of the data and average the prediction error results over them. A special case of this is called K-fold cross validation (CV). In K-fold CV, the data is split into K different subsets and K different analyses are performed. For each analysis, a different set of K-1 subsets are used to build a model, with the remaining subset set aside and used to test its predictive accuracy. The reported prediction error is then obtained by averaging over the predictive accuracies obtained for the K different analyses. It is possible to further improve this estimate by averaging over different random splits into K subsets.

One key point about performing model validation in these high dimensional settings is that it is important to perform external, not internal, cross validation. This means that one should repeat the feature selection process on every subset (external CV), as opposed to the common practice of selecting the predictive features using the entire data set then only repeating the model-fitting process for the different subsets (internal CV). Using internal cross-validation does not deal properly with the multiplicity problem of selecting the features, and can result in strongly optimistically biased assessments of predictive accuracy (37). This point is not well appreciated in the biological literature, as internal cross-validation is widely employed in practice (24).

Validation methods that involve splitting a given data set are useful, but also have limitations because any systematic biases that may be hard-wired into a given data set will be present in both the training and test splits. For example, when sample processing is confounded with group as in case studies 2 and 3 above, classifiers built using a subset of the data will yield impressively low classification errors on the remaining part of the data because of the systematic biases hard-coded into the data by the confounding factor, yet a model based on these data is very likely to have poor predictive accuracy in classifying future data. Thus, it is also beneficial after building a model with one data set to go out and collect some new data in an independent study and see how well the model predicts the outcomes in these data. It may be true that the current proteomic technologies are not consistent enough for this level of validation. If this is the case, then another alternative is to use the high-throughput proteomic technologies to discover proteins that are predictive of outcome, and then subsequently validate these results using more quantitative methods for specific proteins of interest, e.g., using antibodies. These types of assays have the chance to be robust enough to develop clinical applications, so may result in improved translation of proteomic discoveries to clinical settings.

This discussion largely assumed that a feature extraction approach was used for the analysis, as is typical. A feature extraction approach is a two step approach comprised of (1) detecting the "features", e.g. peaks or spots, and (2) analyzing these features to detect which ones are associated with experimental or clinical factors of interest. This approach has almost exclusively used to date, but can miss discoveries if the feature extraction method does not accurately extract all proteomic information from the data. Peaks or spots may fail to be detected, or specific proteins may be masked by adjacent peaks or spots and thus not considered in this analysis. An alternative to the feature extraction approach is a functional data or image-based modeling approach, whereby the proteomic spectrum or image is modeled in its entirety, and then regions of the spectra or gel are flagged as associated with factors of interest.

Wavelet-based functional mixed models (38) are a unified method that models a wide class of functional and image data, and has been applied to MALDI-TOF (39) and 2DGE (40) data. Given a desired fold-change difference and false discovery rate (FDR) threshold, this method can yield a map of the spectrum or image that flags which regions are differentially expressed, while accounting for both the statistical and practical significance. It can simultaneously study the effects of several factors on the image, both factors of interest (like treatment or clinical outcome) and nuisance factors (such as batch or run time).

Using this method, we have found results missed by peak-based methods in MALDI-TOF (39) and spot-based methods for 2DGE (40). For 2DGE, many of the flagged regions were spot-shaped, but corresponded to subregions of abundant spots that were not themselves differentially expressed. These results may indicate that this image-based modeling approach was able to detect differentially expressed minor proteins that were not detected as distinct spots because of co-migration with a more abundant major protein. These methods are more computationally intensive than feature extraction-based approaches, but have automatic, standalone software for their implementation, and represent a promising and exciting novel approach for proteomic analysis moving forward.

6. Conclusion

In this chapter, we have outlined some statistical principles for clinical proteomics studies. These guidelines span all aspects of the study, from sample collection and experimental design to preprocessing to biomarker discovery. While these principles are well known in the world of statistics, some of them have been underappreciated and underutilized in laboratory science, and specifically in clinical proteomics. The incorporation of these principles into your scientific workflow can ensure that the data collected will be able to answer the questions of interest and can prevent the results from being distorted by external biases. While anyone can follow the principles outlined here, we believe there is a strong benefit for a proteomics research team to involve a fully collaborating statistician or other quantitative scientist in all aspects of the study, and to involve them in all phases of the study design and planning process. Their involvement provides an individual whose role is specifically to think through these issues and concerns and ensure that the study has the best chance of fulfilling its intended purpose.

Acknowledgements

JSM's effort was partially supported by a grant from the NCI, R01 CA107304-01. HBG's effort was supported by NIDA grants DA18310 (cell–cell signaling neuroproteomics center). HBG and JSM are supported by AA13888.

References

1. O'Farrell P. H. (1975) High resolution two-dimensional electrophoresis of proteins. *Journal of Biological Chemistry* **250** 4007–4021.
2. Petricoin, E. F., Ardekani, A. M., Hitt, B. A., Levine, P. J., Fusaro, V. A., Steinberg, S. M., Mills, G. B., Simone, C., Fishman, D. A., Kohn, E. C., and Liotta, L. A. (2002). Use of proteomic patterns in serum to identify ovarian cancer. *Lancet* **359** 527–577.
3. Sorace, J. M. and Zhan, M. (2004). A data review and re-assessment of ovarian cancer serum proteomic profiling. *BMC Bioinformatics* **4** 24.
4. Baggerly, K. A., Morris, J. S. and Coombes, K. R. (2004). Reproducibility of SELDI-TOF protein patterns in serum: comparing datasets from different experiments. *Bioinformatics* **20** 777–785.
5. Diamandis, E. P. (2004a). Proteomic patterns to identify ovarian cancer: 3 years on. *Expert Review of Molecular Diagnostics* **4** 575–577.
6. Diamandis, E. P. (2004b). Mass spectrometry as a diagnostic and a cancer biomarker discover tool: opportunities and potential problems. *Molecular and Cellular Proteomics* **3** 367–378.
7. Baggerly K. A., Coombes K. R., and Morris J. S. (2005). Are the NCI/FDA ovarian proteomic data biased? A reply to producers and consumers. *Cancer Informatics* **1(1)** 9–14.
8. Baggerly K. A., Morris J. S., Edmonson S., and Coombes K. R. (2005). Signal in noise: evaluating reported reproducibility of serum proteomic tests for ovarian cancer. *Journal of the National Cancer Institute* **97** 307–309.
9. Zhang, Z., Bast, R. C., Yu, Y., Li, J., Sokoll, L. J., Rai, A. J., Rosenzweig, J. M., Cameron, B., Wang, Y. Y., Meng, X., Berchuck, A., Haaften-Day, C. V., Hacker, N. F., Bruijn, H. W. A., Zee A. G. J., Jacobs, I. J., Fung, E. T., and Chan, D. W. (2004). Three biomarkers identified from serum proteomic analysis for the detection of early stage ovarian cancer. *Cancer Research* **64**, 5882–5890.
10. Hu J., Coombes K. R., Morris J. S., and Baggerly, K. A. (2005). The importance of experimental design in proteomic mass spectrometry experiments: some cautionary tales. *Briefings in Genomics and Proteomics* **3(4)** 322–331.
11. Coombes, K. R., Fritsche, H. A. Jr., Clarke, C., Chen, J. N., Baggerly, K. A., Morris, J. S., Xiao, L. C., Hung, M. C., and Kuerer, H. M. (2003). Quality control and peak finding for proteomics data collected from nipple aspirate fluid by surface-enhanced laser desorption and ionization. *Clinical Chemistry* **49** 1615–1623.
12. Conrads, T. P., Fusaro, V. A., Ross, S., Johann, D., Rajapakse, V., Hitt, B. A., Steinberg, S. M., Kohn, E. C., Fishman, D. A., Whitely, G., Barrett, J. C., Liotta, L. A., Petricoin, E. F. III, Veenstra, T. D. (2004). High-resolution serum proteomic features of ovarian cancer detection. *Endocrine Related Cancer* **11(2)** 163–178.
13. Baggerly K. A., Edmonson S., Morris J. S., and Coombes K. R. (2004). High-resolution serum proteomic patterns for ovarian cancer detection. *Endocrine-Related Cancers* **11(4)** 583–584.

14. Box, G. E. P., Hunter, W. G., and Hunter, J. S. (2005). *Statistics for experimenters: an introduction to design, data analysis, and model building*. 2nd ed., Wiley: New York.
15. Baggerly, K. A., Morris, J. S., Wang, J., Gold, D., Xiao, L. C., and Coombes, K. R. (2003). A comprehensive approach to the analysis of matrix-assisted laser desorption/ionization time of flight proteomics spectra from serum samples. *Proteomics* **3**, 1667–1672.
16. Diamandis, E. P. (2004c). Analysis of serum proteomic patterns for early cancer diagnosis: drawing attention to potential problems. *Journal of the National Cancer Institute* **96(5)** 353–356.
17. Coombes K. R., Morris J. S., Hu J., Edmondson S. R., and Baggerly K. A. (2005) Serum proteomics profiling: a young technology begins to mature. *Nature Biotechnology* **23(3)** 291–292.
18. Coombes, K. R., Tsavachidis, S., Morris, J. S., Baggerly, K. A., Hung, M. C., and Kuerer, H. M. (2005). Improved peak detection and quantification of mass spectrometry data acquired from surface-enhanced laser desorption and ionization by denoising spectra with the undecimated discrete wavelet transform. *Proteomics* **5** 4107–4117.
19. Coombes, K. R., Baggerly, K. A., and Morris, J. S. (2007). Preprocessing mass spectrometry data. In: M. Dubitzky, M. Granzow, and D. Berrar (eds) *Fundamentals of data mining in genomics and proteomics*. Boston: Kluwer, pp 79–99
20. Morris, J. S., Coombes, K. R., Koomen, J. M., Baggerly, K. A., and Kobayashi, R. (2005). Feature extraction and quantification of mass spectrometry data in biomedical applications using the mean spectrum. *Bioinformatics* **21(9)** 1764–1775.
21. Karpievitch, Y. V., Hill, E. G., Morris, J. S., Coombes, K. R., Baggerly, K. A., and Almeida, J. S. (2007). PrepMS. *Bioinformatics* **23(2)** 264–265.
22. Morris, J. S., Clark, B. N., and Gutstein, H. B. (2008). *Pinnacle*: a fast, automatic method for detecting and quantifying protein spots in 2-dimensional gel electrophoresis data. *Bioinformatics* **24(4)** 529–536.
23. Morris, J. S., Clark, B. N., Wei, W., and Gutstein, H. B. (2010). Evaluating the performance of new approaches to spot quantification and differential expression in 2-dimensional gel electrophoresis studies. *Journal of Proteome Research* **9(1)** 595–604.
24. Dupuy A. and Simon R. M. (2007). Critical review of published microarray studies for cancer outcome and guidelines for statistical analysis and reporting. *Journal of the National Cancer Institute* **99(2)** 147–157.
25. Benjamini, Y. and Hochberg, Y. (1995). Controlling the false discovery rate: a practical and powerful approach to multiple testing. *Journal of the Royal Statistical Society, Series B: Methodological* **57** 289–300.
26. Benjamini, Y. and Liu, W. (1999). A step-down multiple hypotheses testing procedure that controls the false discovery rate under independence. *Journal of Statistical Planning and Inference* **82** 163–170.
27. Yekutieli, D. and Benjamini, Y. (1999). Resampling-based false discovery rate controlling multiple test procedures for correlated test statistics. *Journal of Statistical Planning and Inference* **82** 171–196.
28. Storey, J. D. (2002). A direct approach to false discovery rates. *Journal of the Royal Statistical Society, Series B: Statistical Methodology* **64** 479–498.
29. Storey, J. D. (2003). The positive false discovery rate: a Bayesian interpretation and the q-value. *The Annals of Statistics* **31** 2013–2035.
30. Genovese, C. and Wasserman, L. (2002). Operating characteristics and extensions of the false discovery rate procedure. *Journal of the Royal Statistical Society, Series B: Statistical Methodology* **64** 499–517.
31. Ishwaran, H. and Rao, J. S. (2003). Detecting differentially expressed genes in microarrays using Bayesian model selection. *Journal of the American Statistical Association* **98** 438–455.
32. Pounds, S. and Morris, S. W. (2003). Estimating the occurrence of false positives and false negatives in microarray studies by approximating and partitioning the empirical distribution of *p*-values. *Bioinformatics* **19** 1236–1242.
33. Efron, B. (2004). Large-scale simultaneous hypothesis testing: the choice of a null hypothesis. *Journal of the American Statistical Association* **99** 96–104.
34. Newton, M. A. (2004). Detecting differential gene expression with a semiparametric hierarchical mixture method. *Biostatistics (Oxford)* **5** 155–176.
35. Pounds, S. and Cheng, C. (2004). Improving false discovery rate estimation. *Bioinformatics* **20(11)** 1737–1745.
36. Strimmer, K. (2008). Fdrtool: a versitile R package for estimating local and tail area-based false discovery rates. *Bioinformatics* **24** 1461–1462.
37. Lecocke, M. and Hess, K. (2006). An empirical study of univariate and genetic algorithm-based feature selection in binary classification with microarray data. *Cancer Informatics* **2** 313–327.

38. Morris, J. S. and Carroll, R. J. (2006). Wavelet-based functional mixed models. *Journal of the Royal Statistical Society, Series B* **68(2)** 179–199.
39. Morris, J. S., Brown, P. J., Herrick, R. C., Baggerly, K. A., and Coombes, K. R. (2008). Bayesian analysis of mass spectrometry data using wavelet based functional mixed models. *Biometrics* **12** 479–489.
40. Morris, J. S., Baladandayuthapan, V., Herrick, R. C., Sanna, P., and Gutstein, H. B. (2010). Automated analysis of quantitative image data using isomorphic functional mixed models, with application to proteomics data. UT MD Anderson Cancer Center Department of Biostatistics Working Paper Series. Working Paper 56.

Chapter 10

A Review of Experimental Design Best Practices for Proteomics Based Biomarker Discovery: Focus on SELDI-TOF

Rebecca E. Caffrey

Abstract

Surface Enhanced Laser/Desorption Ionization-time of flight (SELDI-TOF) mass spectrometry is a technique uniquely suited to the study of the urine proteome due to its salt tolerance, high-throughput, and small sample requirements. However, due to the extreme sensitivity of the technique, sample collection and storage conditions, as well as instrument protocols and analysis conditions, must be rigorously controlled to ensure that data generated and collected is accurate and free from artifacts. Robust and reproducible data sets can be generated and compared between clinical sites when experimental protocols are carefully standardized. This chapter aims to review known factors that cause irreproducible results so that the experiments may be designed with appropriate sample and process controls for successful biomarker discovery. A suggested protocol follows the review. A number of issues for study design are discussed and these are generally applicable to biomarker discovery experiments.

Key words: SELDI, Proteomics, Urine, Mass spectrometry, Methods, Experimental design, Protocol, Protein profiling

1. Introduction

1.1. SELDI Is Uniquely Suited for High-Throughput Studies of the Urine Proteome

Urine has a long history of providing insight into human health as a diagnostic fluid; however, urine exhibits tremendous biological variability compared to other fluids such as plasma and cerebrospinal fluid. Osmolarity and protein content of urine can vary drastically from patient to patient as a result of a number of factors, including differences in patient hydration and diet, disease processes such as diabetes or renal failure, and other hormonal and/or electrolyte imbalances (1). Normally functioning kidneys can adjust urine osmolarity from 40 to 1,400 mosmol/L (2). As a result of physiological processes, then, urine may be quite dilute

Alex J. Rai (ed.), *The Urinary Proteome: Methods and Protocols*, Methods in Molecular Biology, vol. 641, DOI 10.1007/978-1-60761-711-2_10, © Springer Science+Business Media, LLC 2010

(low in salt content) or concentrated (high in salt content), and may have a varying amount of protein.

Urine normally contains much less protein than blood plasma in a healthy individual. Blood plasma contains approximately 60 mg/mL of protein, whereas urine from healthy individuals contains only micrograms/mL of protein. It has been estimated that healthy adults secrete about 50–70 g of solids per day in their urine; only 50–120 mg of these solids are proteins, and the balance is comprised mostly of salts (3). In fact, 20 mL of urine from a healthy subject yields, on average, only 200 µg of protein (4). Conditions such as renal failure and infections can increase the overall concentration of protein in the urine as well as the absolute numbers of proteins detectable. When a disease process results in urine with higher than normal protein concentration, this is termed "proteinuria." Even when proteinuria is present, however, the urine proteome is still less complex and easier to analyze than serum.

Surface-Enhanced Laser Desorption/Ionization Time of Flight Mass Spectrometry (SELDI-TOF-MS, hereinafter abbreviated as SELDI) is a technique that weds Matrix-Assisted Laser Desorption/Ionization Time of Flight Mass Spectrometry (MALDI) with chromatography, by immobilization of functional groups to the chip surface where proteins are bound. Whereas MALDI uses inert chip surfaces such as glass or gold that do not chromatographically select the proteins bound to them, in the SELDI technique, the chemical moieties on the ProteinChip® array surface actively select which subsets of proteins bind to the chip surface based on their chemical properties. This on-chip retentate chromatography utilizes multiple surface chemistries, such as anion-exchange, cation-exchange, metal affinity-binding, reverse phase, and preactivated surfaces that contain reactive groups capable of forming covalent linkages with primary alcohols or amines (for capturing "bait" molecules such as antibodies, receptors, etc…). Utilization of multiple ProteinChip® array surface chemistries allows for binding and selection of a subset of proteins with common properties on each chip type from a complex mixture such as urine (5). For instance, a reverse-phase chip surface binds hydrophobic regions of proteins by hydrophobic interactions, whereas a cation-exchange or anion-exchange chip surface binds charged protein regions by electrostatic interactions, just as in column chromatography.

SELDI lends itself to study of the urine proteome because it is the only salt tolerant mass spectrometric technique. Other techniques such as 2D-gel electrophoresis (4), electrospray, or MALDI must remove the salts naturally present in urine, plasma, and physiological buffers, but SELDI does not require an additional salt removal step. Additional salt removal steps necessary for other techniques, such as reverse-phase sample clean-up on a chromatographic column or in ZipTips (Millipore, Billerica, MA, USA), can result in sample loss. In SELDI, after the proteins in a urine sample are allowed to bind to the chip surface, the chip is simply washed in deionized water, which elutes any remaining salts while leaving behind bound proteins. Then, the chip is dried, matrix is added, and the bound proteins are analyzed by mass spectrometry.

SELDI is also a valuable technique because it conserves precious samples; as little as 5–10 μL can be used per sample in a profiling experiment, which means an abundance of leftover sample is available for protein identification later. SELDI also has no molecular weight bias; 2D gels are best suited to discovery in the high molecular weight range, and other mass spec techniques may require predigestion of whole proteins with trypsin in order to generate peptides small enough to be detected and identified. SELDI enables biomarker discovery of whole peptides and proteins across a wide range of masses, up to well over 200 kD. Additionally, the authors of numerous studies have noted that SELDI, in contrast to some other techniques, allows for high-throughput protein profiling of multiple urine samples simultaneously (6–10), thus accelerating the discovery process. One drawback to the use of SELDI is that once biomarkers of interest have been discovered, identification of the markers involves a second step.

1.2. The Importance of Appropriate Experimental Design

There is no one protocol to ensure biomarker discovery success in urine with SELDI. The reader must familiarize themselves with the issues discussed herein and plan a protocol based on the clinical question that they are trying to answer. It is the intent of this chapter to inform the reader about a number of factors that should be taken into consideration when planning a SELDI experiment on urine samples so that they can optimize experimental design and maximize chances of success. Ill-conceived experimental designs can be relied upon to produce only two things: results that cannot be replicated, and erroneous conclusions (11–13).

One of the most elementary considerations in the experimental design process is the likely location of the protein biomarkers. Urine contains fluid and solids – some of the solids are cells shed from the transitional epithelial lining of the bladder. Thus, prior to any biomarker work, urine is centrifuged to remove solids. If the researcher is looking for markers of kidney disease, or proteins/peptides that originate in serum but end up in kidney filtrate, then the supernatant is the logical place to begin a biomarker study, and the pellet should be discarded. If the biomarkers sought are markers for bladder cancer, for instance, both the supernatant (14) and the pellet (after cell disruption and solubilization of the proteins) may be useful.

If the research is hypothesis-driven, and a monoclonal or polyclonal exists against the protein family of interest, then questions of fractionation and depletion are largely moot. Similarly, if a non hypothesis-driven profiling strategy is to be employed, consideration should be given to the questions such as fractionation and standardization.

The size of the proteins sought is another parameter of experimental design that should be considered ahead of time. Smaller proteins and peptides require a different matrix and different instrument settings for data collection than larger proteins. Matrix, sometimes abbreviated as EAM (Energy Absorbing Matrix), is the chemical deposited on the chip surface after sample has been bound, and chips washed and dried. The matrix is applied in a volatile liquid, and allowed to crystallize on the chip surface, where the bound proteins are incorporated as imperfections in the matrix crystals (5). When excited by the laser energy in the mass spectrometer, the matrix vaporizes, imparting its energy to the bound proteins, which ionize and fly down the flight tube to the detector. For low molecular weight proteins and peptides (under 20 kD), CHCA (Bio-Rad, Hercules, CA) is generally used; for higher-weight proteins (20 kD and over), SPA (Bio-Rad, Hercules, CA) is the matrix of choice in SELDI. In a profiling experiment where the biomarker size is unknown at the outset, it is wise to run samples in duplicate, binding to sets of ProteinChip® arrays for discovery in both molecular weight ranges, so that different matrix can be applied and data collected at appropriate settings for both large and small proteins. Duplicate sample binding may not be necessary if the molecular weight of the biomarkers of interest is known ahead of time.

Randomization is another aspect of experimental design that is often overlooked by those new to SELDI, but its importance can be understood if we compare SELDI to ELISA. Imagine that there are a large number of control/healthy and a large number of disease/experimental samples that must be examined by ELISA, enough to fill at least two ELISA plates. Imagine that a new laboratory technician thaws out only the control samples (all collected in Washington, DC) and runs them all on one plate along with a standard curve. That same laboratory technician, the following day, thaws the experimental samples (all collected in Japan) and runs them the next day on a separate plate on a different plate reader. Then, the technician calculates levels of biomarker X for both groups based on the standard curve that was run the previous day and finds that the levels of biomarker X vary significantly between groups. Is this finding believable?

When put in the context of ELISA, it should be obvious to the reader that these experimental results would be completely invalid for a number of reasons. Control and experimental samples should have been collected from both US and Japanese populations, thawed simultaneously, randomized over the ELISA plate, and each sample should have been run in duplicate or triplicate. Furthermore, standard curves should have been included on each individual plate, and the plates should have been run and read on the same day, in the same lab, on the same instrument. All of these potential experimental design errors must be considered when designing a SELDI study for the same reason that they are considered when designing any other kind of study.

SELDI is no different from other techniques in that it is only as robust as the care that goes into the experimental design. Samples must also be randomized in SELDI to avoid generation of false artifacts. SELDI ProteinChip® arrays bind together, 12 at a time, in a bioprocessor for high-throughput robotic handling. The footprint of a bioprocessor is the same 96-well footprint familiar from tissue culture plates and ELISAs. Ciphergen Express software allows for randomization of sample groups within and between bioprocessors, and bar-coding on ProteinChip® arrays simplifies the data management. Sample randomization in SELDI is critical because it ensures that minor variance in data collection and biological noise does not become statistically amplified into false conclusions.

Finally, proper instrument qualification and calibration prior to chip reading is critical for optimizing data quality. Mass spectrometers detect and record proteins based on their mass to charge ratio (m/z), and all mass spectrometers drift with time and must be adjusted. Therefore, calibration with proteins or peptides of known molecular weight immediately prior to reading chips on the instrument is essential to the generation of accurate data. This is not a step that can be skipped. Calibrants for peptide or protein mass ranges are available for purchase through Bio-Rad (formerly Ciphergen), as are chips preloaded with insulin or IgG for qualification of instrument resolution and sensitivity, respectively. An instrument that fails to pass QC for sensitivity and resolution should not be used to read chips; however, an instrument failure does not mean that an experiment is lost. Once loaded with matrix, chips can be stored in a dark, humidity-controlled environment for weeks before reading.

1.3. To Standardize or Not to Standardize; That Is the Question

Some researchers choose to standardize their urine samples prior to SELDI analysis by adjusting all samples to a common total protein concentration. This is usually accomplished by measuring total protein, and then diluting more concentrated samples such that they have the same total protein content as the most dilute sample in the group. Whether or not standardization of samples should be performed depends on which disease one is studying, whether one plans a profiling experiment or plans to capture specific proteins of interest, and whether the proteins in question are clinically significant due to their presence or absence, or due to their relative abundance. If relative quantitation of biomarkers is a desired study endpoint, standardization by overall sample protein content may be wise.

Besides total protein, other urinary markers may be used to standardize samples prior to analysis. Urinary creatinine levels have been commonly used to standardize intersample variation, especially when studying proteins that exhibit diurnal variation (15). If one wishes to take advantage of the semiquantitative nature of SELDI to measure relative abundance of a protein biomarker, standardization is a must. SELDI is semiquantitative in the same way that an ELISA is semiquantitative. If a concentration gradient of protein standards is run next to unknowns for either technique, an r^2 value can be calculated that approaches unity, and from this standard curve, the concentration of the unknown can be extrapolated. In an ELISA, of course, color change is being measured. In SELDI, peak height is being measured and compared.

1.4. Depletion, Fractionation, or Enrichment; Yet More Questions

1.4.1. Depletion

The strategy for removal of albumin and/or other abundant species has its proponents who advocate that the removal of the superabundant proteins in any sample allows discovery of less abundant protein species that would otherwise be masked by common proteins at the high end of the dynamic range. However, depletion of common proteins such as albumin also depletes all associated proteins bound to the protein being depleted. In the case of albumin, which is known to be a “sticky” binding protein, so many proteins are bound and thus codepleted on its removal that the term “albuminome” has come into common usage, and some groups have begun to study only this subset of the proteome. Depletion is not a common or necessary strategy in SELDI urine proteomics.

1.4.2. Fractionation

Some researchers conducting serious biomarker searches into the deep urine proteome employ a fractionation strategy like that recommended by Ciphergen Biosystems (Fremont, California) for serum and plasma; others forego fractionation since urine has fewer proteins than serum (5). A fractionation strategy does not involve depletion; rather, it involves disrupting protein–protein interactions in the sample by application of 9 M urea and subsequent fractionation of the urea-diluted sample with appropriate buffers on an anion exchange resin. Usually, this is done on a BioMek robot (Beckman Coulter), for ease of use and rapidity of processing. In the anion exchange fractionation, the original sample is split into six fractions, the first collected being flow-through plus the first fraction at pH 9, the second pH 7, the third pH 5, the fourth pH 4, the fifth pH 3, and the sixth and last being an organic fraction composed of the stripped-off remaining proteins in acetonitrile. Together, this fractionation protocol comprises the first dimension of separation in SELDI (16). The second dimension of separation is the binding of proteins in each fraction to ProteinChip® arrays with different surface chemistries, and the third dimension of separation is achieved in the flight tube of the mass spec itself, when retained proteins are separated by m/z, the mass to charge ratio. Thus, with SELDI studies that employ fractionation, there are three dimensions of separation, and depletion strategies are not needed.

1.4.3. Enrichment

1.4.3.1. Immuno-Affinity Enrichment

Antibodies to a protein family of interest pull down all members of this protein family and all fragments thereof that contain the epitope to which the antibody in question binds. This technique also pulls down proteins that bind to the target protein in the urine. Monoclonals or polyclonals work fine in most cases and can be easily captured in whole urine by sepharose beads coated with protein A or G (7); target proteins and antibody can then be eluted from these beads and analyzed on a variety of chip surfaces. For experiments where the captured proteins will later be digested by trypsin and analyzed by MS/MS for confirmation of identity, monoclonals are preferable. Not as commonly employed as monoclonals, but effective and with the potential to generate less "noise" in MS/MS, are the aptamers, thioaptamers, and other next-generation antibody analogs that are not digested by enzymes such as trypsin. Antibody pull-down of target proteins "on-chip" is also possible by use of ProteinChip® arrays with preactivated surface chemistries that allow covalent coupling of antibody in wet buffer systems. While this works well for "fishing expeditions," it is usually not a suitable technique for binding enough target to do protein sequencing by MS/MS.

1.4.3.2. Equalizer Bead Enrichment

Use of hexameric peptide ligand library-coated beads to bind a large variety of peptides and proteins allows capture and analysis of analytes at the low end of the dynamic range without depletion. Equalizer Beads (developed by Ciphergen Biosystems, currently sold by Bio-Rad, Hercules, California) work via a novel concentration/equalization technique, acting by capturing the whole protein spectra contained in the sample, by drastically reducing the level of the most abundant species, while strongly concentrating the less abundant species (17, 18). Castagna et al. reported that in a control urine sample, 96 unique proteins could be identified. Using the Equalizer Beads, the first bead eluate (in thiourea, urea and CHAPS) allowed identification of 334 proteins, whereas the second eluate (in 9 M urea, pH 3.8) allowed identification of yet another 148 unique proteins. The equalizer bead strategy, by eliminating redundancies and increasing the total number of proteins detected, could revolutionize the proteomic study of urine by allowing discovery of biomarkers at the low end of the dynamic range of protein concentration in urine.

1.5. Sources of Variation in SELDI Experiments: Intrinsic Versus Extrinsic Factors

SELDI is an exquisitely sensitive experimental technique. Most mass specs have been engineered with long flight paths in order to ensure maximum resolution; generally speaking, the better the resolution, the lower the sensitivity, and the fewer low-abundance proteins can be detected. The Ciphergen instruments were deliberately engineered with short flight tubes for maximum sensitivity. Thus, failure of the experimenter to anticipate and adequately control for a range of factors will result in artifacts in the data set produced (11–13). For the purposes of this review, an intrinsic factor is defined as sample-related factor such as collection and storage conditions, contamination, or dilution. An extrinsic factor will be defined as an experiment-related factor such as instrument settings, matrix composition, calibration, data collection settings, etc… When both intrinsic and extrinsic factors are controlled for, it is possible to generate reproducible, concordant results between multiple instruments at multiple sites (6, 19).

1.5.1. Intrinsic Factors

Schaub et al. published an excellent study on the impact of intrinsic and extrinsic factors on urine protein profiling using SELDI. This group found that the presence of blood in the urine (microscopic hematuria), urine dilution, and urine collection parameters (first-void vs. midstream urine) all impacted the quality of the SELDI data (6). This is further described below.

1.5.1.1. Collection Parameters

When urine is collected from patients, instructions usually specify to collect urine "from mid-stream," as opposed to "first-void." For most techniques, this is not a critical distinction. Due to the sensitivity of SELDI, again, this distinction becomes important. When Schaub et al. analyzed first-void vs. mid-stream urine from male and female healthy volunteers, the results indicated that for females, especially, first-void urine had a significantly different ratio of proteins present compared to a mid-stream urine sample. One cluster of unique peaks in female first-void urine shows up as three peaks between 3.3 and 3.5 kD (6). These peaks are consistent with the masses of the alpha-defensins, which are an important component of innate mucosal immunity against microbial infection (20–22). These peaks have been detected by SELDI in urine before (23) and appear at lower intensity or not at all in mid-stream collections. As such, the reader should familiarize themselves with the masses of the alpha-defensins, as they may serve as a useful marker for which collections were truly midstream. In view of this information, it is particularly important to educate the female patient about midstream collection to ensure a clean sample catch. SELDI protein profiles have been shown to yield stable spectra for up to 24 h after clean-catch midstream urine sample collection (6), indicating that significant proteolytic sample degradation is not occurring during at least the first 24 h of refrigerated storage.

This means that in contrast to serum and tissue studies with SELDI, no protease inhibitors need to be added for clean-catch urine.

1.5.1.2. Hematuria

Just as the reader should familiarize themselves with the alpha-defensin masses, another set of peaks that frequently occur in urine samples correspond to the masses for the subunits of hemoglobin. Schaub's group deliberately spiked whole blood into a urine sample in order to demonstrate the single- and double-charged peaks of the alpha- and beta-chains of hemoglobin (M/Z of 15.142 and 15.888 kD for the single-charged species). They further demonstrated that even at a 1:6,400 dilution of blood:urine, hemoglobin peaks remained prominent in the spectra. Also, even when urine was centrifuged to remove the spiked red blood cells, peaks corresponding to the masses of hemoglobin and albumin (66.472 kD MW) were still visible in multiple charge states. Thus, even if urine samples are contaminated with blood at a level invisible to the naked eye, SELDI can still detect aberrant peaks.

Blood can contaminate a sample and influence normal urine protein profiling in a number of different ways. First, tumors or infections of the urinary tract, kidney stones, or other renal disease can result in hematuria. Also, urine samples collected from women during the menstrual cycle may become contaminated with blood. It is important to note that while the protein profile may be more complex with blood contaminant, biomarker discovery may still be possible. Familiarizing one's self with the parent masses of "the usual suspects" helps in deconvoluting the data, and the Ciphergen data analysis software can be set to mark the multiply charged peaks for each parent mass so that one can avoid marking them if one chooses to peak-pick by hand rather than automatically. Also, if only a small number of samples out of the total are contaminated, the contaminant peaks will not be statistically significant in the final analysis.

1.5.1.3. Urine Dilution

The variability of urine osmolarity and protein concentration has already been discussed. Schaub's group studied the quality of SELDI protein profiles obtained from various dilutions of urine, and determined that for a "normal," nondilute urine output of 1–2 L/day, stable protein profiles could be collected on a neutral phase (NP-20) chip. Other chip surfaces that capture proteins via electrostatic or reverse phase interactions are likely to have stable protein profiles at lower dilutions, but this was not studied. They did determine that after a dilution of normal urine with buffer at 1:4 or greater, a steady and continuous loss of detectable peaks occurred at greater dilutions. The most dilute sample that they were able to test had a protein level of 30 mg/L and creatinine at 0.9 mmol/L; this sample was far below the optimum and few peaks were detected. This data should be borne in mind when deciding whether or not to standardize urine samples to creatinine or total protein.

1.5.1.4. Sample Storage: Refrigeration, Freezing, and Thawing

Schaub et al. found that midstream urine samples did not undergo changes in their protein profile when stored for up to 3 days at 4°C; however, this practice is risky as inadvertent sample contamination could lead to bacterial growth and protease activity that could alter the protein profile despite refrigeration. Further, it was noted that urine samples could be subjected to up to five freeze/thaw cycles before significant deterioration in quality of protein profile. However, repeated freeze–thaw cycles should still be avoided by aliquoting samples into smaller volumes whenever possible. Traum et al. reported high concordance between MS profiles within sets of quadruplicate split urine samples frozen at 0, 2, 6, and 24 h after sample collection (9). This group therefore concluded that time elapsed prior to freezer storage did not affect data quality. Thus, once samples are properly collected and frozen down within 24 h, it is likely that they can be safely banked at –80°C without deterioration in quality.

1.5.1.5. Inclusion of Appropriate Samples as Disease Controls

For most biomarker studies, additional control samples beyond "normal" or "not diseased" are needed. Since most disease processes involve inflammatory responses, controlling for inflammatory responses is critical. For example, SELDI proteomic profiling studies that have sought biomarkers for prostate cancer have included biofluids collected from patients with confirmed cancers, patients with no prostatic disease, and patients who suffer from benign prostatic hypertrophy (BPH); the inclusion of BPH samples helps to differentiate markers of generalized inflammation of the prostate from markers specific for cancer (24). When possible, efforts should also be made to collect samples where age and sex matching is balanced between the experimental and control groups.

1.5.1.6. Summary of Best Practices for Minimizing Impact of Intrinsic Factors

Appropriate sample collection and handling will minimize intrinsic factors that contribute to poor data quality. The protocol followed by Khurana et al. and Traum et al., is one of the best examples of good laboratory practice for urine sample handling in the literature (7, 9). These groups collected at least 3 mL from midstream per sample, and immediately centrifuged the samples at 1000 × *g* for 5 min at room temperature. They then divided the supernatant into 1 mL aliquots in dry ice, and stored the aliquots at –80°C until needed for future proteomics studies. By aliquoting, any issues regarding repeated freeze/thaw cycles are avoided altogether, and multiple studies can be performed on the same samples without deterioration in sample quality. It is worth noting that when urine samples are thawed for SELDI analysis, a second centrifugation is necessary to remove precipitate caused by the freezing process. The second centrifugation should be between 1000 × *g* and 3000 × g for 5 min (25).

1.5.2. Extrinsic Factors

The extrinsic factors that most critically influence reproducibility and peak detection in urine protein profiling are matrix composition and instrument settings (6). Additional extrinsic factors include instrument calibration and randomization of control and experimental sample processing order (12, 13), and manual sample and matrix application vs. robotic application in bioprocessors (19). Another extrinsic factor that should be considered is the development of air bubbles in the bottom of the bioprocessor, which can lead to poor or complete lack of sample binding, or improper washing of the chip surface. This is one of the reasons that samples should always be run in duplicate or triplicate for each chip type, randomized across the bioprocessor.

1.5.2.1. Instrument Settings

As previously discussed, prior to data collection, the SELDI instrument must pass qualification for resolution (via insulin chip protocol) and sensitivity (IgG chip protocol). It must also be calibrated using peptide or protein standards for the mass range of the proteins of interest. After this, data collection settings must be optimized for the experiment. These settings include laser and detector settings for intensity and sensitivity, lag time, and setup of spot protocols and chip protocols.

It is beyond the scope of this chapter to restate detailed protocols from the instrument manual or individual papers. The reader is advised to begin by reading the instrument manual as new software updates may confer additional levels of instrument control, and new generations of ProteinChip® instruments may have additional features (this is especially true since Ciphergen's instrumentation division, which produces the instruments and chips, was sold to Bio-Rad in 2006). The optimal settings for data collection from different chip types and molecular weight ranges will vary across time and between instruments due to aging of detectors and lasers; therefore, it is impossible to publish recommended settings in a manual such as this. Once optimal settings for an instrument are determined, it must never be taken for granted that the settings that worked optimally at for one experiment will work optimally for the next.

2. Materials

1. Sterile urine collection containers (Fisher Scientific, Pittsburgh, PA).
2. Cryostorage vials (Fisher Scientific, Pittsburgh, PA).

3. –80°C freezer.
4. Pipettors and sterile tips (Fisher Scientific, Pittsburgh, PA).
5. Ciphergen Express Software (Bio-Rad, Hercules, CA).
6. Ciphergen Biomarker Patterns Software (Bio-Rad, Hercules, CA).
7. Ciphergen ProteinChip® reader mass spectrometer (Bio-Rad, Hercules, CA).
8. ProteinChip® arrays (Bio-Rad, Hercules, CA).
9. Binding and wash buffers specific to each array type (Bio-Rad, Hercules, CA).
10. Bioprocessor lid and assembly (Bio-Rad, Hercules, CA).
11. Preloaded insulin chips (Bio-Rad, Hercules, CA).
12. Preloaded IgG chips (Bio-Rad, Hercules, CA).
13. Low molecular weight calibrants (peptide calibrants) (Bio-Rad, Hercules, CA).
14. High molecular weight calibrants (protein calibrants) (Bio-Rad, Hercules, CA).
15. SPA matrix (Bio-Rad, Hercules, CA).
16. CHCA matrix (Bio-Rad, Hercules, CA).
17. Biomek FX robot (Beckman Coulter).
18. 96-well plates.
19. Deionized water.
20. 9 M urea solution.
21. Acetonitrile.
22. TriFluoroAcetic Acid (TFA).
23. Centrifuge.
24. U9 buffer (9 M urea, 2% CHAPS, in 50 mM Tris pH 9.0).

3. Methods

3.1. Sample Preparation

1. Collect at least 3 mL from midstream per sample.
2. Immediately centrifuge the samples at 1000 × *g* for 5 min at room temperature.
3. Divide the supernatant into 1 mL aliquots in dry ice.
4. Store the aliquots at –80°C immediately.
5. Thaw samples at 4°C.
6. Centrifuge at 1000 × *g* and 3000 × *g* for 5–10 min to remove precipitate.

7. If necessary, measure total protein concentration or creatinine concentration and standardize samples by dilution in appropriate binding buffer.
8. Randomize samples in Ciphergen Express software.
9. Following software-generated sample map, prepare urine samples in a 96-well plate.
10. To each well, add 60 μL of sample and 20 μL of U9 buffer for a total volume of 80 μL, pipetting up and down three times (see Note 2).
11. Place sample plate on orbital shaker at 4°C while preparing arrays for binding.

3.2. ProteinChip® Array Surface Preparation and Sample Binding

1. Assemble arrays into bioprocessor.
2. Set bioprocessor in place on Biomek FX robot.
3. Pretreat arrays according to manufacturer instructions for each specific chip type (see Note 1).
4. Remove bioprocessor from the robot, invert, and spin at 500 × *g* for 1 min to remove any additional liquid.
5. Add 30 μL binding buffer to the arrays (see Note 3).
6. Add 30 μL sample to arrays.
7. Place bioprocessor on orbital shaker for 30 min.
8. Wash once with 100 μL of binding buffer; remove wash buffer to waste.
9. Wash three times with 100 μL of deionized water; remove deionized water to waste.
10. Remove bioprocessor from the robot, invert, and spin at 500 × *g* for 1 min to remove any remaining water.
11. Remove reservoir from bioprocessor.
12. Allow arrays to air dry while preparing matrix.

3.3. Matrix Preparation

1. Prepare solvent for matrix (either CHCA or SPA) with 250 μL acetonitrile and 250 μL deionized water, with 50 μL of 10% TFA added.
2. Powdered matrix is then mixed with 550 μL of the solvent and vortexed vigorously for 2 min.
3. Centrifuge for 3 min to pellet undissolved matrix powder.
4. Add 25 μL saturated supernatant into each well of a 96-well plate to be applied to the dry ProteinChip® arrays (see Note 4).
5. Place the bioprocessor, containing newly dry ProteinChip® arrays, onto the Biomek robot.
6. 1 μL of saturated SPA or CHCA matrix solution is applied to each spot on the chip.

7. Allow matrix to dry.
8. Repeat addition of 1 μL of saturated SPA or CHCA matrix solution to each spot on the chip.
9. Allow matrix to dry at least overnight in a dark, humidity-controlled environment before data collection on ProteinChip reader (see Note 5).
10. Arrays may be stored for weeks prior to reading in a dark, humidity controlled environment with no loss of signal.

3.4. Chip Reading

1. Run instrument checks: insulin chip for resolution and IgG chip for sensitivity. Proceed with chip reading if instrument passes checks.
2. Calibrate instrument using "all in 1 protein standard" mixture (Ciphergen Biosystems Inc.). For the low MW (peptide) size range, 2,000–20,000 Da, the following peptides may be used for calibration: beta-endorphin, bovine insulin, and bovine ubiquitin. For the high MW range (protein) size range, the following proteins may be used: bovine cytochrome C, bovine beta-lactoglobulin A, horse-radish peroxidase, and bovine albumin.
3. Save calibration curves to application to spectra later prior to analysis.
4. Insert chips into cassettes and insert cassettes into protein chip reader.
5. Make test shots to adjust laser intensity and detector settings to optimum.
6. Incorporate laser intensity and detector settings into a spot protocol.
7. Assemble spot protocols into a chip protocol.
8. Assemble chip protocols into a cassette protocol.
9. Run cassette protocols and collect data.

3.5. Data Analysis

1. Normalize data to total ion current, external value of 1.
2. View spectra and for high laser intensities for detection of higher molecular weight species, set a cutoff mass range around 10 kD to exclude noise, and leave high mass range unrestricted or set to 200 kD.
3. View spectra for low laser intensities for detection of low molecular weight species, and set a cutoff for exclusion of matrix noise between 1,500 and 2,000 kD, depending on how the spectra looks.
4. Set noise definitions to desired settings.
5. Perform manual or automatic peak picking.
6. Perform statistical analysis. (see Note 6)

4. Notes

1. Each surface chemistry requires different pretreatment. For example, immobilized metal affinity capture (IMAC) arrays require precharging with 0.1 M $CuSO_4$ solution, and reverse-phase arrays should be washed with acetonitrile to clean off atmospheric hydrocarbons that have bound themselves to the surface prior to sample binding. CM10 (cation exchange) chips are activated by 10 mM HCl for 5 min, followed by rinsing deionized water. Follow the manufacturer's instructions for the chosen array types, and use the pretreatment, binding, and wash buffers that are included in the kits from the manufacturer (Bio-Rad, Hercules, CA).
2. For larger experiments, more total volume can be added to the sample plate, up to 250 μL, as long as the ratio of urine:U9 buffer remains 3:1.
3. Binding buffers should be purchased from the manufacturer, and used according to directions.
4. This step must proceed as quickly as possible due to the volatility of the acetonitrile.
5. Internal optimization studies conducted at Ciphergen's Fremont, California Biomarker Discovery Center indicated that for optimal detection of peaks (highest number and best signal), overnight drying of chips was required. Reading right away did not produce the best results. Chips can be stored for at least 2 weeks prior to reading with no deterioration in signal.
6. Statistical analysis components are included in the Ciphergen software package, particularly in Biomarker Patterns Software. A Student's *t*-test and Analysis of Variance (ANOVA) can be run, as well as Classification and regression tree (CART) analysis and Principle Component analysis (PCA). Since new software functionality may have been added, the reader is advised to read the manual. Also, Ciphergen software allows for the export of data files in other formats for analysis by other software programs.

Acknowledgments

The author would like to thank Dr. Lee O. Lomas of Ciphergen Biosystems, Inc., for his invaluable insights, encouragement, and patience.

References

1. Bockenkamp, B. (2003) Understanding and managing acute fluid and electrolyte disturbances. *Curr Paediatr* **13**(7), 520–528.
2. Subcommittee on the Tenth Edition of the Recommended Dietary Allowances, Food and Nutrition Board, Commission on Life Sciences, National Research Council (1989) Recommended Dietary Allowances, 10th Edition. The National Academies Press, pp. 248–249.
3. Anderson, N.G., Anderson, N.L., Tollaksen, S.L., Hahn, H., Giere, F., and Edwards, J. (1979) Concentration and 2-D gel electrophoretic analysis of human urinary proteins. *Anal Biochem* **95**, 48–61.
4. Kahn, A., and Packer, N.H. (2006) Simple urinary sample preparation for proteomic analysis. *J Proteome Res* **5**, 2824–2838.
5. Fung, E., Diamond, D., Simonsen, A.H., and Weinberger, S.R. (2003) The use of SELDI ProteinChip®® array technology in renal disease research. In: *Renal Disease: Techniques and Protocols,* (Goligorsky, M.S., ed.) Humana Press, Inc., Totowa, NJ, Chap. 20, p. 295.
6. Schaub, S., Wilkins, J., Weiler, T., Sangster, K., and Rush, D. (2004) Urine protein profiling with surface-enhanced laser desorption/ionization time-of-flight mass spectrometry. *Kidney Int* **65**, 323–332.
7. Khurana, M., Traum, A.Z., Aivado, M., Wells, M.P., Guerrero, F.G., Towia, A.L., and Schacter, A.D. (2006) Urine proteomic profiling of pediatric nephrotic syndrome. *Pediatr Nephrol* **21**(9), 1257–1265.
8. Nguyen, M.T., Ross, G.F., Dent, C.L., and Davarajan, P. (2005) Early prediction of acute renal injury using urinary proteomics. *Am J Nephrol* **25**, 318–326.
9. Traum, A.Z., Wells, M.P., Aivado, M., Towia, A.L., Ramoni, M.F., and Schachter A.D. (2006) SELDI-TOF MS of quadruplicate urine and serum samples to evaluate changes related to storage conditions. *Proteomics* **6**(5), 1676–1680.
10. O'Riordan, E., Orlova, T.N., Mei, J.J., Butt, K., Chander, P.M., Rahman, S., Mya, M., Hu, R., Momin, J., Eng, E., Hampel, D.J., Hartman, B., Kretzler, M., Delaney, V., and Goligorsky, M.S. (2004) Bioinformatic analysis of the urine proteome of acute allograft rejection. *J Am Soc Nephrol* **15**, 3240–3248.
11. Petricoin, E.F., Ardekani, A.M., Hitt, B.A., Levine, P.J., Fusaro, V.A., Steinberg, S.M., Mills, G.B., Simone, C., Fishman, D.A., Kohn, E.C., and Liotta L.A. (2002) Use of proteomic patterns in serum to identify ovarian cancer. *Lancet* **359**, 572–577.
12. Baggerly, K.A., Morris, J.S., and Coombes, K.R. (2004) Reproducibility of SELDI-TOF protein patterns in serum: Comparing data sets from different experiments. *Bioinformatics* **20**, 777–785.
13. Baggerly, K.A., Coombes, K.R., and Morris, J.S. (2005) Bias, randomization, and ovarian proteomic data: A reply to "Producers and Consumers". *Cancer Inform* **1**(1), 1–7.
14. Liu, W., Guan, M., Wu, D., Zhang, Y., Wu, Z., Xu, M., and Lu, Y. (2005) Using tree analysis pattern and SELDI-TOF-MS to discriminate transitional cell carcinoma of the bladder cancer from noncancer patients. *Eur Urol* **47**, 456–462.
15. Wolthers, O.D., Heuck, C., and Heickendorff, L. (2001) Diurnal variations in serum and urine markers of type I and type III collagen turnover in children. *Clin Chem* **47**, 1721–1722.
16. Poon, T.C.W., Yip, T-T., Chan, A.T.C., Yip, C., Yip, V., Mok, T.S.K., Lee, C.Y., Leung, W.T., Ho, K.W., and Johnson, P.J. (2003) Comprehensive proteomic profiling identifies serum proteomic signatures for detection of hepatocellular carcinoma and its subtypes. *Clin Chem* **49**, 752–760.
17. Castagna, A., Cecconi, D., Sennels, L., Rappsilber, J., Guerrier, L., Fortis, F., Boschetti, E., Lomas, L., and Righetti, P.G. (2005) Exploring the hidden human urinary proteome via ligand library beads. *J Proteome Res* **4**(6), 1917–1930.
18. Guerrier, L., Thulasiraman, V., Castagna, A., Fortis, F., Lin, S., Lomas, L., Righetti, P.G., and Boschetti, E. (2006) Reducing protein concentration range of biological samples using solid-phase ligand libraries. *J Chromatogr B Analyt Technol Biomed Life Sci* **833**(1), 33–40.
19. Rai, A., Stemmer, P., Zhang, Z., Adam, B-L., Morgan, W., Caffrey, R., Podust, V., Patel, M., Semmes, J., Lim, L., Shipulina, N. (2005) Analysis of HUPO-PPP reference specimens using SELDI-TOF mass spectrometry: Multi-institution correlation of spectra and identification of biomarkers. *Proteomics* **5**, 3467–3474.
20. Ganz, T. (2001) Defensins in the urinary tract and other tissues. *J Infect Dis* **183**(Suppl. 1), S41–S42.
21. Kunin, C.M., Evans, C., Bartholomew, D., and Bates, D.G. (2002) The antimicrobial

defense mechanism of the female urethra: A reassessment. *J Urol* **168**, 413–419.

22. Zhang, L., Yu, W., He, T., Yu, J., Caffrey, R.E., Dalmasso, E.A., Fu, S., Pham, T., Mei, J., Ho, J.J., Zhang, W., Lopez, P., Ho, D. (2003) Contribution of human alpha-defensin 1, 2 and 3 to the anti-HIV-1 activity of CD8 antiviral factor. Science **298**, 995–1000.
23. Hampel, D.J., Sonsome, C., Sha, M., Brodsky, S., Lawson, W.E., and Goligorsky, M.S. (2001) Towards proteomics in uroscopy: Urinary protein profiles after radiocontrast medium administration. *J Am Soc Nephrol* **12**, 1026–1035.
24. Cazares, L.H., Adam, B.L., Ward, M.D., Nasim, S., Schellhammer, P.F., Semmes, O.J., and Wright, Jr, G.L. (2002) Normal, benign, pre-neoplastic, and malignant prostate cells have distinct protein expression profiles resolved by surface-enhanced laser desorption/ionization mass spectrometry. *Clin Cancer Res* **8**, 2541–2552.
25. Dr. Lee Lomas, Director of Research and Development, Ciphergen Biosystems, Inc., (2007) Personal communication.

Chapter 11

Urine Proteomic Profiling for Biomarkers of Acute Renal Transplant Rejection

Shu-Ling Liang and William Clarke

Abstract

Acute allograft rejection is a serious impediment to long-term success in renal transplantation. Early detection of rejection is crucial for treatment of rejection, and can help avoid long-term effects such as chronic rejection or loss of the transplanted organ. The current diagnostic paradigm is a combination of clinical presentation, biochemical measurements (serum creatinine), and needle biopsy. There are significant efforts underway to find alternate biomarkers for early detection of acute rejection, including protein profiling of urine by mass spectrometry. One approach for protein profiling is to use affinity mass spectrometry – we describe a method for this using ProteinChips and SELDI-TOF mass spectrometry.

Key words: Acute allograft rejection, Kidney biomarkers, SELDI-TOF, Protein profiling, Mass spectrometry

1. Introduction

Acute renal allograft rejection is characterized by an increase in serum creatinine that is usually accompanied by other clinical findings such as decreased urine output, hypertension, graft pain/tenderness, and raised body temperature (1). Most episodes of acute rejection occur within the first 6 months after transplantation, with many such episodes occurring early after surgery – some of these are completely asymptomatic. Currently, many centers are achieving the rate of acute rejections below 15%, due in part to the use of different immunosuppressive regimens (2, 3). However, approximately 8% of the patients with functioning grafts would still experience a first episode of rejection after 1 year (4). Early detection along with adequate treatment is extremely crucial for successful management of acute rejection. Diagnostic methods utilized to determine whether a patient is experiencing

Alex J. Rai (ed.), *The Urinary Proteome: Methods and Protocols*, Methods in Molecular Biology, vol. 641,
DOI 10.1007/978-1-60761-711-2_11,

acute rejection include clinical presentation, biochemical parameters, and tissue biopsies. The first two methods are often nonspecific and subjective based on the clinical experience of the transplant physician. Frequent monitoring of the serum creatinine concentration has been strongly recommended in detection of acute rejection episodes, since a rise in serum creatinine is usually the first available indication of allograft dysfunction (5). Unfortunately, this is undermined by the observation that an elevation in the serum creatinine concentration is a relatively late development in the course of a rejection episode and invariably indicates the presence of significant histologic damage (6–8), and therefore this approach is not sensitive enough.

While histologic examination of the percutaneous core needle transplant biopsy remains the gold standard for the diagnosis of acute rejection, less invasive procedures that could diagnose incipient rejection and simultaneously provide mechanistic information on the rejection process (allowing delivery of more tailored therapy) are being sought. To address these problems, a number of alternative diagnostic procedures have been suggested, including duplex Doppler ultrasound assessment, fine-needle aspiration biopsy, urine cytology, urine cytokine analysis, serum cytokine analysis, and cytokine analysis of biopsy material (1) although none of these tests have yet reached wide clinical application. A noninvasive biomarker of rejection would allow not only an early diagnosis but also frequent monitoring of immunosuppressive therapy. Several approaches, such as mRNA measurements in urinary lymphocytes, urine flow cytometry, and measurements of alloreactive peripheral blood lymphocytes, have been attempted but none has reached clinical application either (9).

In recent years, a group of research efforts have been focused on studying urinary proteomics as a novel approach to assess early detection of acute renal transplant rejection (10–13). Multivariate analysis of urine profiles of different disease states suggests that there is an ensemble of subtle changes, comprising a proteomic signature of acute rejection at an early stage, a more detailed evaluation of which might provide novel opportunities for the diagnosis of acute rejection. In particular, the application of surface-enhanced laser desorption/ionization-time of flight-mass spectrometry (SELDI-TOF-MS) offers a novel, noninvasive, sensitive, highly predictive, reproducible, rapid method for the prediction of acute renal injury. SELDI-TOF-MS is a protein analysis tool capable of detecting protein profile differences between biological samples. Since proteins are excreted from the urinary tract, protein-based urinalysis may offer insights into renal disease diagnosis. Although the day-to-day proteomic analysis using this approach is far from being routine, requiring technical precision followed by exhaustive bioinformatic analysis, the main thrust of the analyses is to select an optimal number of the most valuable variables – protein peaks – which will have the highest discriminative and predictive power, and then develop analyte-specific assays for detection of renal allograft rejection.

Several factors have led to the choice urine as the subject for analysis and biomarker discovery: collection is risk-free and non-invasive, there is no discomfort to the patient, and there are few collection restraints. In contrast to biopsies, urine allows for sampling of the entire organ and therefore should more accurately reflect its physiologic state and level of function. Urine represents a rich source of information related to the functioning of many internal organs including the kidney, bladder, and prostate. Protein/peptide composition of the urine is determined by the function of the glomerular filtration apparatus, proximal tubular absorption of ultrafiltered proteins, and the capacity of the brush border and lysosomal proteolytic machinery to degrade filtered proteins (14). In addition, the appearance of certain "pathologic" proteins in the blood (for instance, when a distant tumor synthesizes them), or generation of certain proteins by the diseased kidney or the lower urinary tract, may result in their appearance in the urine either in the intact form or, more likely, as peptide fragments. In either case, detection of a specific protein/polypeptide or their ensemble should provide a signature of a particular pathological process. Moreover, such aberrations may be detectable even prior to any clinical manifestation of a disease process, thus offering potential prophylactic utility (14).

2. Materials

2.1. Urine Collection and Processing

1. 15 mL conical tubes.
2. Fixed angle centrifuge.
3. 1.5 mL micro-centrifuge tubes.

2.2. Hydrophobic SELDI

1. 50% Acetonitrile (ACN) solution: Mix equal parts water and ACN; store at 4°C.
2. 20% ACN solution; store at 4°C.
3. H50 SELDI chips, 8-spot (Bio-Rad, Hercules, CA).
4. 96-well bioprocessor (Bio-Rad).
5. Saturated solution (in 50% ACN, 0.5% TFA) of sinapinic acid (SPA).

2.3. Immobilized Metal Affinity SELDI

1. 100 mM $CuSO_4$ solution: add 1.6 g of anhydrous $CuSO_4$ to 100 mL of deionized water.

2. 10× phosphate buffered saline (PBS), pH 7.4.
3. IMC30 SELDI chips, 8-spot (Bio-Rad).
4. 96-well bioprocessor (Bio-Rad).
5. Saturated solution (in 50% ACN, 0.5% TFA) of sinapinic acid (SPA).

3. Methods

3.1. Urine Collection and Processing

1. Urine specimens are collected into a 15-mL conical tube from renal transplantation patients at multiple time points post-transplantation; no specimens are collected less than 4 days post-transplant (see Note 1).
2. Specimens are immediately centrifuged for 5 min at 1,000 ×*g* to remove sediment (see Note 2).
3. Supernatant is divided into 1 mL aliquots and stored frozen at –80°C until analysis (see Note 3).

3.2. Urine Proteomic Profiles by Hydrophobic SELDI

1. H50 ProteinChips are pretreated with 50% ACN wash buffer for 30 min; wash solution is then discarded.
2. 50 μL of urine specimen (disease, control, QC) is added to each well of the bioprocessor.
3. Incubate for 30 min with continuous mixing.
4. Excess specimen is discarded from the bioprocessor.
5. Add 50 μL of 20% ACN wash buffer to each sample well.
6. Incubate for 5 min with continuous mixing to remove non-specific binding components; wash solution is then discarded.
7. Add 50 μL of deionized water to each sample well.
8. Incubate for 5 min with continuous mixing; water is then discarded.
9. Add 1 μL of matrix solution (SPA) and let air dry.
10. Obtain mass spectra from ProteinChips.
11. Peaks from spectra should be normalized using urine biochemical parameters (e.g., creatinine).
12. Normalized spectra are submitted to bioinformatic/statistical analysis to identify biomarker candidates.

3.3. Urine Proteomic Profiles by Immobilized Metal Affinity SELDI

1. IMC30 ProteinChips are pretreated with 50 μL of 100 mM $CuSO_4$ for 30 min; $CuSO_4$ solution is then discarded.
2. ProteinChips should be briefly washed with deionized water to remove excess copper.

3. Add 200 μL of PBS to each well.
4. Incubate for 5 min with continuous mixing; discard PBS solution.
5. Add 50 μl of urine specimen (disease, control, QC) to each well of the bioprocessor.
6. Incubate for 30 min with continuous mixing; excess specimen is discarded.
7. Add 100 μL of PBS to each well.
8. Incubate for 5 min with continuous mixing; PBS solution is discarded.
9. Briefly rinse ProteinChips with deionized water, let air dry.
10. Add 1 μL of matrix solution (SPA) and let air dry.
11. Obtain mass spectra from ProteinChips.
12. Peaks from spectra should be normalized using urine biochemical parameters (e.g., creatinine).
13. Normalized spectra are submitted to bioinformatic/statistical analysis to identify biomarker candidates.

4. Notes

1. Specimens must be collected no sooner than 4 days post-transplant to avoid excessive inflammatory response proteins from surgery.
2. If specimens cannot be immediately centrifuged, they can be stored briefly at 4°C prior to processing. They should not be stored overnight without processing.
3. Specimens should be aliquoted before any freezing because recent reports in the literature have shown that multiple freeze–thaw cycles can introduce many extra peaks into the mass spectrum (15).
4. It is recommended that when an aliquot is removed for MS analysis, a separate aliquot is removed for biochemical analysis (creatinine, specific gravity, total protein). Normalization is important due to protein concentration variation that is not rejection related (e.g., patient hydration state, time of collection, due to postural changes).
5. All specimens should be run on the same day to avoid bias stemming from day-to-day variability in the analytical system.
6. Specimens should be run in triplicate or more to identify differences in protein peaks that stem from random or analytical system variability.

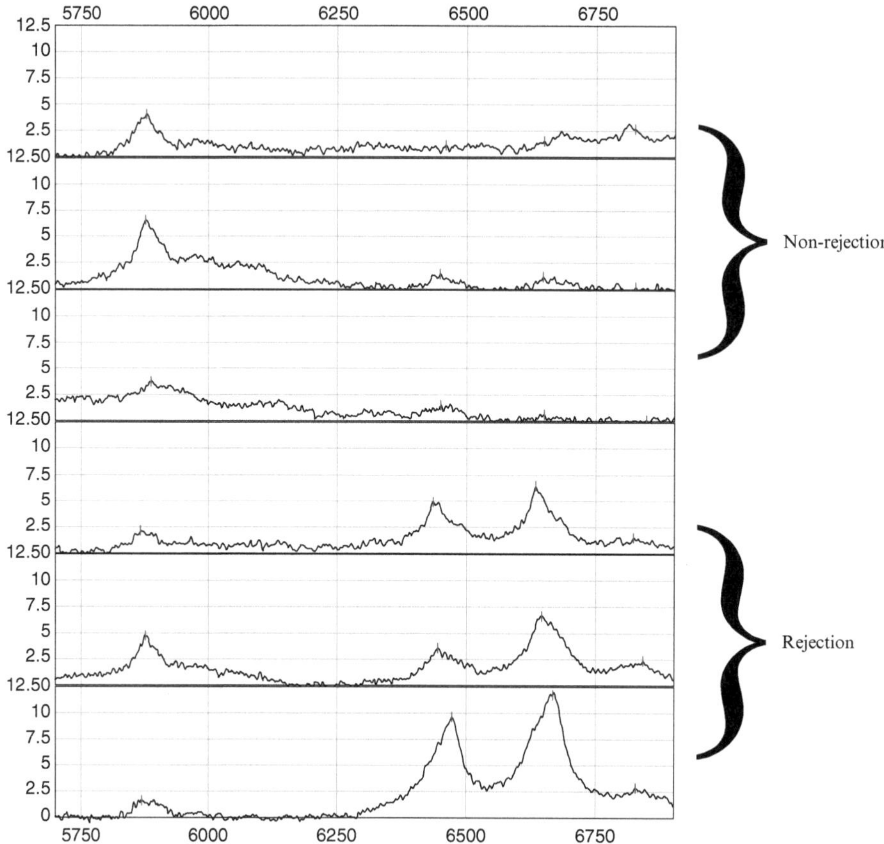

Fig. 1. Protein profiles of three nonrejection and three rejection samples using SELDI-TOF mass spectrometry.

7. The order that specimens are applied to the ProteinChips should be randomized to minimize bias that may stem from chip variability or positional variability.
8. Multiple control specimens (QC) should be included in the each biomarker discovery experiment; these specimens should be made from a pool of the experimental urine specimens. One spot per chip should be reserved for a QC specimen.
9. Sample spectra can be seen in Fig. 1.

References

1. Chandraker, A. (1999) Diagnostic techniques in the work-up of renal allograft dysfunction-an update. Curr Opin Nephrol Hypertens. **8**, 723–728.
2. Wiland, A.M., Fink, J.C., Weir, M.R., Philosophe, B., Blahut, S., Weir, M.R. Jr., Copenhaver, B., Bartlett, S.T. (2004) Should living-unrelated renal transplant recipients receive antibody induction? Results of a clinical experience trial. Transplantation. 77, 422–425.
3. Port, F.K., Dykstra, D.M., Merion R.M., Wolfe R.A. (2005) Trends and results for organ donation and transplantation in the United States. Am J Transplant **5**, 843–849.
4. Burke, J.F., Jr., Pirsch, J.D., Ramos, E.L., Salomon, D.R., Stablein, D.M., Van Buren, D.H., West, J.C. (1994) Long-term efficacy

and safety of cyclosporine in renal-transplant recipients. N Engl J Med **331**:,358–363.

5. Li, B., Hartono, C., Ding, R., Sharma, V.K., Ramaswamy, R., Qian, B., Serur, D., Mouradian, J., Schwartz, J.E., Suthanthiran, M. (2001) Noninvasive diagnosis of renal-allograft rejection by measurement of messenger RNA for perforin and granzyme B in urine. N Engl J Med **344**, 947–954.
6. Roberti, I., Reisman, L. (2001) Serial evaluation of cell surface markers for immune activation after acute renal allograft rejection by urine flow cytometry – correlation with clinical outcome. Transplantation **71**, 1317–1320.
7. Woodle, E.S., Cronin, D., Newell, K.A., Millis, J.M., Bruce, D.S., Piper, J.B., Haas, M., Josephson, M.A., Thistlethwaite, J.R. (1996) Tacrolimus therapy for refractory acute renal allograft rejection. Transplantation **62**, 906–910.
8. Beckingham, I.J., Nicholson, M.L., Bell, P.R. (1994) Analysis of factors associated complications following renal transplant needle core biopsy. Br J Urol **73**, 13–15.
9. Gonzalez-Buitrago, J., Ferreira, L., and Lorenzo, I. (2007) Urinary proteomics. Clin Chem Acta **375**, 49–56.
10. Clarke, W., Silverman, B.C., Zhang, Z., Chan, D., Klein, A., Molmenti, E. (2003) Characterization of renal allograft rejection by urinary proteomic analysis. Ann Surg **237**, 660–665.
11. Schaub, S., Rush, D., Wilkins, J., Gibson, I.W., Weiler, T., Sangster, K., Nicolle, L., Karpinski, M., Jeffery, J., Nickerson, P. (2004) Proteomic-based detection of urine proteins associated with acute renal allograft rejection. J Am Soc Nephrol **15**, 219–227.
12. Schaub, S., Wilkins, J.A., Antonovici, M., Krokhin, O., Weiler, T., Rush, D., Nickerson, P. (2005) Proteomic-based identification of cleaved urinary beta 2-microglobulin as a potential marker for acute tubular injury in renal allografts. Am J Transplant **5**, 729–738.
13. O'Riordan, E., Orlova, T.N., Mei, J.J., Butt, K., Chander, P.M., Rahman, S., Mya, M., Hu, R., Momin, J., Eng, E.W., Hampel, J., Hartman, B., Kretzler, M., Delaney, V., Goligorsky, M.S. (2004) Bioinformatic analysis of the urine proteome of acute allograft rejection. J Am Soc Nephrol **15**, 3240–3248.
14. O'Riordan, E. and Goligorsky, M. (2005) Emerging studies of the urinary proteome: the end of the beginning? Curr Opin Nephrol Hypertens **14**, 579–585.
15. Fiedler, G.M., Baumann, S., Leichtle, A., Oltmann, A., Kase, J., Theiry, J., Ceglarek, U. (2007) Standardized peptidome profiling of human urine by magnetic bead separation and matrix-assisted laser desorption/ionization time-of-flight mass spectrometry **53**, 421–428.

Chapter 12

Surface Plasmon Resonance Biosensorics in Urine Proteomics

Peter B. Luppa, Jochen Metzger, and Heike Schneider

Abstract

Surface plasmon resonance (SPR) is a novel biophysical detection method. In combination with sophisticated surface chemistries and sensing instrumentations, SPR biosensors are approved as tools for molecular interaction studies. SPR plays also a role in interaction proteomics. Once being detected in urine, SPR helps to unravel the functions of new proteins. Due to its outstanding analytical characteristics, SPR also moves more and more into the realm of quantitative analyses in the clinical laboratory. Complex urine determinations of proteins and/or metabolites will bring the SPR biosensor both to the core lab and to point-of-care-testing.

This review delineates first the optical phenomena of SPR near to the gold surface, and also the main features of bioconjugation chemistry on a solid-state surface. Then the kinetic calculation of molecular interaction analysis using SPR is introduced. In order to portray the capability of the method, new applications in urine proteomics and proteinuria diagnostics are finally described in detail.

Key words: Surface plasmon resonance, Biosensor, Kinetic analysis, Bioconjugation, Protein array, Proteinuria

1. Introduction

1.1. Biosensors: Definition and Applications

1.1.1. Terms and Definitions

In the following chapter, we will stringently use the word *analyte* for an interesting protein, peptide or other macromolecule, which should be analyzed in a biological sample in terms of mass concentration or biochemical function by use of an SPR biosensor. In contrast, the term *ligand* will be used as natural interaction partner for the *analyte*. The *ligand* molecule is attached to the biosensor surface and enables the biospecific interaction with the *analyte*.

A biosensor integrates a *surface-immobilized biological element*, enabling a reversible biospecific interaction with the analyte,

Alex J. Rai (ed.), *The Urinary Proteome: Methods and Protocols*, Methods in Molecular Biology, vol. 641, DOI 10.1007/978-1-60761-711-2_12,

and a *signal transducer*. Compared to conventional analytical instruments, biosensors are characterized by an integrated structure of these two pivot components. The biological element is an attached layer of macromolecules qualified for biorecognition, such as enzymes, receptors, peptides, single-stranded DNA or RNA. Even whole cells or other complex systems are applicable for the biological element (1–7).

In general, there are two different devices: biocatalytic and bioaffinity-based biosensors.

- The *biocatalytic biosensor* uses enzymes as the biological compound, catalyzing a signaling biochemical reaction. Probably the most successful commercialization of biosensors is the near patient measurement of capillary glucose using various handheld systems with disposable reagent cartridges (8).
- The *bioaffinity-based biosensor*, designed to monitor the specific binding event with a ligand molecule, uses specific binding proteins, lectins, receptors, nucleic acids, membranes, whole cells, antibodies, or antibody-related substances for biomolecular recognition.

Many devices are connected with a flow-through cell, enabling a flow-injection analysis mode of operation. Biosensors combine high analytical specificity with the processing power of modern electronics to achieve highly sensitive detection systems. Up to now, they are already extensively used as diagnostic tools in point-of-care testing (9).

1.1.2. Measuring Principles

The general biosensor design is depicted in Fig. 1 (2). There are four types of biosensor detection modes: *electrochemical* (potentiometric, amperometric, or conductometric/capacitative), *micro-gravimetric*, *thermometric*, and *optical*. The latter detection mode is most widespread in use. All biosensor types may either be run direct (nonlabeled) or indirect (labeled). The direct sensors are able to detect the physical changes during the biospecific interaction, whereas the indirect sensors use signal-generating labels, which allow more sensitive and versatile detection modes in many cases. Although the latter are highly sensitive due to the analytical characteristics of the label applied, the concept of a direct sensor device is still fascinating. When using an optical detection system, such a biosensor holds multiple advantages due to its simplicity and makes the device progressive and future directed.

1.2. Surface Plasmon Resonance

Biosensors applying the surface plasmon resonance (SPR) principle for signaling feature the advantages of a label-free optical detection mode. The following advantages compared to other bioanalytical methods are worth mentioning:

- Molecular interaction analysis without modification of the analyte molecules.

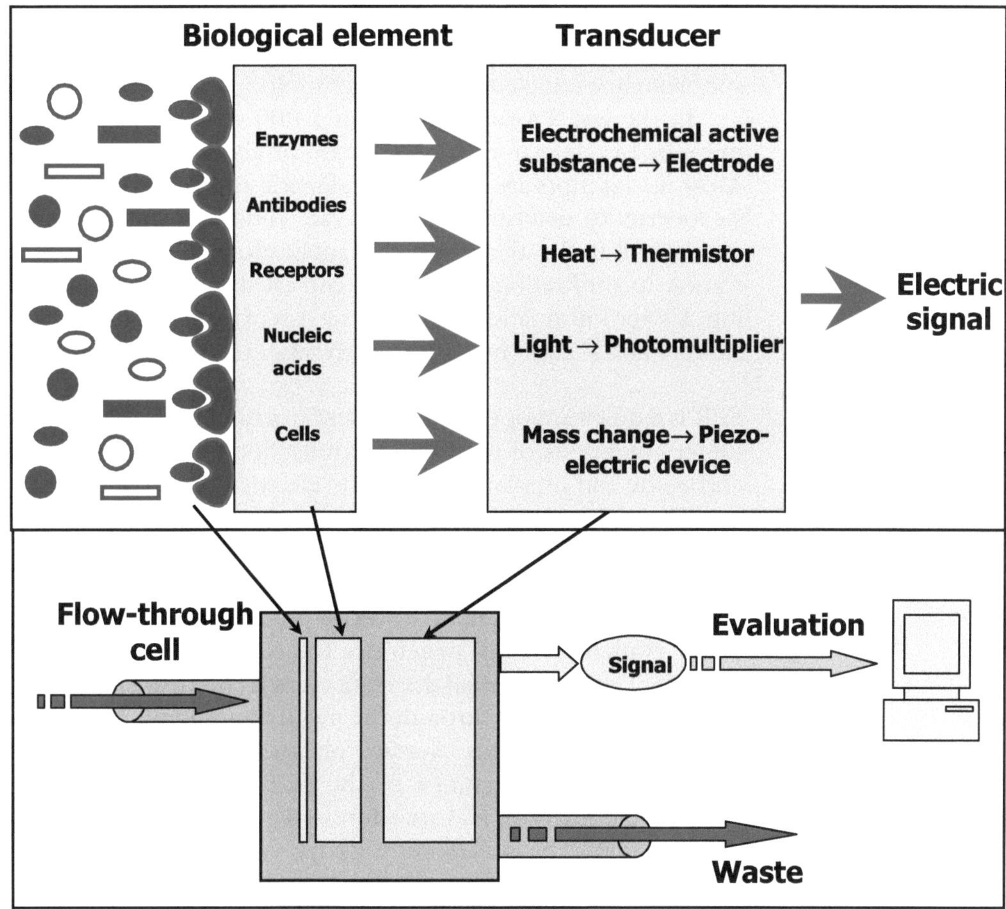

Fig. 1. General biosensor design depicting the intimate integration of biochemical recognition at the solid-state surface and the signal transduction.

- Universal approach to assay development due to the possibility to use native interaction partners without being influenced by the physicochemical properties of a label.
- Measurement of kinetic parameters without need of complex competition or displacement assays

1.2.1. General Application Fields

Applications of optical biosensors in life science comprise food, veterinary, pharmaceutical, and environment investigations, as well as numerous clinical applications in research and diagnostics. In the latter fields, the SPR technology was proven to be useful in a variety of tasks such as ligand fishing, high-throughput small molecule screening, epitope mapping, and others (9–12). Analytes are proteins in general, particularly enzymes, peptides, antibodies, and receptors, as well as oligonucleotides. But also viruses, whole cells, and other complex systems are investigated. Worth mentioning are in particular the improvements in mass spectrometry-coupled biosensor technology.

In the last 3 years, more than 1,000 research articles were published each year with an increase of about 10% per annum. Most manuscripts are dealing with kinetic and equilibrium analysis formats to determine reaction rate constants and interaction affinities. But also the number of applications in analyte concentration quantification, notably in clinical diagnostics, is increasing. Closer information in this context is presented by the review surveys 2002–2005 by Rich and Myszka (13–16).

1.2.2. SPR Optical Phenomenon

SPR is a quantum optical phenomenon arising in thin metal films under conditions of total internal reflection of incoming monochromatic and p-polarized (i.e., the electric vector component is parallel to the plane of incidence) light at the interface of two transparent media of different refractive indices (RI) (17). At a certain angle, the incident light is completely reflected even though an electromagnetic field component of the light, the so-called evanescent field, penetrates the substance with a low RI (Fig. 2). That exponential decaying evanescence wave is then produced and penetrates through the metal film leading to the oscillation of free electrons. Surface plasmons are electromagnetic surface density fluctuations of the oscillating electrons at the metal-liquid interface and are addressed quantum mechanically as quasi particle. The evanescent wave has components in all spacial orientations. It penetrates and "interrogates" even thicker biological layers up to 300 nm, regardless of a potential sample turbidity. In response to the dissipation of energy, there is a sharp dip in the intensity of reflected light. The corresponding light angle is called SPR angle. Its position depends on the amount of surface-bound molecules and is detected by measuring the reflected light intensity. The SPR angle is highly sensitive to changes in the RI of a thin layer adjacent to the metal surface. These changes

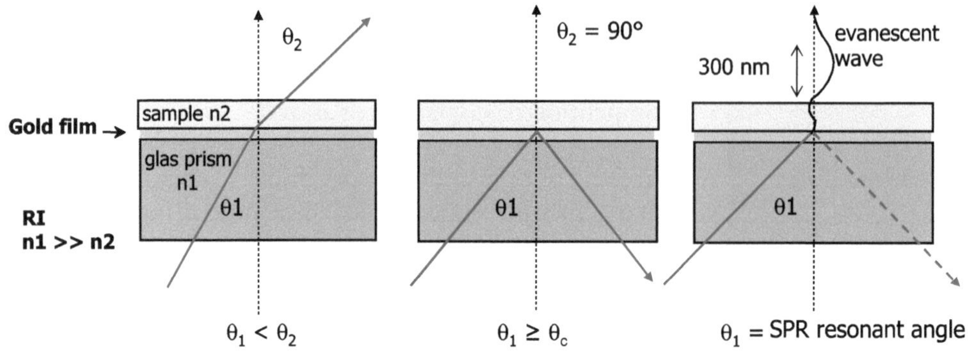

Fig. 2. Optical principle of SPR technology.

correspond to the evanescent wave probing distance when a volume, defined by the size of the illuminated area and the penetration depth, is investigated. For example, when biomolecules interact with already immobilized molecules in this volume, an increase in the surface concentration occurs and the resonant angle shifts to greater values. The magnitude of the angular shift, defined as the SPR response depends on the mean RI change due to the biomolecular interaction in the probed volume.

Gold (Au) is most practical to be used as metal film. It produces a strong, easy to measure SPR signal in the near infrared region. On one side, Au is resistant to oxidation and other contaminants. On the other side, this metal (in case when freshly spattered on a glass or plastic support) is reactive to a series of organic compounds in order to be immobilized onto the metal surface. The SPR generating surface is usually composed of approximately 50 nm thick layer of Au deposited. Other metals are applicable for the generation of plasmons but are not as practical. Indium is very expensive, copper, and aluminium have a too expanded SPR response, and silver is susceptible to oxidation.

In general, there are three configurations of biosensor devices that are suited to generate SPR resonance:

- Prism-based systems (see Fig. 3, configuration applied for most Biacore systems),
- Grating-based systems and
- Optical waveguide-based SPR systems.

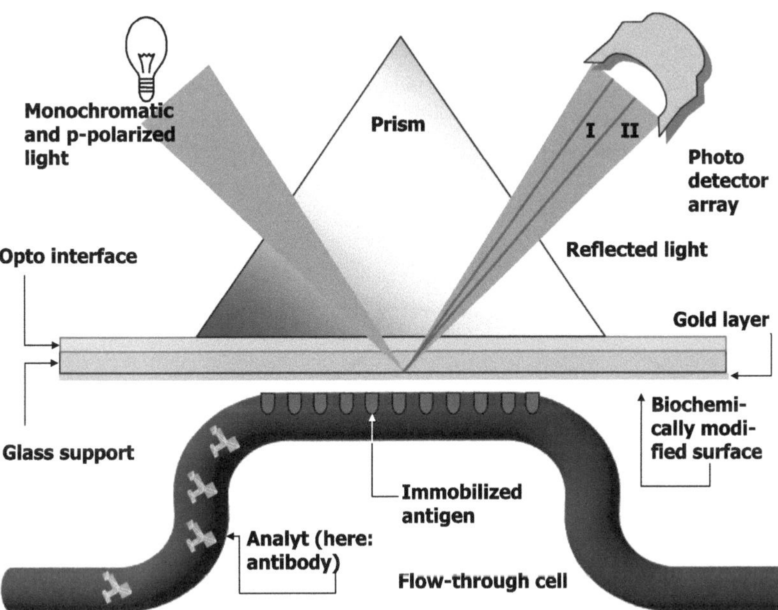

Fig. 3. Design of a typical SPR biosensor showing how the biological interactions occur in the flow cell on the functionalized solid-state surface of the sensor.

In a typical biosensor run, one of the interaction partners is coupled to the sensor surface, whereas the dissolved opponent passes over the surface under continuous flow conditions. The SPR signal as a measure of the SPR angle is generated by a change in mass concentration over the biosensor chip as the soluble macromolecular analyte binds to or dissociates from the immobilized ligand. There is a linear relationship between the amount of bound material and the shift of the SPR angle. Scanning mirror biosensors measure the SPR angle shift in millidegree as a response unit to quantify the binding of macromolecules to the sensor surface. The response also depends on the refractive index of the bulk solution.

Binding is expressed in an increase of arbitrary resonance units (RU) and plotted in form of a sensorgram (Fig. 4). The sensorgram is the transformation of the detector array output to the RU as a function of time. The absolute value of the signal is not relevant to biosensing. It is the change in signal that is important. The response of 1,000 RU corresponds to a shift of 120 millidegree in the resonant angle position, which represents a protein concentration of about 1.0 ng/mm^2 or in a bulk RI change of about 10^{-3}.

Recently, the development of grating- and waveguide-based systems enables protein arrays, which use the "SPR imaging technique" (18) or a multichannel SPR sensor based on spectroscopy of plasmons and wavelength division multiplexing of sensing channels (19). SPR spectroscopic imaging was also established as "SPR wavelength interrogation-based sensor" (20).

The detection principle limits the size of the analytes in solution, which are bound to the chip surface. If the molecular weight of the compound is below 200 Da, the change in refractive index during the biospecific interaction is too low to be detected directly. Therefore, SPR proved to be more sensitive for larger analytes

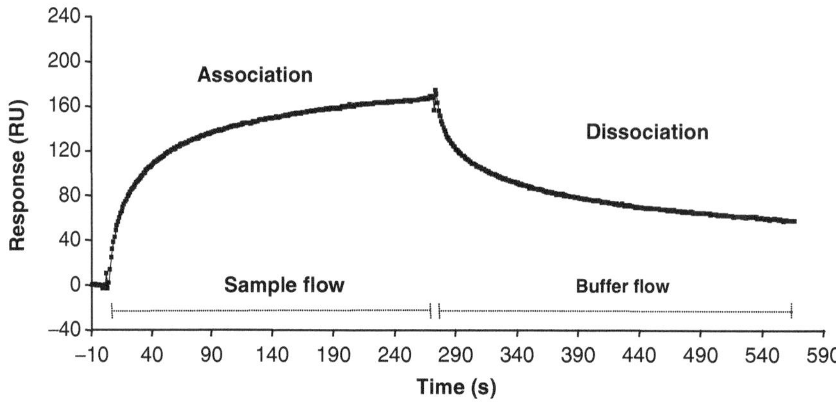

Fig. 4. Typical SPR sensorgram showing association of analyte with the ligand during sample injection and dissociation of ligand-bound analyte after the subsequent switch to running buffer.

than for small molecules. This is an evident disadvantage of the SPR technology, which is, however, controllable by skilled scientists. The penetration depth of the evanescent wave of 200–300 nm also determines the size of macromolecules or particles that can be studied. Particles larger than 400 nm cannot be measured totally. As a result, the signal is not linear related to the amount of bound particles. Under these circumstances, a quantitative or kinetic analysis cannot be performed – but it is possible to study the binding qualitatively.

1.3. Real-Time Biospecific Interaction Analysis Using SPR Sensorics

Biomolecular interactions can be directly monitored using so-called evanescent field sensors. Conventionally, biomolecular interactions are studied using techniques as immunoassays (ELISA or RIA), equilibrium dialysis, affinity chromatography, and spectroscopy. SPR gives two main advantages over these techniques: The binding events are monitored in real-time and it is not necessary to label the interacting biomolecules. Thus, a kinetic evaluation of these interactions is possible. In principle, absolute reaction rate constants can be measured.

1.3.1. Kinetic Measurements Using Biosensor Sensorgrams

The fundamentals for the kinetic evaluation of SPR biosensor data were compiled in the 1990s (21–23). Various commercially or freely available evaluation programs are able to calculate the association and dissociation data (see below). The evaluation algorithms perform a nonlinear regression analysis of the measured sensorgrams.

The analysis of binding events can follow the respective interaction models:

- 1:1 binding;
- 1:1 binding with mass transfer;
- Bivalent analyte binding;
- Heterogeneous analyte or ligand binding and
- Two-state reaction with conformational change.

The most appropriate model will be chosen by a fitting analysis based on the residual plots with low χ^2 values, which were obtained from analysis of the sensorgrams, the χ^2 value being a standard statistical measure of the closeness of fit.

Exemplarily, this is depicted for a Langmuir 1:1-binding: If a soluble analyte molecule A reacts with a ligand B (e.g., f_{ab} fragment with an immobilized epitope), this reaction is calculated as a kinetic pseudo-first-order. The rate constant of the complex formation is described by the association constant k_{a}, the rate of the dissociation being defined by the dissociation rate constant k_{d}.

$$A + B \underset{k_{\mathrm{d}}}{\overset{k_{\mathrm{a}}}{\rightleftharpoons}} \mathrm{AB}$$

These rate constants are defined as follows:

$$d[AB]/dt = k_a[A][B]$$

$$-d[AB]/dt = k_d[AB]$$

The net formation rate of the complex formation (*association phase*) in the SPR sensorgram is defined as sum of the complex formation and simultaneous dissociation:

$$d[AB]/dt = k_a[A][B] - k_d[AB]$$

where

- $[AB]$ is the amount of formed complex, corresponding to the measured parameter R.
- $[A]$ is the concentration C of the analyte protein in the sample solution.
- $[B]$ is the current available amount of nonbound ligand $(R_{max} - R)$.

For the *association kinetics*, it follows from the above that:

$$dR/dt = k_a C(R_{max} - R) - k_d R$$

This can be transformed into:

$$dR/dt = k_a C R_{max} - (k_a C + k_d)R \tag{1}$$

In a plot of dR/dt versus R, both k_a and k_d can be determined from a single association sensorgram in case R_{max} is known. If this is not the case, the determination of sensorgrams at various analyte concentrations is advisable.

The transformation of Eq. 1 under the assumption that C is constant produces:

$$R = k_a C R_{max} / (k_a C + k_d) \times \left(1 - e^{-(k_a C + k_d)t}\right) \tag{2}$$

Equation 2 can be used for a nonlinear regression analysis of the sensorgrams cohort.

Dissociation kinetics: After the injection phase, when the surface is rinsed with buffer and merely the complex dissociation process takes place, the following equation is valid:

$$-dR/dt = k_d R$$

This zero-order reaction can be described by

$$R_t = R_0 \times e^{-k_d(t - t_0)} \quad \text{or} \quad \ln R_0 / R_t = k_d(t - t_0) \tag{3}$$

Equation 3 can now also be used for nonlinear regression analysis of the dissociation phases of the sensorgrams cohort in the dR/dt versus R plot according to the transformation to:

$$dR/dt = k_d R_0 \times e^{-k_d(t-t_0)}$$

1.3.2. Implications from the Kinetic Measurements

According to these formulas, it is possible to determine the rate constants for association and dissociation from the measured binding curves. The premise hereof is simply the knowledge of the molar concentration of the analyte. The kinetic evaluation process can be done by use of an adequate software programme. Examples are:

- BiaEvaluation (Biacore, http://www.biacore.com) (24)
- SPRevaluation (patented by Maureen D. O'Connor-McCourt) (25) or
- SCRUBBER-2 (http://www.cores.utah.edu/interaction/software.html)

There are, however, only few interactions recorded on SPR biosensors, which can be correctly described by the simple 1:1 interaction model. Several other more complicated binding mechanisms are possible:

- Simple bivalent
- Two-state or three-state conformational change
- Surface heterogeneity

Such kinetic reactions of higher order are best described by multiple rate equations. But the prerequisite should be considered that fitting data concerning the association and dissociation phase are to be collected for a concentration series of the analyte. This so-called global fitting analysis discriminates between the above mentioned reaction types.

There is also a series of other factors affecting the binding response. One part are simple artifacts, and most important are, however, avidity effects, mass transport, or rebinding effects (22).

Mass transport phenomena can be reduced by working at low densities of immobilized binding partners on the sensor surface. The mass transport coefficient k_m, given in Fig. 5, is dependent on the diffusion coefficient of the analyte, the dimensions of the flow cell and the flow rate. By incorporating a mass-transport step into the mathematical binding model, it enables to calculate reaction rate constants (24).

Under steady-state conditions (association of the analyte with the surface is balanced by the simultaneous dissociation of the surface-bound complex), the so-called equilibrium analysis is able to derive the thermodynamically relevant association constant:

$$dR/dt = k_a C(R_{max} - R_{eq}) - k_d R_{eq} = 0 \quad \text{at equilibrium.}$$

The equation gives:

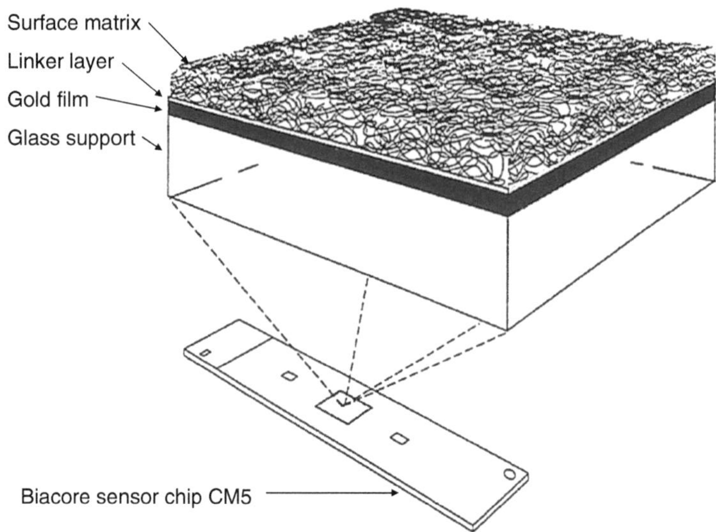

Fig. 5. Interactions at the biosensor surface under special consideration of mass transfer phenomena.

$$k_a C(R_{max} - R_{eq}) = k_d R_{eq}.$$

This can be transformed to:

$$k_a / k_d = \text{Association constant } K_A = R_{eq} / C(R_{max} - R_{eq})$$

or

$$R_{eq} / C = K_A R_{max} - K_A R_{eq}$$

The Scatchard plot R_{eq}/C versus R_{eq} gives a straight line from which R_{max} and K_A can be calculated.

1.4. Technical Aspects

1.4.1. Flow Cell and Microfluidics

Apart from cuvette systems, e.g., applied in the IAsys device, SPR measurements are usually performed in flow injection analysis (FIA) systems. Biacore devices use such flow cells on the biosensor surface for interaction analysis. These flow cells are formed when a microfluidic flow system is brought into contact with a sensor surface on which one interacting partner is immobilized. The second interaction partner is delivered in the solution by being pumped into the FIA cell.

Several flow system designs have evolved:

- Serial flow cell systems
- Independent flow cell systems
- Hydrodynamic addressing flow cell systems

One major advantage of the flow cell design is that there are at least two flow cells (FC) of which one may act as a reference cell when being loaded with a reference ligand in order to subtract unspecific binding phenomena. In the clinical laboratory, this is an imperative prerequisite for analysis when applying (un)diluted serum or urine samples. Thus, the gross signal can be subtracted from the reference signal to gain a specific sensorgram.

There are, however, problems when using such FIA systems. Often, an interdiffusion of sample and of flow buffer is encountered during the transport process from the inlet to the interior of the flow cell. To minimize this problem, Biacore designed an integrated μ-fluidic cartridge (IFC) system that brings the sample loops and valves close to the reaction site on the biosensor surface. This IFC consists of a series of precision cast channels built-in a silicon polymer plate, forming sample loops and flow channels. The unit is controlled by a series of pneumatically driven diaphragmatic valves directing the solvent flow through a chosen loop to the solid-state surface. The material properties of the silicon polymer and the precision of the docking mechanism prevent leakages from the channels.

The traditional flow cell technology, which uses a small number of channels to flow the analyte sequentially in a single direction, limits flexibility, throughput, and optimization of experimental conditions. The *Flexchip* from Biacore uses the "single pass, multi-target flow cell system" for an array format. Potential ligands are spotted onto a gold-coated surface of an array chip. A sealed window with inlet/outlet valves is then positioned over the array to form a flow cell and inserted in the instrument. A solution containing the interacting analyte is injected over the array and interactions on each spot are detected in real time via grating-coupled SPR. Up to 400 interactions can be monitored in a single experiment, which is beneficial for proteomics applications (26).

Also, the new *ProteOn XPR36* from Bio-Rad Laboratories expands the flow cell approach to SPR by providing multiple flow channels. A multichannel module forms six channels on the sensor chip surface, and up to six different ligand samples can be immobilized in each of the channels. A second set of channels is superposed orthogonally, creating a crisscross pattern of two sets of flow channels. This results in a 6×6 interaction array. The response at each interaction spot generates 36 sensorgrams corresponding to six analytes interacting with six ligands.

The ProteOn XPR36 utilizes a synchronized sequential scanning illumination system coupled with an imaging device to detect the SPR response for each point on the sensor chip. The measurement uses a total of 78 spots, 36 of which represent the interaction data. The additional 42 spots are called interspots and are used for normalization.

1.4.2. The Biosensor Surface

One of the most striking advantages of SPR analysis is the wide range of different proteins that can be measured in a quantitative way without major nonspecific binding effects. The SPR response is a measure of the RI changes of a chip surface, which occur during binding of an analyte protein to a specific surface matrix. But also the sample bulk solution itself plays a role in the RI change. The instrument and the experimental design, however, can eliminate this signal.

Therefore, it can be stated that the SPR response primarily reflects mass concentration changes of biomolecules on the sensor chip in the flow cell. By use of radiolabeled proteins, Stenberg et al. were able to correlate closely the absolute surface concentration of protein to the SPR response in a classical paper (27). They applied antibodies, transferrin, and human chymotrypsinogen as radiolabeled proteins. Responses were linear correlated over the range of 2–50 ng/mm^2 surface concentrations. Interestingly, there was no significant difference in specific responses for the different proteins tested. This close agreement in RU response makes the SPR sensor a very appropriate tool for urinary protein determinations. The smallest response, which can be reliably measured in different SPR devices, is around 10 RU, corresponding to a potential lower limit of detection of 10 pg/mm^2.

In the commonly used biosensor chip formats, a surface matrix of noncrosslinked carboxymethyl dextran (CMD) is applied for covalent attachment of ligands (28). The main reasons for its broad application are:

- High ligand binding capacity
- Resistance to alkaline and acidic treatment
- Provision of a hydrophilic environment
- Applicability to a wide range of coupling chemistries

It should be stated, however, that partially due to the hydrogel-like nature of the CMD matrix, nonspecific binding phenomena and mass-transport limitations may occur, which lead to misrepresentation of true binding kinetics. Therefore, it is sometimes more preferable to use planar surfaces such as self-assembling monolayer (SAM) composed of long chain *n*-alkylthiols or polyethylene glycol block copolymers as functionalized surface coatings. Another advantage for the use of a SAM rather than a CMD layer is the possibility to construct biosensor supports with various reactive groups allowing immobilization of combinations of two or more ligands.

1.4.2.1. CMD-Coated Biosensor Surface

The commercially available CMD-coated biosensor supports (e.g., CM5 sensor chips from Biacore (Fig. 6)) are sufficiently eligible for the majority of biomolecular interaction analyses. In cases with special demands on CMD chain length, molecular structure or carboxymethyl composition, dextran coatings can be efficiently

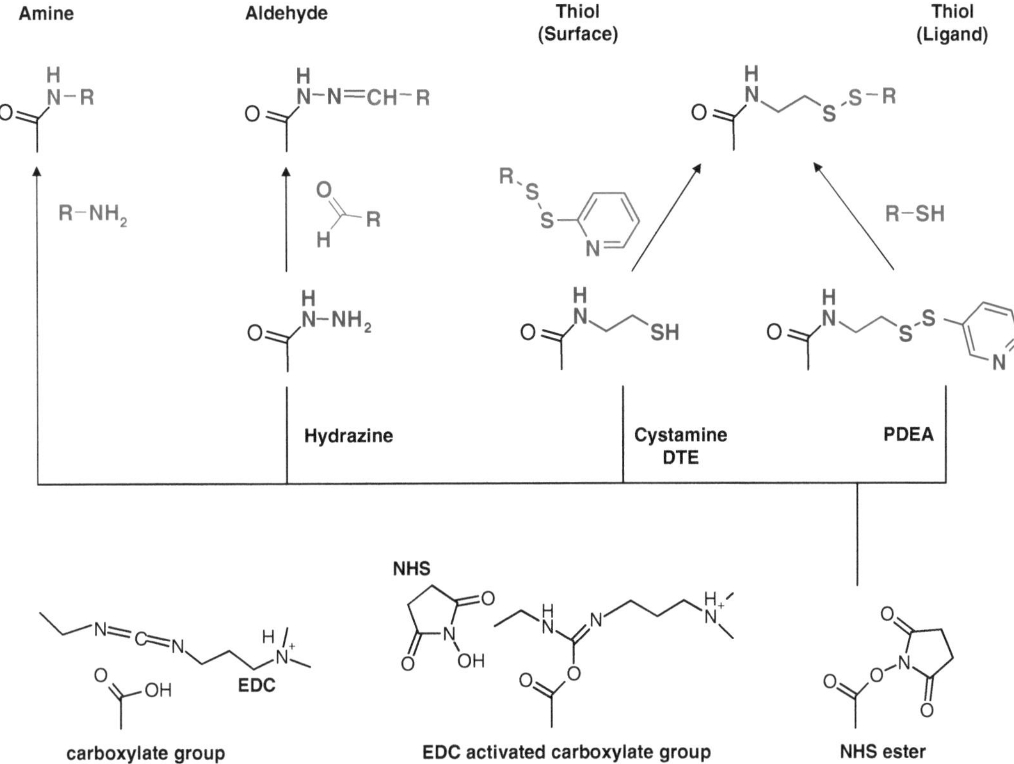

Fig. 6. Design of a SPR biosensor chip.

prepared after formation of a 16-mercapto-hexadecane-1-ol linker layer on the gold surface followed by an epichlorohydrin treatment. The introduced epoxide functions readily react with the dextran, which can subsequently be carboxymethylated to different degrees of substitution using bromoacetic acid (29).

1.4.2.2. SAM-Coated Biosensor Surface

SAM formation is initiated by the interaction of sulfur of *n*-alkylthiol head groups with gold atoms. After initial chemisorption, which takes only a few minutes, a higher degree of conformational order is reached after several hours by the subsequent reorganization of the hydrophobic *n*-alkyl chains. This late crystallization-like process is driven by lateral interchain interactions, such as van der Waals, steric, repulsive, and electrostatic forces, and leads to a densely packed rigid arrangement of the molecules on the gold surface (for review see (30)).

Due to these slow rearrangement processes, it is necessary to expose the gold-coated surfaces to the *n*-alkylthiols for an extended period of time. There are other factors, however, that affect the stability of the SAM, like total *n*-alkylthiol concentration and alkyl chain length. Generally, ethanol solutions containing 1 mM of 6- to 16-C long *n*-alkylthiols at incubation times of more than 12 h are recommended for optimal results in SAM formation.

1.4.2.3. PEG-Coated Biosensor Surfaces

Polyethylene glycol block copolymers are starting materials for the preparation of high-density grafted polymer brushes. Due to their resistance to nonspecific adsorption, they play an important role as surface modifiers of silica substrates, i.e., for the production of medical implants and nonbiofouling devices. Preparation and functionalization rely on a complex and sophisticated chemistry due to the fact that polymerisation is initiated directly on the surface. This restricts their usefulness in SPR biosensorics to certain special applications. A method that combines self-assembly of *n*-alkanethiols on gold and surface-confined atom transfer radical polymerization of poly(2-vinylpyridine) to determine thermodynamic adsorption properties of fibronectin adhesion-promoting peptide is given by Li et al. (31).

1.4.3. The Optointerface

Another pivotal component for the generation of the SPR signal at the chip surface is the optointerface enabling an optimal optical coupling. The gold film is usually deposited on a thin glass slide, which is tightly coupled to the glass (or moulded plastics) prism by use of an optically matched soft polymer. This coupling technique is comparable to the use of immersion oils for enhancing the optical resolution in light microscopy. Instead of oily liquids, however, the soft polymer is applied (see also Fig. 3) (21).

2. Materials

2.1. Commercially Available Systems

The majority of researchers use Biacore SPR systems. Also, in literature, the vast majority of work is performed using Biacore platforms (13–16). In the last 2 years, however, the quote of mentionable manufacturers increased significantly. Table 1 shows a comprehensive (but surely not complete) survey of commercially available SPR biosensor devices.

Because of the large number of applications, in particular for medical laboratory diagnostics, we focus this survey on Biacore platforms and chips, but many aspects may be applied also to other available SPR instruments.

2.2. Buffer Solutions

All buffers for biosensor measurements should be sterile-filtered using 0.2 μm FP CA-S filters (Whatman) and exhaustively helium-degassed before application.

Table 1
Survey of commercially available SPR Biosensor instrumentations (status as of December, 2006)

Biosensor type	Company	Website
Autolab Esprit/Springle	Eco Chemie BV, Utrecht, The Netherlands	www.ecochemie.nl
Biacore 2000, 3000, A100, C, X, Q, J, T100, Flexchip[a]	Biacore AB, Uppsala, Schweden (newly acquired by GE Healthcare)	www.biacore.com
BIOS-1	Artificial Sensing Instruments, Zürich, Switzerland	www.microvacuum.com
Biosuplar	Analytical µ-Systems GmbH, Sinzing, Germany	www.biosuplar.de
FT-SPR 100	GWC Technologies, Madison, WI, USA	www.gwctechnologies.com
IAsys[b]	NeoSensors, Sedgefield, UK	www.neosensors.com
MultiSPRinter	Toyobo Co., Ltd., Fukui, Japan	www.toyobo.co.jp
OWLS	MicroVacuum Ltd, Budapest, Hungary	www.microvacuum.com
Plasmonic	Hofmann Sensorsysteme, Wallenfels, Germany	www.plasmonic.de
ProteOn XPR36	Bio-Rad Laboratories, Hercules, CA, USA	www.biorad.com
SensiQ	Nomadics, Inc., Oklahoma City, OK, USA	www.nomadics.com
SPR 670	Rikei Corporation, Tokyo, Japan	www.rikei.com
SPRi Array	GenOptics, Orsay Cedex, France	www.genoptics-spr.com
Spreeta TSPR2KXY Biosensor	Texas Instruments, Dallas, TX, USA	www.spreeta.com
SR7000 Surface Plasmon Resonance Instrument	Reichert Analytical Instruments, Depew, NY, USA	www.reichert.com
VirChip	Vir A/S, Taastrup, Denmark	www.vir.dk

[a]Formerly "ABI 8500 Affinity Chip Analyzer" from Applied Biosystems, Foster City, CA, USA
[b]Formerly product of Affinity Sensors

For Biacore sensor systems, HEPES-buffered saline (HBS) is preferably used as general running buffer.

HBS-N: 10 mM HEPES pH7.4, 0.15 M NaCl.

HBS-EP: 10 mM HEPES pH7.4, 0.15 M NaCl, 3.4 mM EDTA, 0.005% v/v surfactant P20.

HBS-P: 10 mM HEPES pH7.4, 0.15 M NaCl, 0.005% v/v surfactant P20.

These buffers may be modified individually in order to suppress possibly occurring nonspecific binding phenomena. These modifications include the addition of various compounds such as bovine serum albumin, human serum albumin, Tween 20, etc. at different concentrations.

Ligand immobilizations in a biosensor system are best performed by using buffer systems with low ionic strength to favor covalent binding of the ligand to the chemically preactivated biosensor matrix by electrostatic attraction. Additionally, the pH of the buffer should be below the isoelectric point of the ligand. Recommended buffers are 10 mM formiate for pH3.0–4.5; 10 mM acetate for pH4.0–5.5; 5 mM malate or 20 mM 2-(*N*-morpholino)-ethanesulfonic acid for pH5.5–6.5; 20 mM 4-(2-hydroxyethyl)-1-piperazineethanesulfonic acid (HEPES) for pH6.5–8.0, and 10 mM borate for pH8.0–10.0.

2.3. Preparation of Functionalized Chip Surfaces

2.3.1. Preparation of a CMD Biosensor Surface

For the preparation of CMD biosensor surfaces, the gold slides have first to be coated with hydroxyalkylthiols, such as 11-mercaptoundecan-1-ol or 16-mercaptohexadecan-1-ol. Adsorption of the alkylthiol compounds on the gold is readily achieved by contacting the slides with a 1 mM ethanolic solution for 18–20 h. This and all subsequent reactions are best performed at room temperature. After extensive washing with 96% ethanol and distilled water to remove unbound material, the gold surface is treated with a 0.6 M solution of epichlorohydrin in a 1:1 mixture of 0.4 M sodium hydroxide and bis-2-methoxyethyl ether for 4 h. After an additional washing with distilled water and 96% v/v ethanol, the slides are incubated for 20 h in a 30% (wt/vol) basic dextran solution (Sigma), which was prepared in 100 mM sodium hydroxide. For the substitution with carboxymethyl groups, the CMD-coated slides are finally incubated with 1 M bromoacetic acid in 2 M sodium hydroxide for 16 h. After rinsing with 96% ethanol and drying under a stream of argon, the CMD-coated slides are mounted onto a chip carrier and inserted into the biosensor flow system.

2.3.2. Preparation of n-Alkylthiol-Based SAM

Functionalized SAM surfaces are prepared by immersing gold slides in a 1 mM ethanolic solution of the appropriate *n*-alkylthiols for 18–20 h at room temperature. SAM-coated gold surfaces are subsequently rinsed with 96% v/v ethanol, dried under an argon stream, assembled onto a chip carrier, and inserted into the biosensor flow system.

3. Methods

3.1. Biochemical Immobilization Strategies

The choice of the appropriate immobilization technique depends on the chemical properties of the ligand. Due to the requirement to retain analyte-binding activity over a series of measurements and postanalytical regeneration steps, covalent coupling is generally favored over noncovalent capturing. However, there are certain instances, where affinity tags and adaptor proteins are more amenable for ligand binding.

- First, in the case of biomolecules with a low isoelectric point and high electrostatic repulsion from the hydrophilic biosensor surface, such as nucleic acids, polysaccharides, and acidic proteins, conjugation or fusion to affinity tags often represents the only viable option to achieve stable attachment under physiological salt and pH conditions.
- Secondly, affinity-mediated capturing instead of rigid chemical coupling to the biosensor surfaces better preserves full analyte binding activity of labile proteins. In this context, it is worth mentioning that for affinity capturing, the protein structure can be better maintained by the addition of stabilizing agents and cosolvents, whereas this is in most cases not applicable for covalent coupling due to the risk of chemical side reactions.
- Thirdly, affinity tag-mediated ligand immobilization allows site-directed orientation, which often improves the accessibility of the analyte to its binding regions. The latter property is important in particular for the immobilization of synthetic peptides or other small molecules where reactive groups are lacking or reserved for biological function.

3.1.1. Strategies for Covalent Ligand Immobilization

CMD and SAM biosensor surfaces both allow the application of multiple chemistries for ligand immobilization. Covalent attachment of natural or synthetic ligands can be achieved by derivatization of the functionalized biosensor surface, for example, with amine- or thiol-reactive cross-linker containing *N*-hydroxysuccinimidyl or maleimidyl groups. The standard chemistries commonly used for the coupling of amino acid or carbohydrate moieties are summarized in Fig. 7.

A central role in all chemical coupling reactions play the water-soluble amine-reactive cross-linker *N*-ethyl-*N′*-(3-diethylaminopropyl) carbodiimide (EDC) and *N*-hydroxysuccinimide (NHS). A combination of these two reagents can be used for rapid transformation of carboxylate into active ester intermediates, which in turn are susceptible to the attack of primary amines and other nucleophilic groups. Since all proteins contain amine functions at

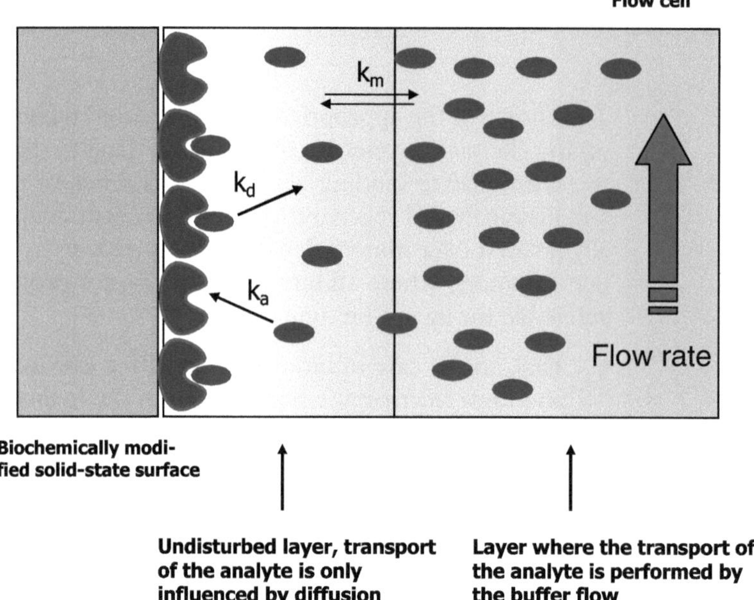

Fig. 7. Overview of approved immobilization chemistries for covalent ligand coupling on SPR biosensor matrices (abbreviation: *PDEA* poly(2-(diethylamino) ethyl methacrylate)).

their *N*-terminus and in ε-position of lysine side chains amine linkage is recommended as first choice for the covalent attachment of proteins. However, the carbodiimide reaction requires uncharged amino groups. Hence, for ligands with an isoelectric point less than 3.5, other coupling chemistries have to be applied. One alternative method is coupling by thiol-disulfide exchange. The introduction of disulfides can either be performed on the chip by injection of cysteamin over an EDC/NHS-activated surface followed by dithioerythritol-mediated reduction of free sulfhydryls or on the ligand by thiolation of carboxylate or amine functions with pyridyl disulfide containing heterobifunctional cross-linker, such as 2-(2-pyridinyl-dithio)ethaneamine or 4-succinimidyloxycarbonyl-α-methyl-α-(2-pyridyldithio) toluene.

In situations where polysaccharides and glycoproteins serve as ligands, immobilization can be carried out by selective periodate oxidation of *cis*-diols and conjugation of the resulting aldehyde groups on a hydrazide preactivated support. One representative example for the use of the aldehyde chemistry in protein attachment is the oriented immobilization of IgGs via their Fc-localized carbohydrate moieties (32).

3.1.2. Affinity-Based Strategies for Ligand Binding

The major challenge in the development of an affinity-based coupling strategy is the selection of an appropriate combination of affinity tag and surface immobilized capture agent. Especially,

Table 2
Affinity-tags and capture molecules for SPR biosensor applications

Tag	Capture	Mode	Comments
Fc portion of Igs	Protein G Protein A Protein M	Fc-receptor binding	– Oriented immobilization of mAbs, i.e., for epitope mapping – Selection of Fc-receptor depends on Ig isotype
Fc portion of Igs	Anti-human Fc RAMFc	Anti-Ig binding	– Oriented immobilization of mAbs – Recommended for the capture of Fc fusion proteins – Anti-Immunoglobulins must be affinity pure
Biotin	SA Avidin Neutravidin	Biotinylation/ SA-binding	– Extremely stable affinity interaction (10^{15}/M) – Coupling of oligonucleotides and peptides – Small molecules require the introduction of a linker sequence – Other Biotin/SA-Systems are available, i.e., on the basis of SA-binding peptides
Poly-His	Ni^{2+}-Chelate	Poly-amino acid fusion/Nickel-binding	– Short recognition motif – Weak binding may lead to gradual loss of ligand – Binding is affected by EDTA, imidazole and acids – Anti-His mAbs can be used as an alternative capture reagent
GST	anti-GST mAbs	Protein fusion/ Ab-binding	– GST allows simple one-step purification of fusion proteins prior SPR analysis – High affinity anti-GST mAbs are available that resist harsh regeneration conditions
hAGT	Benzylguanine	Protein fusion/ substrate linkage	– Highly stable due to the formation of a thioether bond – No further purification of fusion proteins needed

hAGT O6-alkylguanine-DNA-alkyltransferase, *EDTA* ethylene diamine tetraacetic acid, *GST* glutathione-S-transferase, *Ig* immunoglobulin, *mAb* monoclonal antibody, *RAMFc* rabbit anti-mouse Fc, *SA* strepavidin

the ability to mediate tight interaction under a broad variety of experimental conditions is of particular importance in this context. The decision which affinity labels should be applied crucially depends on the chemical and physical characteristics of the ligand. Table 2 outlines the most accepted affinity methods for biosensor applications. With exception of antigen capturing by surface-immobilized specific antibodies, these capturing methods require modification of the target protein by either chemical reaction, i.e., with biotinylation reagents, or by genetic fusion to small affinity protein domains or polyamino acid sequences.

To immobilize capture antibodies in a site-directed manner on the biosensor support, proteins are available, which specifically react with the Fc portion of immunoglobulins. These include the staphylococcal Fc-receptor proteins A and G as well as anti-immunoglobulins from animals being immunized with species-specific Fc-fragments, most typically of mouse or human origin. It should also be mentioned that, besides capturing of xenoantibodies, antiimmunoglobulins can be applied for the immobilization of recombinant Fc fusion proteins.

The *streptavidin-biotin system* provides a convenient method for the capture of DNA oligonucleotides and peptides since the conjugation of biotin can be performed directly during automated synthesis. Another important area of application concerns immobilization of acidic proteins, polysaccharides, and glycoconjugates due to the fact that the streptavidin–biotin interaction can overcome electrostatic repulsion of these anionic compounds from the biosensor surface. A variety of biotinylation reagents directed against different functional groups are commercially available. The most important are NHS-biotin for primary amines, biotin hydrazide for carboxylic acids and aldehydes, iodoacetyl-biotin, and biotin maleimide for thiol compounds, and photo-activatable biotin derivatives for nucleic acids.

In recent years, a series of *fusion tags* have been developed for applications such as protein purification and microarray preparation. Some of them were found to be very useful in biosensor applications. The most widely used among these are capture of polyhistidine-tagged proteins on nickel coated surfaces and binding of gluthatione-S-transferase (GST)-fused proteins to surface-bound GST-specific monoclonal antibodies.

These affinity tag-mediated immobilization techniques have the collective attribute to allow oriented immobilization under mild conditions so that antigenicity and biological activity is retained. They suffer, however, from the drawback that the interaction is not stable enough and results in a drift of the baseline signal and in a lack of resistance to the harsh regeneration conditions necessary for efficient reconstitution of the biosensor surface after analysis. A new affinity capture strategy – based on the irreversible reaction of the human DNA repair enzyme O_6-alkylguanine-DNA-alkyltransferase (hAGT) with O_6-benzylguanine (BG) – has recently been developed (33): It enables both covalent as well as site-directed attachment of proteins. For immobilization of hAGT-fusion proteins, a BG derivative with an aminopolyethylene glycol substitution is previously coupled via EDC/NHS-chemistry to a carboxylated biosensor matrix. hAGT reacts with the BG head group to form a stable thioether linkage after injection of the hAGT-fusion protein into the biosensor flow system. Due to the high specificity of the hAGT–BG interaction, hAGT-fusion proteins can be immobilized directly from cell culture lysates without the need of prior purification.

3.1.3. Analyte Measurement and Chip Regeneration

In a typical analytical run, a solution with the analyte is injected under continuous flow conditions over the biosensor surface. This period defines the association phase. It is followed by a switch to running buffer for the detection of analyte dissociation. At the end of the dissociation phase, the ligand surface is treated with regeneration solutions containing, for example, acids, bases, or organic solvents.

It is generally recommended for biosensor measurements to simultaneously evaluate binding of the analyte to an irrelevant reference surface. This allows a satisfying compensation of nonspecific binding, bulk refractive index background, and matrix effects. Ideally, the reference surface should be prepared by coupling of a control antigen with similar biochemical properties.

For the reliable determination of binding affinities and rate constants, the binding reactions must be conducted under optimal experimental conditions. It is worth mentioning that there are a number of sources of error that have to be considered in the design of the experimental protocol in SPR biosensor analysis. The major causes are nonspecific binding, matrix effects, mass transport limitations, rebinding of the analyte, and gradual loss of ligand binding activity, the latter being caused mainly by the accumulation of bound material on the chip surface due to incomplete regeneration (see also Subheading 1.3.2).

Nonspecific binding can be prevented depending on its chemical nature either by an increase of the ionic strength or by the addition of suitable additives to the analyte dilution and running buffer. Recommended blocking agents include nonionic surfactants, such as Tween-20, and irrelevant proteins, such as bovine or human serum albumin. The addition of soluble CMD to the buffers is an effective way to reduce nonspecific binding in cases where compounds in the sample exhibit affinity to a dextran hydrogel matrix.

Matrix effects arise from changes in the extension of the surface matrix in response to differences in the pH or ionic strength of sample and running buffer. They can be minimized by adjusting the composition of the running buffer and the analyte solution.

In order to prevent *mass transport limitations*, it is advisable to perform kinetic measurements at low ligand immobilization levels and relatively high flow rates. Conversely, low flow rates and high surface binding capacities were found to be more appropriate for concentration measurements.

Rebinding occurs in situations when the analyte binds with only low affinity to the immobilized ligand. Rebinding can be assumed when an increase in the flow rate results in higher dissociation rates. Another indication of rebinding phenomena is when a slight dissociation is visible in the sensorgrams at the end of the analyte injection phase. Rebinding can be prevented by the use of higher flow rates, by immobilization of ligand at low densities, or by injection of soluble ligand in the dissociation phase.

After each cycle of measurement, the sensor chip has to be regenerated in such a way that all bound materials from the sample are completely removed. Although there are some general guidelines available in literature, for example, removal of antibodies by repeated short injections of hydrochloric acid or glycine buffer at pH 2.5, the optimum regeneration conditions have to be evaluated empirically in most cases. An adequate strategy to find the best regeneration conditions on the basis of six multivariate stock solutions is presented by Andersson et al. (34). During this phase of method optimization, it is important to ascertain that the rate of analyte binding is stable over a sufficient number of sample injections and does not decrease due to ligand inactivation. Alternatively to generally used regeneration methods, a new approach for the collection of kinetic binding data is the so-called "kinetic titration" (35). This method involves sequentially injecting an analyte concentration series without regeneration.

3.2. Exemplified: SPR Measurements and Kinetic Evaluation

3.2.1. General Recommendations for Chemical Coupling Reactions

HBS as running buffer and a constant flow rate between 2 and 5 μl/min is generally recommended for all chemical coupling strategies. Since efficient ligand immobilization to the biosensor matrix requires electrostatic preconcentration, dilution buffers should be adjusted to a pH below the pI of the ligand molecule.

3.2.1.1. Amine Coupling

The carboxyl groups of CMD or SAM biosensor chips are activated in one flow cell (FC) by a 35 μl-injection of an aqueous solution of 200 mM EDC and 50 mM NHS. After ligand immobilization, nonconverted active ester groups are blocked by a 35 μl-pulse of a 1.0 M ethanolamine hydrochloride solution, pH 8.5.

3.2.1.2. Aldehyde Coupling

After activation of the CMD or SAM biosensor surface with EDC and NHS (see Subheading "Amine Coupling"), 35 μl of 5 mM hydrazine or carbohydrazine in distilled water are injected to generate reactive hydrazine groups. Next, nonconverted NHS-esters are inactivated with a 1.0 M ethanolamine solution, pH 8.5. After injection of the aldehyde, containing ligand, the reactive hydrazide groups on the biosensor surface spontaneously react with the aldehyde functions to yield hydrazone bonds. These can be reduced with sodium cyanoborohydride to form an imine bond, which is in contrast to the hydrazone bond completely stable under acidic conditions. For complete reduction of hydrazone bonds, a volume of 40 μl of 0.1 M sodium cyanoborohydride in 0.1 M acetate buffer, pH 4.0, is injected over the biosensor surface at a flow rate of 2 μl/min.

3.2.1.3. Ligand Thiol Coupling

In the ligand thiol coupling procedure, the ligand of interest contains free thiols to bind to reactive disulfide groups on the biosensor surface. After activation of the CMD or SAM biosensor surface with EDC and NHS, sulfhydryl-reactive groups are introduced on the biosensor surface by injection of 20 μl of 80 mM 2-(2-pyridinyldithio)-ethaneamine hydrochloride in 0.1 M borate buffer, pH 8.5. This is immediately followed by the addition of the ligand at a concentration of 10–200 μg/ml. For maximum coupling efficiency, it is recommended to dissolve the ligand in dilution buffers with moderate acidic pH (4.0–5.0). Finally, residual disulfide groups are blocked on the biosensor surface by a 20 μl-injection of 50 mM l-cysteine and 1 M NaCl in 0.1 M formate buffer, pH 4.3.

3.2.1.4. Surface Thiol Coupling

In the surface thiol coupling technique, the ligand of interest is modified to contain active disulfide groups, which in turn can be coupled to thiol groups on the biosensor surface. Firstly, the CMD or SAM biosensor is activated by EDC and NHS injection. This is followed by a 15 μl-injection of 40 mM cysteamine in 0.1 M borate buffer, pH 8.0, to introduce disulfide groups on the biosensor surface. Subsequently, the cystamine disulfides are reduced to thiols by injection of 15 μl of 0.1 M dithioerythritol or dithiothreitol. Next, the ligand of interest, which was prepared to contain reactive disulfide groups by exposure to the sulfhydryl-containing cross-linker poly(2-(diethylamino) ethyl methacrylate (PDEA) or *N*-succinimidyl 3-(2-pyridyldithio)propionate, is injected over the thiolated biosensor matrix at a concentration of 10–200 μg/ml. After the end of the ligand coupling reaction, nonconverted thiol groups are blocked on the biosensor surface by a 20 μl-pulse of 20 mM PDEA and 1 M NaCl in 0.1 M formate buffer, pH 4.3.

3.2.2. Selected Examples for Surface Ligand Capturing

3.2.2.1. Immobilization of Capture Antibodies

For covalent immobilization of capture antibodies, such as rabbit anti-mouse IgG, goat anti-human IgG, goat anti-GST (Biacore), or the His-tag specific Penta-His antibody (Qiagen, Hilden, Germany), a volume of 70 μl of a 30 μg/ml antibody solution in 10 mM sodium acetate, pH 5.0, is injected over an EDC/NHS-activated CMD or SAM biosensor surface. After deactivation of remaining ester groups with 1.0 M ethanolamine hydrochloride, pH 8.5, the high density antibody coated biosensor chips can be tethered in the specific FC with monoclonal antibodies or recombinant fusion proteins at amounts in the range of 1–100 µg. In contrast, the reference FC can be loaded with an irrelevant antibody or antigen control, allowing its use for the elimination of matrix effects in the resulting Fc2-Fc1 sensorgram.

3.2.2.2. Immobilization of hAGT-Fusion Proteins

For covalent attachment of the benzylguanine derivative BG-PEG-NH_2 as an amine-terminated hAGT substrate onto the solid state of a CMD or SAM biosensor surface, a 3.1 µM solution in 1 ml HBS buffer is prepared. After subsequent centrifugation at 20,000×g for 20 min to remove all insoluble material from solution, a 70 µl-aliquot is applied over the EDC/NHS-activated biosensor surface resulting in an approximately 2,000 RU stable increase of the SPR baseline signal. All nonconverted NHS-activated ester groups are then blocked by a 35 µl-pulse of 1.0 M ethanolamine hydrochloride, pH 8.5. After an at least 15 min period of buffer flow, the hAGT-fusion protein can be immobilized by a 35 µl-injection of 100 µg/ml of purified protein at a flow rate of 1 µl/min. Due to the extraordinarily high specificity of the hAGT–BG interaction, however, it is also possible to alternatively inject whole cell extracts from the bacterial expression system at a total protein concentration of 10 mg/ml. hAGT without a fusion partner can be immobilized in the reference FC to allow the subtraction of matrix effects in a Fc2-Fc1 sensorgram.

3.2.3. Kinetic Measurements of Immobilized Ligand and Its Receptor in Solution

Kinetic analysis is exemplarily described for the interaction of the sexual hormone binding globulin (SHBG) to 17α- and 1α-aminoalkyl dihydrotestosterone (DHT) derivatives (36). Measurements were performed on the BIAcore X device at 25°C. Commercially available B1-chips that contain a CMD matrix with a very low-degree of carboxymethylation were chosen to ensure a low steroid surface density in the amine coupling step. For immobilization of the steroidal ligands in one flow cell, the carboxyl groups of the dextran layer were activated with a 35 µl-injection of a 0.2 M EDC/0.05 M NHS solution followed by a 30 µl-injection of a mixture of 100 µM aminoalkyl DHT derivative and 200 µM of ethanolamine in 100 mM borate buffer (pH 9.0) at a flow-rate of 5 µl/min. After the signal reached approx. 200–350 RU, the nonconverted active ester groups were blocked by a 35 µl-pulse of 1.0 M ethanolamine hydrochloride (pH 8.5). The second flow cell was treated in a similar fashion but without the addition of the steroid/ethanolamine mixture so as to allow its use as a reference surface. In order to obtain a homogeneous analyte preparation, affinity-purified SHBG was *N*-deglycosylated using PNGase F and subjected to size-exclusion chromatography. 45 µl-aliquots of SHBG were injected at concentrations of 50, 75, 100, 150, 200, 300, and 400 nM into the biosensor flow system. The total sensorgram recording times were 1,050 s with an association phase of 450 s and a dissociation phase of 600 s. At the end of the dissociation phase, 100 mM H_3PO_4 was added for regenerating the chip surface.

Analysis of the sensorgram data for the calculation of kinetic rate constants was done with the BIAevaluation program from Biacore AB. The evaluation algorithms perform a nonlinear regression analysis of the measured sensorgrams. The most appropriate model describing the SHBG–steroid interaction was determined to be the bivalent binding model. This was evaluated by a global fitting analysis based on the residual plots with low χ^2 values. The χ^2 value is a standard statistical measure of the closeness of fit and can be obtained from analyzing the sensorgrams. The bivalent binding mode is supported by the structure of SHBG as a homodimer. The association constants derived from the biosensor studies were found to be in good agreement with those determined by Scatchard analysis under equilibrium conditions.

4. Notes

1. SPR is a useful completion of protein analysis in proteome research. This chapter will portray the capability of the SPR method by the description of new applications in urine proteomics and proteinuria diagnostics. The applications are due to the main advantage of SPR: The ability to study interactions of proteins and peptides in their native forms with sizes ranging a few hundreds of daltons to large complexes such as cells (37).
2. It is estimated that the urine proteome contains more than 1,500 proteins. Provided most of proteins are digested to peptides on their way to excretion, one will find more than 8,000 peptides in urine (38). State-of-the-art peptide analysis methods are still unable to cover and characterize all peptides in urine samples from patients with end-stage nephritic syndrome in a manageable time at low costs. To cover as much proteins and peptides as possible, one has to combine different separation or capture approaches such as 2D-PAGE, capillary electrophoresis, free flow electrophoresis, or different selecting surfaces for liquid chromatography and chip technology (SELDI). This has to be combined with the detection methods of mass spectrometry or specific dye labeling (39). A worthwhile task for SPR biosensorics is the identification and kinetic description of protein/ligand binding events. A promising approach offers the highly selective SPR-analysis combined with mass spectrometry (MS), called biospecific interaction analysis/MS (BIA/MS). This approach is

not only able to identify the captured protein or peptide but also determines the respective binding affinities. A fully automated system for ligand fishing SPR combined to MALDI-TOF MS was recently established. This system is able to perform ligand capture, recovery, deposition onto a MALDI target, and the sample preparation including tryptic digestion (40). Apart from the fact that SPR is used to analyze cell extracts or other body fluids, there are already a series of promising examples of urine proteomics applications. Nedelkov et al. studied protein–protein interactions using BIA/MS. The glomerular filtration marker cystatin C was isolated from urine samples by its interaction with papain. Whereas no novel protein–protein interactions could be discovered, some general aspects of urine SPR analysis were elucidated in this publication (41). For example, the signal-to-noise ratio in urine samples was found to be favorably compared to serum/plasma samples. Human urine samples from apparently healthy subjects are characterized by low protein concentrations but might have high salt load. Requisite sample dilutions for the SPR analysis with buffers used (e.g., HBS-EP at pH 7.4) bring the salt content to an appropriate level and prevent degradation of small proteins in urine by adjusting the pH value. Proteins are usually degraded in urine with low pH values below 6.5. Additionally, unspecific binding to the chip is prevented by a tenfold dilution (42).

3. For comprehensive proteinuria diagnostics, it is valuable to use various techniques that cover as many proteins and peptides as possible. In this context, the combination BIA/MS helps to detect novel disease specific markers or complete a marker profile as mentioned above. But for diagnosis in routine clinical chemistry laboratories, these methods are too laborious even if data interpretation is automated. To make a laboratory diagnosis, one has to select a limited number of proteins for urine analysis. Most SPR protein analyses are carried out in individual reactions rather than multiplex systems, but there are some publications with parallel or sequential multityping. The novel Flexchip and ProteON XPR36 devices (see Subheading 2.3.2) will facilitate these multiarray approaches.

   Nedelkov and Nelson (6) characterized the tubular dysfunction markers retinol-binding protein, urinary protein 1, β2-microglobulin, and cystatin C simultaneously by applying BIA/MS (42). A fundamental benefit of using MALDI-TOF is that the amount of truncated forms of proteins may be quantified. Many proteins are truncated in particular when cells in the renal tubular system are damaged. A fundamental drawback of the BIA/MS system is that the sensor chip can only be used singular, caused by the irreversible shuttering through the matrix. Furthermore, the technique is costly and extremely complex.

Another approach is to use a sequential SPR analysis. Its feasibility was demonstrated with model substances and two clinical chemistry parameter in human urine samples. Chung et al. (43) analyzed human chorionic gonadotrophin (hCG) and albumin (hA) in urine by coupling anti-hCG and anti-hA to the SPR surface. The combination of hCG and hA is useful to monitor a pregnancy at risk, in particular for women with type I diabetes. The concentration range for sequential analysis of hCG and hA was 415–46,100 mIU/ml and 20–200 μg/ml, respectively. The SPR assay uses secondary detection antibodies in a sandwich system for signal generation.

Urinary N-telopeptide (NTx), a biochemical marker for bone resorption, is used to monitor the therapeutic management of primary osteoporosis. Lung et al. (44) established a SPR-based assay to quantify NTx, which is a degradation product from bony type I collagen and excreted in urine. The authors compared the SPR assay with conventional ELISA and found similar results in identifying subjects with significantly increased bone loss. Advantageous for the SPR-based assay is the fact that it works faster than the ELISA and that the SPR method is reusable manifold.

4. Due to the fact that SPR biosensorics can provide affinity and kinetic data, unique features such as protein–peptide interaction analysis will be fruitful also for urine proteomics. Therefore, the urine matrix poses no special challenge for SPR. The integration of SPR into protein array technology will accelerate discovery in existing applications and will also play a significant role in driving the development of urine proteomics (45). Moreover, chip data processing has the potential to create a new quality in the field of proteinuria diagnostics. What is emerging by applying the SPR technology is the discovery and characterization of new diagnostic protein markers in their native state in urine.

References

1. Ekins, R. P. (1999) Immunoassay and other ligand assays: from isotopes to luminescence. *J. Clin. Ligand Assay* **22**, 61–77
2. Luppa, P. B., Sokoll, L. J., and Chan, D. W. (2001) Immunosensors – Principles and applications to clinical chemistry. *Clin. Chim. Acta* **314**, 1–26
3. Mullett, W., Lai, E. P., and Yeung, J. M. (2000) Surface plasmon resonance-based immunoassays. *Methods* **22**, 77–91
4. Pearson, J. E., Gill, A., Vadgama, P. (2000) Analytical aspects of biosensors. *Ann. Clin. Biochem.* **37**, 119–145
5. Rogers, K. R. (2000) Principles of affinity-based biosensors. *Mol. Biotechnol.* **14**, 109–129
6. Ng, J. H., Ilag, L. L. (2003) Biochips beyond DNA: technologies and applications. *Biotechnol. Annu. Rev.* **9**, 1–149
7. Marquette, C. A. and Blum, L. J. (2006) State of the art and recent advances in

immunoanalytical systems. *Biosens. Bioelectron* **21**, 1424–1433
8. Newman, J. D. and Setford, S. J. (2006) Enzymatic biosensors. *Mol. Biotechnol.* **32**, 249–268
9. Price, C. P., St. John, A., and Hicks, J. M., eds. (2004) Point of Care Testing, 2nd edition. *AACC Press*, Washington, DC.
10. Karlsson, R. (2004) SPR for molecular interaction analysis: a review of emerging application areas. *J. Mol. Recognit.* **17**, 151–161
11. Homola, J., Yee, S, S., and Gauglitz, G. (1999) Surface plasmon resonance sensors: review. *Sens. Actuators B Chem.* **54**, 3–15
12. Homola, J. (2003) Present and future of surface plasmon resonance biosensors. *Anal. Bioanal. Chem.* **377**, 528–539
13. Rich, R. L. and Myszka, D. G. (2005) Survey of the year 2003 commercial optical biosensor literature. *J. Mol. Recognit.* **18**, 1–39
14. Rich, R. L. and Myszka, D. G. (2005) Survey of the year 2004 commercial optical biosensor literature. *J. Mol. Recognit.* **18**, 431–478
15. Rich, R. L. and Myszka, D. G. (2006) Survey of the year 2005 commercial optical biosensor literature. *J. Mol. Recognit.* **19**, 478–534
16. Rich, R. L. and Myszka, D. G. (2003) A survey of the year 2002 commercial optical biosensor literature. *J. Mol. Recognit.* **16**, 351–382
17. Liedberg, B., Nylander, C., and Lundström, I. (1995) Biosensing with surface plasmon resonance – how it all started. *Biosens. Bioelectron* **10**, i–ix
18. Kyo, M., Usui-Aoki, K., and Koga, H. (2005) Label-free detection of proteins in crude cell lysate with antibody arrays by a surface plasmon resonance imaging technique. *Anal. Chem.* 77, 7115–7121
19. Homola, J., Vaisocherová, H., Dostálek, J., and Piliarik, M. (2005) Multi-analyte surface plasmon resonance biosensing. *Methods* **37**, 26–36
20. Yi, S. J., Yuk, J. S., Jung, S. H., Zhavnerko, G. K., Kim, Y. M., and Ha, K. S. (2003) Investigation of selective protein immobilization on charged protein array by wavelength interrogation-based SPR sensor. *Mol. Cells* **15**, 333–340
21. Jönsson, U. and Malmqvist, M. (1992) Real time biospecific interaction analysis. The integration of surface plasmon resonance detection, general biospecific interface chemistry and microfluidics into one analytical system. *Adv. Biosens.* **2**, 291–336
22. Myszka, D. G. (1997) Kinetic analysis of macromolecular interactions using surface plasmon resonance biosensors. *Curr. Opin. Biotechnol.* **8**, 50–57
23. Myszka, D. G., Morton, T. A., Doyle, M. L., and Chaiken, I. M. (1997) Kinetic analysis of a protein antigen-antibody interaction limited by mass transport on an optical biosensor. *Biophys. Chem.* **64**, 127–137
24. Khalifa, M. B., Choulier, L., Lortat-Jacob, H., Altschuh, D., and Vernet, T. (2001) Biacore data processing: an evaluation of the global fitting procedure. *Anal. Biochem.* **293**, 194–203
25. De Crescenzo, G., Pham, P. L., Durocher, Y., and O'Connor-McCourt, M. D. (2003) Transforming growth factor-beta (TGF-β) binding to the extracellular domain of the type II TGF-β receptor: receptor capture on a biosensor surface using a new coiled-coil capture system demonstrates that avidity contributes significantly to high affinity binding. *J. Mol. Biol.* **328**, 1173–1183
26. Usui-Aoki, K., Shimada, K., Nagano, M., Kawai, M., and Koga, H. (2005) A novel approach to protein expression profiling using antibody microarrays combined with surface plasmon resonance technology. *Proteomics* **5**, 2396–2401
27. Stenberg, E., Persson, B., Roos, H., and Urbanisczky, C. (1991) Quantitative determination of surface concentration of protein with surface plasmon resonance using radiolabeled proteins. *J. Colloid Interface Sci.* **143**, 513–526
28. Lofas, S. and Johnsson, B. (1990) A novel hydrogel matrix on gold surfaces in surface plasmon resonance sensors for fast and efficient covalent immobilization of ligands. *J. Chem. Soc. Chem. Commun.* **21**, 1526–1528
29. Lofas, S. (1995) Dextran modified self-assembled monolayer surfaces for use in biointeraction analysis with surface plasmon resonance. *Pure Appl. Chem.* **67**, 829–834
30. Ulman, A. (1996) Formation and structure of self-assembled monolayers. *Chem. Rev.* **96**, 1533–1554
31. Li, X., Wei, X., and Husson, S. M. (2004) Thermodynamic studies on the adsorption of fibronectin adhesion-promoting peptide on nanothin films of poly(2-vinylpyridine) by SPR. *Biomacromolecules* **5**, 869–876
32. Bilkova, Z., Mazurova, J., Churacek, J., Horak, D., and Turkova, J. (1999) Oriented immobilization of chymotrypsin by use of suitable antibodies coupled to a nonporous solid support. *J. Chromatogr. A* **852**, 141–149
33. Kindermann, M., George, N., Johnsson, N., and Johnsson, K. (2003) Covalent and

selective immobilization of fusion proteins. *J. Am. Chem. Soc.* **125**, 7810–7811

34. Andersson, K., Hamalainen, M., and Malmqvist, M. (1999) Identification and optimization of regeneration conditions for affinity-based biosensor assays. A multivariate cocktail approach. *Anal. Chem.* **71**, 2475–2481
35. Karlsson, R., Katsamba, P. S., Nordin, H., Pol, E., and Myszka, D. G. (2006) Analyzing a kinetic titration series using affnity biosensors. *Anal. Biochem.* **349**, 136–147
36. Metzger, J., Schnitzbauer, A., Meyer, M., Söder, M., Cuilleron, C. Y., and Luppa, P. B. (2003) Binding analysis of 1alpha- and 17alpha-dihydrotestosternone derivatives to homodimeric sex hormone-binding globulin. *Biochemistry* **42**, 13735–13745
37. McDonnell, J. M. (2001) Surface plasmon resonance: towards an understanding of the mechanisms of biological molecular recognition. *Curr Opin Chem Biol.* **5**, 572–577.
38. Adachi, J., Kumar, C., Zhang, Y., Olsen, J. V., and Mann, M. (2006) The human urinary proteome contains more than 1500 proteins, including a large proportion of membrane proteins. *Genome Biol.* 7, R80.1–R80.16
39. Nedelkov, D., Kiernan, U. A., Niederkofler, E. E., Tubbs, K. A., and Nelson, R. W. (2006) Population proteomics: the concept, attributes, and potential for cancer biomarker research. *Mol. Cell Proteomics* **5**, 1811–1818
40. Zhukov, A., Schurenberg, M., Jansson, O., Areskoug, D., and Buijs, J. (2004) Integration of surface plasmon resonance with mass spectrometry: automated ligand fishing and sample preparation for MALDI MS using a Biacore 3000 biosensor. *J. Biomol. Tech.* **15**, 112–119
41. Nedelkov, D. and Nelson, R. W. (2003) Delineating protein-protein interactions via biomolecular interaction analysis-mass spectrometry. *J. Mol. Recognit.* **16**, 9–14
42. Nedelkov, D. and Nelson, R. W. (2001) Analysis of human urine protein biomarkers via biomolecular interaction analysis mass spectrometry. *Am. J. Kidney Dis.* **38**, 481–487
43. Chung, J. W., Bernhardt, R., and Pyun, J. C. (2006) Sequential analysis of multiple analytes using a surface plasmon resonance (SPR) biosensor. *J. Immunol. Methods* **311**, 178–188
44. Lung, F. D., Chen, H. Y., and Lin, H. T. (2003) Monitoring bone loss using ELISA and surface plasmon resonance (SPR) technology. *Protein Pept. Lett.* **10**, 313–319
45. Boozer, C., Kim, G., Cong, S., Guan, H., and Londergan, T. (2006) Looking towards label-free biomolecular interaction analysis in a high-throughput format: a review of new surface plasmon resonance technologies. *Curr. Opin. Biotechnol.* **17**, 400–405

Chapter 13

Urinary Proteins for the Diagnosis of Obstructive Sleep Apnea Syndrome

Ayelet Snow, David Gozal, Roland Valdes Jr., and Saeed A. Jortani

Abstract

Approximately 2–3% of all children in the United States suffer from obstructive sleep apnea (OSA). This condition is characterized by repeated events of partial or complete obstruction of the upper airways during sleep leading to recurring episodes of hypercapnia, hypoxemia, and arousal throughout the night as well as snoring, which afflicts 7–10% of all children. Since clinical history and physical examination are unreliable in the differentiation between children with OSA and children with primary snoring (PS) who have no apparent alteration in sleep architecture, current diagnostic approaches for OSA require an overnight sleep study (ONP). ONP is onerous, relatively unavailable, labor intensive, and inconvenient, leading to long waiting periods and unnecessary delays in diagnosis and treatment. Development of noninvasive biomarker(s) capable of reliably distinguishing children with PS from those with OSA would greatly facilitate timely screening and diagnosis of OSA in children. Therefore, we hypothesized that proteomic strategies in the urine may permit the identification of biomarker(s) that reliably screen for OSA. In this study, time-of-flight mass spectrometry was used to profile proteins in the first morning void urines from children. We discovered that urocortins are increased in OSA and provide a noninvasive approach for quick and convenient diagnosis otf OSA in snoring children.

Key words: Obstructive sleep apnea syndrome, Biomarkers, Urine protein profiling

1. Introduction

1.1. OSAS in Children

Obstructive sleep apnea syndrome (OSAS) in children is a fairly common condition affecting between 2 and 3% of all children and in recent years has been shown to impose substantial adverse consequences encompassing physiological, neurological, and cognitive. Nevertheless, awareness to OSAS is still relatively low among both public and medical professional, a fact that may further be compounded by the unavailability of inexpensive and rapid diagnostic tests.

Alex J. Rai (ed.), *The Urinary Proteome: Methods and Protocols*, Methods in Molecular Biology, vol. 641, DOI 10.1007/978-1-60761-711-2_13, © Springer Science+Business Media, LLC 2010

1.2. Characteristics

OSAS symptoms can be divided into those primarily occurring during the night and daytime manifestations. The main nocturnal symptom and usually the presenting one is habitual snoring. Additional sleep-related symptoms include respiratory pauses (usually ending with a snort), noisy breathing sounds, paradoxical breathing movements, cyanosis, restless sleep, excessive diaphoresis, and sleep enuresis. Diurnal symptoms may include difficulties waking up, daytime sleepiness, hyperactivity or other behavioral problems, mouth breathing, poor growth and mood, poor school performance, and morning headaches (1–7).

OSAS is characterized by recurrent events of partial or complete upper airway obstruction during sleep. These events are associated with gas exchange abnormalities such as hypoxemia and hypercapnia as well as with sleep fragmentation due to the occurrence of multiple arousals (8). OSAS encompasses the severe end of the clinical spectrum of obstructive sleep disordered breathing. At the other, less severe end, there is the condition called primary or habitual snoring, which is considered to be a more benign form of upper airway resistance. Habitual snoring occurs in the absence of apnea, gas exchange abnormalities, or with no evidence of disruption of sleep architecture. An intermediate form of sleep disordered breathing is a condition termed upper airway resistance syndrome (UARS). UARS is associated with normal gas exchange patterns but with increased frequency of respiratory- and snore-related arousals and consequent sleep fragmentation.

1.3. Etiology

It is believed that the majority of childhood OSAS is due to enlarged adenoids and tonsils. The high prevalence of OSAS among 2–8 year olds might be due to a disproportionate lymphatic tissue growth in the upper airway which may be triggered in response to a variety of factors like respiratory viral illnesses, allergens, passive cigarette smoking, and many other environmental airborne irritants (9–12). It has also been suggested that the upper airway of children with OSAS is more collapsible in comparison to nonsnoring children due to possible decreased upper airway muscle dilator recruitment, increased upper airway compliance, and excessive inspiratory driving pressures caused by proximal airway narrowing (13–15). It has also been reported that both anatomical (severe nasal obstruction, orthodontic, and craniofacial abnormalities) and functional factors (neuromuscular diseases, obesity) may predispose for the presence of OSAS. Prevalence is also higher within families with positive history of OSAS, and therefore the occurrence of OSAS represents most likely the end result of a combination of anatomical and neuronal alterations influenced by environmental and genetic factors that occur in the upper airways.

1.4. Prevalence

Seven to ten percent of all school age children will present with snoring during the majority of their night and are called habitual snorers; however, only 1–3% of all children will have some degree of OSAS (2, 3) such that only 1 in every 4–6 habitual snorers will have what is now perceived as disease (i.e., OSAS). Peak incidence is highest among preschool children and will usually decline after the age of 8–9 years, possibly reflecting the increased airway size and decreased respiratory viral burden that promotes increased proliferation of the upper airway lymphadenoid tissues.

OSAS is more prevalent among African Americans, among obese children, children with asthma and recurrent sinus infections, as well as those born prematurely (16).

1.5. Consequences and Current Diagnostic Procedures

Untreated OSAS may lead to serious neurobehavioral and cardiovascular consequences. Among the first, daytime sleepiness, irritability, depression, ADHD-like behaviors, and poor cognitive function have all been reported. Indeed it has now become clear that OSAS in children strongly and dose-dependently correlates with emotional problems, impaired school performance, hyperactivity, and aggressive and withdrawal behaviors (5, 17–19).

OSAS in adults is causally associated with hypertension. This issue has been recently reviewed in children (20). Marcus et al. (21) measured serial BP during overnight polysomnography (PSG) in 67 children. They found that OSAS children had higher sleep and wake diastolic blood pressure compared with habitually snoring children. Kohyama et al. (22) also measured blood pressure (BP) values during PSG in 32 children and reported that among children with an apnea–hypopnea index (AHI; used as a common index of respiratory disturbance) ≥10/h of sleep, both systolic and diastolic blood pressures were elevated during REM sleep, but not during non-REM sleep. In the TuCASA study (23), increased blood pressure was found among 239 children with OSAS, while in a study performed by Amin et al. (24), ambulatory BP measurements in 60 snoring children revealed significantly greater mean BP variability and smaller nocturnal BP dipping in OSAS subjects. Surprisingly, in contrary to previous reports, they found lower diastolic BP during waking in the group with OSAS. Abnormal systolic and diastolic blood pressures were also reported among habitual snorers (25).

Changes in right ventricular stroke volume, left ventricular mass, and insulin resistance have been reported to be associated with OSAS (4, 26, 27) with insulin resistance being more prevalent in obese children with OSAS (27). The connection between OSAS and systemic inflammation as evidenced by elevated serum C-reactive protein (CRP) level, a serum marker with major implications on cardiovascular morbidity, was studied by Tauman and collaborators in our laboratory (28). This study found a significantly positive correlation between CRP level and AHI, which was later confirmed in a pre-posttreatment investigation (29). Furthermore, a link between OSAS, CRP, and cognitive morbidity has been recently reported by our laboratory (30). Of note, a study in Greece found no evidence for such correlation between CRP and OSAS (31).

OSAS has been linked to impaired growth, as well as with gastroesophageal reflux (GER), the latter being most probably related to the large negative intrathoracic pressure swings accompanying the upper airway obstructive events (7, 32).

Finally, severe and long-standing cases of OSAS may be associated with pulmonary hypertension, cor pulmonale, and sudden unexpected death, similar to those cases described in adults with OSAS.

Clinical evaluation accompanied by a sleep questionnaire such as the "Pediatric Sleep Questionnaire" (33) or the "Children Sleep Habits Questionnaire" (34) has low sensitivity and specificity. Therefore, these tools will not be diagnostic for sleep disordered breathing, but rather can be employed to identify the need for further investigation and intervention (8, 35–37). There are no diagnostic blood tests for OSAS, although this condition may be associated with polycythemia (38), elevated serum CRP (28), insulin resistance (39), abnormal lipid profile (40), abnormal thyroid function (39), and elevated serum VEGF levels (41). When suspecting OSAS a lateral soft tissue neck X-ray may be useful in identifying the extent of adenoidal hypertrophy.

The diagnosis of OSAS traditionally and thus far optimally requires an overnight PSG. Home monitoring with or without videotaping has been used as well as ambulatory cardiorespiratory monitoring including a single channel of oximetry. These monitoring techniques are able to detect the presence of drops in oxygen saturation (SaO_2), apneas, and potentially apneas and hypopneas and may therefore recognize severe SDB. Nevertheless, these techniques are not useful in the discrimination of more subtle forms of OSAS or the presence of other sleep disorders. Therefore, a negative home test result does not rule out the diagnosis of OSAS and must be confirmed by PSG. However, a positive finding may lead to earlier diagnosis and referral for treatment.

As mentioned, the gold standard tool for the diagnosis of OSAS consists in an overnight PSG in a sleep laboratory. This approach is both time-consuming, labor intensive, and onerous. It includes monitoring of sleep/wake states through electroencephalography (EEG), electro-oculography, chin and leg electromyography, electrocardiography, body position, and appropriate monitoring of breathing-related parameters. In the pediatric population, use of nasal cannula-pressure transducer, oral thermistor, end-tidal capnography, chest and abdominal respiratory excursion sensors, a sound recording microphone, and pulse oximetry are all recommended to allow for optimal diagnostic yield.

Additionally, although the most accurate method to differentiate obstructive respiratory events from central events is esophageal pressure monitoring, this method is usually less well tolerated in children, and therefore reserved for only special circumstances (42, 43).

2. Need for Biomarkers for OSA

2.1. Importance of Early Diagnosis

As we have alluded earlier, OSAS imposes substantial adverse consequences on children. We further know that some of these consequences can be reversed upon timely treatment (44–47) but may be long-lasting if treatment is delayed (48, 49). In addition, effective treatment of OSAS in adult patients will prevent cardiovascular complications, although this aspect has yet to be examined thoroughly in children. Children with OSAS may further grow up to become adults suffering from OSAS and exhibit substantial increased risk for complications such as occurrence of myocardial infarction and stroke. Early diagnosis of OSAS and treatment with adenoidectomy and tonsillectomy will for example improve school performance, such that early diagnosis and treatment of OSAS in children are tantamount to favorable outcomes.

2.2. Sleep Study and Its Limitations

The gold standard diagnostic tool for diagnosis of OSAS consists of a full night-time polysomnographic recording in a sleep laboratory. The major obstacle to conducting such studies in all children with suspected OSAS, that is, fourfold to sixfold the number of actual patients with the disease, is accessibility to a sleep center. Indeed, the number of pediatric sleep laboratories in the USA and around the world is small, and therefore, wait lists of more than 1 year are common. In fact, this seems to be situation for all ages since a recent survey clearly indicated that the current numbers of diagnostic sleep facilities could not meet the demand (50). Furthermore, the test itself is a very time-consuming, laborious, and expensive process, which requires from both children and their parents to spend a night away from their home along with the inherent discomfort of wearing multiple electrodes while asleep. As with all biological studies, sleep studies are not exempt from night-to-night variability. However, Katz et al. have shown that for diagnostic purposes, the intraindividual variability does not affect the diagnostic outcome (51).

2.3. Value of a Noninvasive Biomarker for Diagnosis of OSA in Children

Currently, there is a critical need for a cheaper, quicker, and child-friendly tool for the diagnosis of OSAS. The unique characteristics of OSAS including gas exchange abnormalities accompanied by sleep fragmentation and deprivation open a new opportunity to identify specific biomarkers that characteristically accompany the presence of OSAS. Such biological markers may not only be present in the patient serum, but could also be measured in urine. Development of a urine test that enables the diagnosis of OSAS is very appealing, particularly considering that this is a very easy and noninvasive way to test children. Furthermore, these noninvasive biological markers may also provide accurate estimates of the risk for OSAS-induced morbidity.

The possibility of finding urinary biomarkers unique for OSAS is supported by increasing evidence suggesting that alterations in the expression of proteins occur as a result of OSAS.

For example, CRP has been identified as a novel and sensitive marker for detection of progression in atherogenesis (52). CRP is synthesized in the liver and its expression is primarily regulated by cytokines (53). CRP levels are generally stable in the same individual across the circadian cycle (54) such that circulating levels reflect the intensity of inflammatory response, and also provide a reliable estimate for the risk of atherosclerosis. Interestingly in adults with OSAS, elevated levels of CRP were associated with the severity of OSAS (55). Plasma CRP levels were also increased among children with OSAS and were correlated with AHI, arterial oxygen saturation nadir, and arousal index measures. In addition, CRP levels were found to be higher among OSAS children with cognitive deficits, and tended to improve after adenotonsillectomy, the usual first line of treatment for pediatric OSAS (28–30). Endothelial growth factor (VEGF) levels were also studied among OSAS patients and were found to be elevated (41). Taken together, elevation of specific serum proteins in the context of OSAS may serve as potential candidates for biomarkers of the disease.

3. Urinary Proteins as Biomarkers for the Diagnosis of Obstructive Sleep Apnea

3.1. Patients Characteristics

In an attempt to identify potential biomarkers for OSAS, snoring children were recruited from the pediatric sleep medicine clinic at Kosair Children's Hospital and the University of Louisville if they were referred for evaluation of habitual snoring and after undergoing an overnight polysomnographic evaluation. A standard overnight multichannel polysomnographic evaluation was performed with no drugs or sleep deprivation being used to induce sleep, and in the presence of a parent or caretaker. Children were studied for at least 8 h in a quiet, darkened room with an ambient temperature of 24°C. The following parameters were measured: chest and abdominal wall movement by respiratory impedance or inductance plethysmography, heart rate by ECG, air flow monitored with a sidestream end-tidal capnograph that also provided breath-by-breath assessment of end-tidal carbon dioxide levels ($PetCO_2$; Pryon), as well as a nasal pressure transducer

(Braebon, Canada) and/or a thermistor. Arterial oxygen saturation (SaO_2) was assessed by pulse oximetry (Nellcor N 100; Nellcor Inc., Hayward, CA), with simultaneous recording of the pulse waveform. The bilateral electro-oculogram (EOG), eight channels of electroencephalogram (EEG), chin, bilateral anterior tibial and forearm electromyograms (EMG), and analog output from a body position sensor were also monitored. All measures were digitized using a commercially available PSG system (REMBrandt, Embla, Amsterdam). Tracheal sound was monitored with a microphone sensor and a digital time-synchronized video recording. The following parameters were quantified (1) Sleep architecture, sleep latency (time from lights out to sleep onset), percentage of sleep time spent in all sleep stages, and arousals. (2) Obstructive apnea was defined as the presence of continued chest and abdominal motion and the absence of airflow. Since children have higher respiratory rates than adults and frequently desaturate during short apnea episodes, all obstructive apnea events greater than or equal to two-breath duration were counted. The obstructive apnea index was defined as the number of obstructive apneas per hour of total sleep. Mixed apneic events (apnea with both central and obstructive components) were included in the apnea index. (3) Hypopneas were defined as a decrease of greater than or equal to 50% in oronasal flow which was associated with a reduction of greater than or equal to 4% in the SaO_2. The obstructive apnea/hypopnea index (OAHI) was defined as the average number of obstructive apnea and hypopnea per hour of sleep. (4) Arterial oxygen saturation (SaO_2) was assessed by means of the nadir SaO_2, mean SaO_2 and the percentage of total sleep time during which SaO_2 was less than 92%. In addition, a cumulative desaturation index was calculated based on percentage of total sleep time spent at each of 5% oxyhemoglobin saturation steps (100–95, 95–90, 90–85%, etc.) multiplied by a factor of 1, 2, 4, etc. (5) End-tidal carbon dioxide tension ($PetO_2$; the mean and peak $PetCO_2$) was determined. The percentage of total sleep time during which $PetCO_2$ was greater than or equal to 50 mmHg was calculated. (6) The severity of OSAS was scored as (a) None (zero) apneic episodes less than 1/h TST associated with oxygen desaturation, and/or mild hypoventilation, less than 10% TST with $PetCO_2$ greater than 50 mmHg. (b) Mild apnea index (AI) > 2 and <5/h TST. (c) Moderate AI > 5 but < 10/h TST, and (d) severe AI>10/h TST (56).

Based upon sleep study, subjects were categorized into either OSAS (AHI of more than 5/h, $n = 30$) with the group of habitual snorers (HS; apnea–hypopnea index less than 1/h, $n = 25$) serving as controls (Table 1). As shown in Table 1, age, gender, ethnicity, maternal education level, and maternal smoking prevalence were not significantly different between the groups. Table 2 summarizes the polysomnographic findings for the two groups.

Table 1
Demographic data for enrolled children

	OSAS (*n*=30)	HS (*n*=25)
Age (years)	6.6±0.7	6.7±0.4
Gender (*n* male)	16	12
Ethnicity		
African American	8	7
White Non-Hispanic	22	18
Maternal educational attainment		
College or higher	19	16
High school or lower	11	9
Maternal smoking		
Yes	12	11
No	18	14

Table 2
Overnight sleep study findings in 30 children with OSA and 25 children with habitual snoring (HS)

	OSA (*n*=30)	HS (*n*=25)
Sleep efficiency (%)	90.6±9.0	88.6±7.9
Sleep latency (min)	17.3±17.8*	24.7±27.8
REM latency (min)	122.7±49.1	131.1±57.7
Stage 1 (%TST)	10.6±9.2	8.3±4.9
Stage 2 (%TST)	43.9±7.6	45.4±9.6
SWS (%TST)	23.1±7.0	21.7±6.4
REM (%TST)	22.4±6.1	24.6±7.1
Spontaneous arousal index	4.4±4.4**	8.2±2.8
Respiratory arousal index	10.4±10.1***	1.5±0.8
Total arousal index	14.8±7.1***	9.7±3.2
AHI (/hour TST)	10.4±7.6***	0.8±0.4
Mean SpO_2	95.7±1.9**	97.8±0.7
SpO_2 nadir	75.6±10.3***	92.8±1.9

*$p<0.05$; **$p<0.01$; ***$p<0.001$

A first morning void urine specimen was obtained from each patient in the morning after the sleep study. Proteins in the specimen were profiled using time-of-flight mass spectrometry in an attempt to recognize potential biomarkers.

3.2. Protein Profiling

A combination of time-of-flight mass spectrometry and linear discriminative analysis is an appropriate and convenient strategy for the screening of potential biomarkers in biological fluids such as serum, plasma, or urine. One of these approaches specifically involves the use of surface-enhanced laser desorption ionization technology (SELDI) (Bio-Rad Laboratories, Hercules, CA). This proteomic methodology enables selective protein retention on ProteinChip® Array surfaces by means of distinct chromatographic or bioaffinity surfaces. The SELDI process can be described in three parts: Crude biological samples such as urine can be applied directly to the ProteinChip Arrays. Application can be done manually by pipetting or by employing a robot-assisted automation station. To reduce the quenching phenomenon (larger proteins or those with greater concentrations interfering with the ionization and/or movement of other proteins in the mass spectrometer), the sample is also fractionated using a strong anion exchange resin. The fractionation of various proteins is achieved by eluting them during the process of decreasing pH in a step-wise manner. When analyzing urine, samples are generally analyzed without prior fractionation. After a short incubation period, the unbound proteins are washed off the surface of the ProteinChip Array. Only proteins interacting with the chemistry of the array surface are then retained for analysis. After several washes, an energy-absorbing molecule (EAM) is applied to the array as the final step. ProteinChip Arrays are then subjected to analysis in the PBS-II Reader, which basically consists of a time-of-flight (TOF) mass spectrometer. The mass values and signal intensities for detected proteins and peptides can be viewed in several formats, and then the information can be transferred to software programs designed for further in-depth analysis, including clustering, machine-based classification, and data mining.

There are two major approaches for analysis of biological fluids including urine using SELDI-TOF proteomic methodologies. For serum or plasma, the manufacturer generally recommends a pre-fractionation step, which involves running the sample through a strong anion resin. With sequential elutions using decreasing pH, various proteins are subsequently collected and analyzed on multiple chip types or conditions to profile the proteome in the fluid of interest. However, many biomarker discovery reports using SELDI have not fractionated the samples. In fact, based on our survey of the literature on SELDI applications for all biological fluids, >60% of studies of various biological fluids (mainly serum) have been performed on unfractionated samples. For the present study, we profiled proteins using unfractionated urine samples. Due to the reported variability in the intensities of peaks of interest by SELDI, and considering the imprecision issues associated with the molecular weight determination of proteins, we analyzed each urine sample in triplicate. It was predicted that in an 8-spot SELDI chip array marked as A through H, the first and the last spot (i.e., A and H) would demonstrate the greatest degree of imprecision. Therefore, we did not use these spots for proteomic profiling. This allowed urine from two subjects to be analyzed in triplicate on the remaining six spots on the chip array. The mean peak intensities were calculated for each spot and used for the data mining and biomarker discovery analysis. The urine samples for OSAS and HS children were assigned in random to the various chip locations. Each urine sample was analyzed on four different chip types: weak cation exchange (WCX) with low stringency (pH 4), metal binding (IMAC-Cu^{2+}), strong cation exchange (SAX), and hydrophobic (H4). Chip arrays were placed in a 96-well Bioprocessor with new (disposable) reservoir and gasket in place.

3.2.1. Chip Activation

IMAC chips need to be activated by addition of $CuSO_4$. This was done in a 96-well formatted Bioprocessor using 50 μL of 100 mM $CuSO_4$ for 5 min at room temperature. The unbound Cu^{2+} is then removed; chips are rinsed by dH_2O and 100 mM Na-acetate (pH 4.0). Other chip types do not require activation.

3.2.2. Chip Equilibration

All chip types were equilibrated with their respective binding/washing buffers according to the manufacturer instructions. Appropriate buffers for a given chip type (150 μL) were added to each chip in a 96-well Bioprocessor and incubated for 5 min. Then, the buffer was removed and dumped in collection trays.

3.2.3. Chip Binding

In our routine analysis of various sample types (serum, tissues, etc.), we generally load between 5 and 20 μg of total protein per SELDI-TOF chip spot. Our preliminary analysis of total protein (95) in the urine samples of enrolled children (after acetone precipitation and ultracentrifugation) resulted in a protein concentration range of 0.8–1.5 mg/mL. Therefore, a normalized volume of urine to achieve 10 μg of total protein per spot and 40 μL of binding/washing buffer were added in an assembled 96-well Bioprocessor containing the appropriate chip types. Following incubation for 5 min at room temperature and removal of the buffer, the chips were washed by twice adding and discarding dH_2O (200 μL).

3.2.4. EAM Addition

Alpha-cyano-4-hydroxy cinnamicacid (CHCA) and sinapinic acid (SPA) are two typical EAM for SELDI analysis. EAMs are added prior to mass spectrometry for transferring a proton to the proteins to be desorbed from the surface of the chip. In a study recently published by Schaub et al. (57) maximum number of protein peaks in the range of 2–25 kD were detected when CHCA was used. CHCA also resulted in greater peak intensities for proteins with molecular weights of 8–10 kD. In contrast, peak intensities for proteins >8–10 kD were greatest when SPA was the EAM. We expected the enrolled children to have adequate kidney function, and the majority of their urinary proteins to be in the lower molecular weight range. Therefore, we used CHCA as the EAM for processing of all SELDI chip types. CHCA was prepared by adding the contents supplied by the manufacturer to 100 μL of acetonitrile (ACN) and 100 μL of 1% trifluoroacetic acid (TFA) followed by 5 min of vortexing at room temperature.

3.2.5. Chip Reading

We used two protocols developed in our laboratory for low and high mass focusing was used to read each chip. The low mass focusing encompasses a mass range from 1,000 to 30,000 Da with optimization at 3,000–10,000 Da. The laser intensity is set at 285 Å, and the instrument sensitivity at 9. For the high mass focusing, we set the high threshold at 200,000 Da with an optimization range of 10,000–60,000 Da and a laser intensity of 290 Å, with a sensitivity setting of 9.

3.2.6. Peak Detection/ Normalization

For the pediatric urine sample studies described herein, all spectra were initially visually inspected. If the matrix peaks were missing or depressed below 30% log normalized value, the spectra were eliminated from final data analysis. Peaks were detected with first and second pass signal-to-noise ratio of 5 and 2 using Biomarker Wizard Software (Ciphergen Biosystems, Fremont, CA). All spectra were normalized according to the manufacturer's instructions using the Wizard which also allowed for generation of ".csv" output for each experiment mass spectrum. The supervised qualitative inspection of peaks was performed at this stage. The ".csv" files for the three replicates for each urine sample were merged and mean intensities were calculated. Then, the ".csv" files were imported into Biomarker Pattern Software (BPS) for linear format data mining analysis.

3.3. Discovery of Discriminatory Pattern

Using the supervised linear discriminative analysis of the normalized peaks obtained from urine protein profiling, two distinct proteins among the OSAS group were discovered, characterized by molecular weights of 4,382 and 4,640 Da. This pattern had been captured by H4 chip when analyzed with a low molecular weight focus. The mining parameters included the cross-validation model which resulted in relative cost of 0.113. This value is considered to be highly discriminatory. In fact in the learning phase, the pattern had perfect performance, that is, not missing any cases in their appropriate group designation. In the test phase, a diagnostic specificity of 97% and a sensitivity of 93% were achieved. We further challenged this new pattern by randomly holding back 50% of the cases for the testing phase. The 4,640 Da protein emerged as the most discriminatory in the learning phase with a 100% correct performance. The test phase of randomly held back cases resulted in diagnostic specificity of 87% and sensitivity of 100%.

We further interrogated a protein database for preliminary identification of the 4,640 Da putative protein. Following this search using ExPASy Molecular Biology Server, three candidate proteins with rather similar molecular weight ranges were found. The most plausible and relevant preliminary identification of the putative biomarker pointed toward the urocortins (UCNs).

4. Urocortins

The urocortin family consists of three peptides – UCN1, UCN2, and UCN3. Their expression, roles, and possible connection to OSAS are further elaborated below.

4.1. Urocortins in Stress Response

4.1.1. The CRF Family

Corticotropin releasing factor (CRF) was isolated for the first time from bovine hypothalamus in 1981(58) and is presently known to be the hypothalamic activator of the pituitary–adrenal axis during stress, thus activating neuroendocrine, autonomic, and behavioral responses to stress (59, 60). This peptide is the prototype of an important emerging family – the corticotrophin releasing hormone-related peptides – otherwise also known as urocortins. The first member of the UCN family to be identified was UCN1, which was cloned from rat midbrain in 1995 (61–63). It was referred to as a “urocortin” because of its high homology to urotensin1 (the fish homologues of CRF) and CRF itself. The UCN family expanded after the identification of UCNs 2 and 3 in 2001 from mouse genomic libraries and from human libraries (64–67).

4.1.2. The CRF Receptors

The CRF-related peptides bind to two types of receptors: CRF Receptor 1 (CRFR1) and CRF Receptor 2 (CRFR2). The two receptors exhibit significant heterogeneity in their tissue expression, ligand affinity, and pharmacology. All CRF receptor isoforms conform to the classic G-protein-coupled seven hydrophobic transmembrane structures (68, 69).

The CRFR1 is expressed as eight subtypes named a through h, while the CRFR2 gene only has three isoforms named α, β, and γ. Upon activation, these receptors elicit an increase in cAMP (70, 71). There is a variety of G protein species coupled to the different receptor subtypes, and thus presumably are associated with multiple different downstream cascades.

The CRFR1 is expressed primarily in the brain (cortex, cerebellum, hippocampus, amygdala, olfactory bulb) and pituitary where it has the most pronounced effect. It is also expressed at low levels in the skin, adipose tissue, ovary, testis, and adrenal. CRFR1 is activated by UCN1 and CRF but displays a 100-fold higher binding affinity for UCN1 than for CRF itself (61). By binding to CRF or UCN1, CRFR1 mediates ACTH response to stress. UCN1 is also thought to have some immunologic regulations. It was demonstrated to mimic some of CRF influences on T and B lymphocytes proliferation (72). Administration of UCN1 can also suppress inflammation (73). It has a higher affinity for the centrally located CRFR1 found in Edinger–Westphal nucleus, hippocampus, and basal ganglia (74, 75).

CRFR2 is present in both central and peripheral tissues with high density demonstrated in cardiac myocardium and blood vessels (76, 77). CRFR2 is actually the only receptor subtype found in cardiac tissue (78). It is also found in skeletal muscle, skin, adipose tissue, lungs, ovaries, pituitary, and GIT (76, 77). Central expression of CRFR2 was found in the limbic system, hypothalamus, brainstem, cortex, cerebellum, and retina. UCNs 2 and 3 are reported to bind exclusively and with high selectivity to CRFR2 (64, 66). By doing so, they mediate stress-coping responses.

In summary, UCN1 binds strongly to both CRF receptors whereas UCNs 2 and 3 are reported to be highly selective for binding to CRFR2 receptor. In fact, CRFR2 appears to counter-regulate the stress responses mediated by CRFR1. Activation of CRFR2 by all three UCNs regulates anxiety, depression, arousal, appetite and gastric emptying, improves skeleto-muscular mass, and influences learning, energy balance, and immune responses.

The UCNs' cardiovascular influences have been recently the focus of great attention due to their potential stress response and stress-coping influences. UCN1 and UCN2 influence the heart and are implicated in hypotension, increased cardiac contractility, and stroke volume output (61, 62, 78–80). Administration of both UCN1 and UCN2 in an experimental heart failure model has shown to improve hemodynamic status in association with significant suppression of multiple vasoconstrictor hormone systems and augmentation of renal function (81, 82). This has been suggested to be through a reduction in arginine vasopressin release as demonstrated in an ovine model of heart failure (81, 82). Interestingly, UCN1 was observed to have a less desirable hypertrophic effect on the heart through which it might lead to heart failure (83, 84). This hypertrophy is not caused by activation of p42/44 kinase pathways, but rather by the activation of P13 Akt (84). Therefore, analogs of UCN1 may be designed to produce the selective activation of the protective pathway without concurrent hypertrophy response. In addition, UCN1 has been shown to modulate L-type calcium channels (85) and decrease the inward Ca^{2+} current correlating with enhanced cell survival (85). Altogether, these channels may represent novel a mechanism through which cardio protective effects are exerted.

Recently, UCN3 has emerged as an intriguing molecule, which in addition to its UCN2-like effects and targets, shows major differences in its affinity for the microdomains of the receptor (86). In addition, UCN3 produces more pronounced anti-apoptotic, cardioprotective (87), and natriuretic peptide secretory activity in rat cardiomyocytes compared to the other UCNs (88). UCN3 has been located in high concentrations in the myocardium of the human heart (65, 67, 89, 90). Having such a prominent presence in the cardiac tissue suggests a role in increasing response to stress (87). It has also been shown that UCN3 produces potent vasodilatation of rat and human arteries in vitro (77, 91). When administered intracerebroventricularly, UCN3 elevates BP and HR in the rat (92). UCN3 has also been examined in heart failure in a sheep model. When incremental doses were given IV before and after induced heart failure, prominent dose-dependent improvements in cardiac output emerged. Furthermore, UCN3 reduced total peripheral resistance and left atrial pressure, slightly increased mean arterial pressure, significantly suppressed multiple hormonal systems such as arginine vasopressin, endothelin1, rennin-angiotensin-II-aldosterone, epinephrine, and improved renal function (81, 82). The reduced peripheral vascular resistance is consistent with the vasodilator effect of UCN3 as demonstrated in vivo in arteries of both rat (91) and human (77). Left atrial pressure was normalized in heart failure by UCN3 administration, probably due to the increased cardiac output as previously reported (79, 93). It is noteworthy that the aforementioned hemodynamic influences of UCN3 have also been reported for the other UCNs (77, 79, 82, 94). Nevertheless, in the sheep model, there were differences in the latency to onset of action, with the most rapid being UCN3 followed by UCN2 and UCN1. The duration of action was inversed with UCN1 having the longest half-life followed by UCN2 and UCN3. Thus UCN3 is the most rapid and shortest acting member. These attributes are most probably due to its smaller volume of distribution and faster clearance rate.

In summary, UCNs are associated with stress conditions in general including the pathological cardiovascular states. Their roles are in stress response as well as stress coping processes.

4.2. Potential Value as a New Tool for Diagnosis of OSAS

OSAS involves respiratory events that impose significant cardiovascular short-term and long-term effects. Intermittent hypoxia is accompanied by sympathetic nervous system surges, and these processes may in turn trigger the release and expression of UCNs. If this is indeed the case, the concentrations of UCNs in serum and/or in urine should be greater in OSAS patients, and may serve as a novel diagnostic tool. Considering the fact that habitual snoring does not cause any gas exchange abnormalities or sleep disruption, and thus does not recruit as extensively stress responses, greater UCN concentrations may differentiate OSAS patients from those with the same symptoms but without the disease, that is, habitual snoring.

UCNs may also have therapeutic implications. A positive inotropic and vasodilator influence may be beneficial for improving vasoconstriction.

5. Summary

The current methods for screening OSAS are laborious, expensive, and relatively scarce, thereby imposing unnecessary delays in the recognition of OSAS among habitually snoring children. We propose that use of proteomic approaches such as those described herein may serve as the basis for the development of a noninvasive and rapid diagnostic tool for OSAS, and ultimately lead to improved outcomes.

References

1. Findley, L.J., Barth, J.T., Powers, D.C., Wilhoit, S.C., Boyd, D.G., Suratt, P.M. (1986) Cognitive impairment in patients with obstructive sleep apnea and associated hypoxemia. *Chest* **90**(5), 686–90.
2. Gozal, D. (1998) Sleep-disordered breathing and school performance in children. *Pediatrics* **102**(3 Pt 1), 616–20.
3. Punjabi, N.M., Ahmed, M.M., Polotsky, V.Y., Beamer, B.A., O'Donnell, C.P. (2003) Sleep-disordered breathing, glucose intolerance, and insulin resistance. *Respir Physiol Neurobiol* **136**(2–3), 167–78.
4. Brouillette, R.T., Fernbach, S.K., Hunt, C.E. (1982) Obstructive sleep apnea in infants and children. *J Pediatr* **100**(1), 31–40.
5. Brooks, L.J., Topol, H.I. (2003) Enuresis in children with sleep apnea. *J Pediatr* **142**(5), 515–8.
6. Sakai, J., Hebert, F. (2000) Secondary enuresis associated with obstructive sleep apnea. *J Am Acad Child Adolesc Psychiatry* **39**(2), 140–1.
7. Marcus, C.L., Carroll, J.L., Koerner, C.B., Hamer, A., Lutz, J., Loughlin, G.M. (1994) Determinants of growth in children with the obstructive sleep apnea syndrome. *J Pediatr* **125**(4), 556–62.
8. American Thoracic Society. (1996) Standards and indications for cardiopulmonary sleep studies in children. *Am J Respir Crit Care Med* **153**(2), 866–78.
9. Jeans, W.D., Fernando, D.C., Maw, A.R., Leighton, B.C. (1981) A longitudinal study of the growth of the nasopharynx and its contents in normal children. *Br J Radiol* **54**(638), 117–21.
10. Arens, R., McDonough, J.M., Corbin, A.M., Hernandez, M.E., Maislin, G., Schwab, R.J., et al. (2002) Linear dimensions of the upper airway structure during development: assessment by magnetic resonance imaging. *Am J Respir Crit Care Med* **165**(1), 117–22.
11. Arens, R., McDonough, J.M., Costarino, A.T., Mahboubi, S., Tayag-Kier, C.E., Maislin, G., et al. (2001) Magnetic resonance imaging

of the upper airway structure of children with obstructive sleep apnea syndrome. *Am J Respir Crit Care Med* **164**(4), 698–703.
12. Arens, R., Sin, S., McDonough, J.M., Palmer, J.M., Dominguez, T., Meyer, H., et al. (2005) Changes in upper airway size during tidal breathing in children with obstructive sleep apnea syndrome. *Am J Respir Crit Care Med* **171**(11), 1298–304.
13. Marcus, C.L., McColley, S.A., Carroll, J.L., Loughlin, G.M., Smith, P.L., Schwartz, A.R. (1994) Upper airway collapsibility in children with obstructive sleep apnea syndrome. *J Appl Physiol* 77(2), 918–24.
14. Gozal, D., Burnside, M.M. (2004) Increased upper airway collapsibility in children with obstructive sleep apnea during wakefulness. *Am J Respir Crit Care Med* **169**(2), 163–7.
15. Isono, S., Shimada, A., Utsugi, M., Konno, A., Nishino, T. (1998) Comparison of static mechanical properties of the passive pharynx between normal children and children with sleep-disordered breathing. *Am J Respir Crit Care Med* **157**(4 Pt 1), 1204–12.
16. Redline, S., Tishler, P.V., Schluchter, M., Aylor, J., Clark, K., Graham, G. (1999) Risk factors for sleep-disordered breathing in children. Associations with obesity, race, and respiratory problems. *Am J Respir Crit Care Med* **159**(5 Pt 1), 1527–32.
17. Ali, N.J., Pitson, D.J., Stradling, J.R. (1993) Snoring, sleep disturbance, and behaviour in 4–5 year olds. *Arch Dis Child* **68**(3), 360–6.
18. Bass, J.L., Corwin, M., Gozal, D., Moore, C., Nishida, H., Parker, S., et al. (2004) The effect of chronic or intermittent hypoxia on cognition in childhood: a review of the evidence. *Pediatrics* **114**(3), 805–16.
19. Mulvaney, S.A., Goodwin, J.L., Morgan, W.J., Rosen, G.R., Quan, S.F., Kaemingk, K.L. (2006) Behavior problems associated with sleep disordered breathing in school-aged children – the Tucson children's assessment of sleep apnea study. *J Pediatr Psychol* **31**(3), 322–30.
20. Zintzaras, E., Kaditis, A.G. (2007) Sleep-disordered breathing and blood pressure in children: a meta-analysis. *Arch Pediatr Adolesc Med* **161**(2), 172–8.
21. Marcus, C.L., Greene, M.G., Carroll, J.L. (1998) Blood pressure in children with obstructive sleep apnea. *Am J Respir Crit Care Med* **157**(4 Pt 1), 1098–103.
22. Kohyama, J., Ohinata, J.S., Hasegawa, T. (2003) Blood pressure in sleep disordered breathing. *Arch Dis Child* **88**(2), 139–42.
23. Enright, P.L., Goodwin, J.L., Sherrill, D.L., Quan, J.R., Quan, S.F. (2003) Blood pressure elevation associated with sleep-related breathing disorder in a community sample of white and Hispanic children: the Tucson Children's Assessment of Sleep Apnea study. *Arch Pediatr Adolesc Med* **157**(9), 901–4.
24. Amin, R.S., Carroll, J.L., Jeffries, J.L., Grone, C., Bean, J.A., Chini, B., et al. (2004) Twenty-four-hour ambulatory blood pressure in children with sleep-disordered breathing. *Am J Respir Crit Care Med* **169**(8), 950–6.
25. Kwok, K.L., Ng, D.K., Cheung, Y.F. (2003) BP and arterial distensibility in children with primary snoring. *Chest* **123**(5), 1561–6.
26. Tal, A., Leiberman, A., Margulis, G., Sofer, S. (1988) Ventricular dysfunction in children with obstructive sleep apnea: radionuclide assessment. *Pediatr Pulmonol* **4**(3), 139–43.
27. de la Eva, R.C., Baur, L.A., Donaghue, K.C., Waters, K.A. (2002) Metabolic correlates with obstructive sleep apnea in obese subjects. *J Pediatr* **140**(6), 654–9.
28. Tauman, R., Ivanenko, A., O'brien, L.M., Gozal, D. (2004) Plasma C-reactive protein levels among children with sleep-disordered breathing. *Pediatrics* **113**(6), e564–9.
29. Kheirandish-Gozal, L., Capdevila, O.S., Tauman, R., Gozal, D. (2006) Plasma C-reactive protein in nonobese children with obstructive sleep apnea before and after adenotonsillectomy. *J Clin Sleep Med* **2**(3), 301–4.
30. Gozal, D., McLaughlin, C.V., Sans, C.O., Witcher, L.A., Kheirandish-Gozal, L. (2007) C reactive protein, obstructive sleep apnea, and cognitive dysfunction in school-aged children. *Am J Respir Crit Care Med* **176**(2), 188–93.
31. Kaditis, A.G., Alexopoulos, E.I., Kalampouka, E., Kostadima, E., Germenis, A., Zintzaras, E., et al. (2005) Morning levels of C-reactive protein in children with obstructive sleep-disordered breathing. *Am J Respir Crit Care Med* **171**(3), 282–6.
32. Buts, J.P., Barudi, C., Moulin, D., Claus, D., Cornu, G., Otte, J.B. (1986) Prevalence and treatment of silent gastro-oesophageal reflux in children with recurrent respiratory disorders. *Eur J Pediatr* **145**(5), 396–400.
33. Chervin, R.D., Hedger, K., Dillon, J.E., Pituch, K.J. (2000) Pediatric sleep questionnaire (PSQ): validity and reliability of scales for sleep-disordered breathing, snoring, sleepiness, and behavioral problems. *Sleep Med* **1**(1), 21–32.

34. Owens, J.A., Spirito, A., McGuinn, M. (2000) The Children's Sleep Habits Questionnaire (CSHQ): psychometric properties of a survey instrument for school-aged children. *Sleep* **23**(8), 1043–51.
35. Clinical practice guideline: diagnosis and management of childhood obstructive sleep apnea syndrome. (2002) *Pediatrics* **109**(4), 704–12.
36. Carroll, J.L., McColley, S.A., Marcus, C.L., Curtis, S., Loughlin, G.M. (1995) Inability of clinical history to distinguish primary snoring from obstructive sleep apnea syndrome in children. *Chest* **108**(3), 610–8.
37. American Thoracic Society. (1999) Cardiorespiratory sleep studies in children. Establishment of normative data and polysomnographic predictors of morbidity. *Am J Respir Crit Care Med* **160**(4), 1381–7.
38. Nasser, S., Rees, P.J. (1992) Sleep apnoea: causes, consequences and treatment. *Br J Clin Pract* **46**(1), 39–43.
39. Resta, O., Pannacciulli, N., Di, G.G., Stefano, A., Barbaro, M.P., De, P.G. (2004) High prevalence of previously unknown subclinical hypothyroidism in obese patients referred to a sleep clinic for sleep disordered breathing. *Nutr Metab Cardiovasc Dis* **14**(5), 248–53.
40. Tresaco, B., Bueno, G., Pineda, I., Moreno, L.A., Garagorri, J.M., Bueno, M. (2005) Homeostatic model assessment (HOMA) index cut-off values to identify the metabolic syndrome in children. *J Physiol Biochem* **61**(2), 381–8.
41. Gozal, D., Lipton, A.J., Jones, K.L. (2002) Circulating vascular endothelial growth factor levels in patients with obstructive sleep apnea. *Sleep* **25**(1), 59–65.
42. Stoohs, R., Guilleminault, C. (1991) Snoring during NREM sleep: respiratory timing, esophageal pressure and EEG arousal. *Respir Physiol* **85**(2), 151–67.
43. Guilleminault, C., Stoohs, R., Clerk, A., Cetel, M., Maistros, P. (1993) A cause of excessive daytime sleepiness. The upper airway resistance syndrome. *Chest* **104**(3), 781–7.
44. Gozal, D., Kheirandish-Gozal, L. (2006) Sleep apnea in children – treatment considerations. *Paediatr Respir Rev* 7(Suppl 1), S58–S61.
45. Kheirandish, L., Gozal, D. (2006) Neurocognitive dysfunction in children with sleep disorders. *Dev Sci* **9**(4), 388–99.
46. Montgomery-Downs, H.E., Crabtree, V.M., Gozal, D. (2005) Cognition, sleep and respiration in at-risk children treated for obstructive sleep apnoea. *Eur Respir J* **25**(2), 336–42.
47. Friedman, B.C., Hendeles-Amitai, A., Kozminsky, E., Leiberman, A., Friger, M., Tarasiuk, A., et al. (2003) Adenotonsillectomy improves neurocognitive function in children with obstructive sleep apnea syndrome. *Sleep* **26**(8), 999–1005.
48. Gozal, D., Pope, D.W., Jr. (2001) Snoring during early childhood and academic performance at ages thirteen to fourteen years. *Pediatrics* **107**(6), 1394–9.
49. Kheirandish, L., Gozal, D., Pequignot, J.M., Pequignot, J., Row, B.W. (2005) Intermittent hypoxia during development induces long-term alterations in spatial working memory, monoamines, and dendritic branching in rat frontal cortex. *Pediatr Res* **58**(3), 594–9.
50. Flemons, W.W., Douglas, N.J., Kuna, S.T., Rodenstein, D.O., Wheatley, J. (2004) Access to diagnosis and treatment of patients with suspected sleep apnea. *Am J Respir Crit Care Med* **169**(6), 668–72.
51. Katz, E.S., Greene, M.G., Carson, K.A., Galster, P., Loughlin, G.M., Carroll, J., et al. (2002) Night-to-night variability of polysomnography in children with suspected obstructive sleep apnea. *J Pediatr* **140**(5), 589–94.
52. Ridker, P.M., Stampfer, M.J., Rifai, N. (2001) Novel risk factors for systemic atherosclerosis: a comparison of C-reactive protein, fibrinogen, homocysteine, lipoprotein(a), and standard cholesterol screening as predictors of peripheral arterial disease. *JAMA* **285**(19), 2481–5.
53. Castell, J.V., Gomez-Lechon, M.J., David, M., Fabra, R., Trullenque, R., Heinrich, P.C. (1990) Acute-phase response of human hepatocytes: regulation of acute-phase protein synthesis by interleukin-6. *Hepatology* **12**(5), 1179–86.
54. Meier-Ewert, H.K., Ridker, P.M., Rifai, N., Price, N., Dinges, D.F., Mullington, J.M. (2001) Absence of diurnal variation of C-reactive protein concentrations in healthy human subjects. *Clin Chem* **47**(3), 426–30.
55. Shamsuzzaman, A.S., Winnicki, M., Lanfranchi, P., Wolk, R., Kara, T., Accurso, V., et al. (2002) Elevated C-reactive protein in patients with obstructive sleep apnea. *Circulation* **105**(21), 2462–4.
56. Montgomery-Downs, H.E., O'brien, L.M., Gulliver, T.E., Gozal, D. (2006) Polysomnographic characteristics in normal preschool and early school-aged children. *Pediatrics* **117**(3), 741–53.

57. Schaub, S., Wilkins, J., Weiler, T., Sangster, K., Rush, D., Nickerson, P. (2004) Urine protein profiling with surface-enhanced laser-desorption/ionization time-of-flight mass spectrometry. *Kidney Int* **65**(1), 323–32.
58. Vale, W., Spiess, J., Rivier, C., Rivier, J. (1981) Characterization of a 41-residue ovine hypothalamic peptide that stimulates secretion of corticotropin and beta-endorphin. *Science* **213**(4514), 1394–7.
59. Habib, K.E., Gold, P.W., Chrousos, G.P. (2001) Neuroendocrinology of stress. *Endocrinol Metab Clin North Am* **30**(3), 695–728.
60. Tsigos, C., Chrousos, G.P. (2002) Hypothalamic-pituitary-adrenal axis, neuroendocrine factors and stress. *J Psychosom Res* **53**(4), 865–71.
61. Vaughan, J., Donaldson, C., Bittencourt, J., Perrin, M.H., Lewis, K., Sutton, S., et al. (1995) Urocortin, a mammalian neuropeptide related to fish urotensin I and to corticotropin-releasing factor. *Nature* **378**(6554), 287–92.
62. Parkes, D.G., Vaughan, J., Rivier, J., Vale, W., May, C.N. (1997) Cardiac inotropic actions of urocortin in conscious sheep. *Am J Physiol* **272**(5 Pt 2), H2115–22.
63. Latchman, D.S. (2002) Urocortin. *Int J Biochem Cell Biol* **34**(8), 907–10.
64. Reyes, T.M., Lewis, K., Perrin, M.H., Kunitake, K.S., Vaughan, J., Arias, C.A., et al. (2001) Urocortin II: a member of the corticotropin-releasing factor (CRF) neuropeptide family that is selectively bound by type 2 CRF receptors. *Proc Natl Acad Sci* **98**(5), 2843–8.
65. Lewis, K., Li, C., Perrin, M.H., Blount, A., Kunitake, K., Donaldson, C., et al. (2001) Identification of urocortin III, an additional member of the corticotropin-releasing factor (CRF) family with high affinity for the CRF2 receptor. *Proc Natl Acad Sci* **98**(13), 7570–5.
66. Donaldson, C.J., Sutton, S.W., Perrin, M.H., Corrigan, A.Z., Lewis, K.A., Rivier, J.E., et al. (1996) Cloning and characterization of human urocortin. *Endocrinology* **137**(5), 2167–70.
67. Hsu, S.Y., Hsueh, A.J. (2001) Human stresscopin and stresscopin-related peptide are selective ligands for the type 2 corticotropin-releasing hormone receptor. *Nat Med* 7(5), 605–11.
68. Perrin, M.H., Donaldson, C.J., Chen, R., Lewis, K.A., Vale, W.W. (1993) Cloning and functional expression of a rat brain corticotropin releasing factor (CRF) receptor. *Endocrinology* **133**(6), 3058–61.
69. Chen, R., Lewis, K.A., Perrin, M.H., Vale, W.W. (1993) Expression cloning of a human corticotropin-releasing-factor receptor. *Proc Natl Acad Sci* **90**(19), 8967–71.
70. Blank, T., Nijholt, I., Grammatopoulos, D.K., Randeva, H.S., Hillhouse, E.W., Spiess, J. (2003) Corticotropin-releasing factor receptors couple to multiple G-proteins to activate diverse intracellular signaling pathways in mouse hippocampus: role in neuronal excitability and associative learning. *J Neurosci* **23**(2), 700–7.
71. Karteris, E., Grammatopoulos, D., Randeva, H., Hillhouse, E.W. (2000) Signal transduction characteristics of the corticotropin-releasing hormone receptors in the feto-placental unit. *J Clin Endocrinol Metab* **85**(5), 1989–96.
72. Spina, M., Merlo-Pich, E., Chan, R.K., Basso, A.M., Rivier, J., Vale, W., et al. (1996) Appetite-suppressing effects of urocortin, a CRF-related neuropeptide. *Science* **273**(5281), 1561–4.
73. Turnbull, A.V., Vale, W., Rivier, C. (1996) Urocortin, a corticotropin-releasing factor-related mammalian peptide, inhibits edema due to thermal injury in rats. *Eur J Pharmacol* **303**(3), 213–6.
74. Kozicz, T., Yanaihara, H., Arimura, A. (1998) Distribution of urocortin-like immunoreactivity in the central nervous system of the rat. *J Comp Neurol* **391**(1), 1–10.
75. Kozicz, T., Arimura, A., Maderdrut, J.L., Lazar, G. (2002) Distribution of urocortin-like immunoreactivity in the central nervous system of the frog *Rana esculenta*. *J Comp Neurol* **453**(2), 185–98.
76. Wiley, K.E., Davenport, A.P. (2004) CRF2 receptors are highly expressed in the human cardiovascular system and their cognate ligands urocortins 2 and 3 are potent vasodilators. *Br J Pharmacol* **143**(4), 508–14.
77. Kimura, Y., Takahashi, K., Totsune, K., Muramatsu, Y., Kaneko, C., Darnel, A.D., et al. (2002) Expression of urocortin and corticotropin-releasing factor receptor subtypes in the human heart. *J Clin Endocrinol Metab* **87**(1), 340–6.
78. Takahashi, K., Totsune, K., Murakami, O., Shibahara, S. (2004) Urocortins as cardiovascular peptides. *Peptides* **25**(10), 1723–31.
79. Bale, T.L., Hoshijima, M., Gu, Y., Dalton, N., Anderson, K.R., Lee, K.F., et al. (2004) The cardiovascular physiologic actions of urocortin II: acute effects in murine heart failure. *Proc Natl Acad Sci* **101**(10), 3697–702.
80. Coste, S.C., Kesterson, R.A., Heldwein, K.A., Stevens, S.L., Heard, A.D., Hollis, J.H., et al.

(2000) Abnormal adaptations to stress and impaired cardiovascular function in mice lacking corticotropin-releasing hormone receptor-2. *Nat Genet* **24**(4), 403–9.

81. Rademaker, M.T., Charles, C.J., Espiner, E.A., Fisher, S., Frampton, C.M., Kirkpatrick, C.M., et al. (2002) Beneficial hemodynamic, endocrine, and renal effects of urocortin in experimental heart failure: comparison with normal sheep. *J Am Coll Cardiol* **40**(8), 1495–505.
82. Rademaker, M.T., Cameron, V.A., Charles, C.J., Richards, A.M. (2005) Integrated hemodynamic, hormonal, and renal actions of urocortin 2 in normal and paced sheep: beneficial effects in heart failure. *Circulation* **112**(23), 3624–32.
83. Latchman, D.S. (1999) Cardiotrophin-1 (CT-1): a novel hypertrophic and cardioprotective agent. *Int J Exp Pathol* **80**(4), 189–96.
84. Railson, J., Lawrence, K., Stephanou, A., Brar, B., Pennica, D., Latchman, D. (2000) Cardiotrophin-1 reduces stress-induced heat shock protein production in cardiac myocytes. *Cytokine* **12**(11), 1741–4.
85. Tao, J., Xu, H., Yang, C., Liu, C.N., Li, S. (2004) Effect of urocortin on L-type calcium currents in adult rat ventricular myocytes. *Pharmacol Res* **50**(5), 471–6.
86. Perrin, M.H., DiGruccio, M.R., Koerber, S.C., Rivier, J.E., Kunitake, K.S., Bain, D.L., et al. (2003) A soluble form of the first extracellular domain of mouse type 2beta corticotropin-releasing factor receptor reveals differential ligand specificity. *J Biol Chem* **278**(18), 15595–600.
87. Chanalaris, A., Lawrence, K.M., Stephanou, A., Knight, R.D., Hsu, S.Y., Hsueh, A.J., et al. (2003) Protective effects of the urocortin homologues stresscopin (SCP) and stresscopin-related peptide (SRP) against hypoxia/reoxygenation injury in rat neonatal cardiomyocytes. *J Mol Cell Cardiol* **35**(10), 1295–305.
88. Chanalaris, A., Lawrence, K.M., Townsend, P.A., Davidson, S., Jamshidi, Y., Stephanou, A., et al. (2005) Hypertrophic effects of urocortin homologous peptides are mediated via activation of the Akt pathway. *Biochem Biophys Res Commun* **328**(2), 442–8.
89. Fukuda, T., Takahashi, K., Suzuki, T., Saruta, M., Watanabe, M., Nakata, T., et al. (2005) Urocortin 1, urocortin 3/stresscopin, and corticotropin-releasing factor receptors in human adrenal and its disorders. *J Clin Endocrinol Metab* **90**(8), 4671–8.
90. Takahashi, K., Totsune, K., Murakami, O., Saruta, M., Nakabayashi, M., Suzuki, T., et al. (2004) Expression of urocortin III/stresscopin in human heart and kidney. *J Clin Endocrinol Metab* **89**(4), 1897–903.
91. Kageyama, K., Furukawa, K., Miki, I., Terui, K., Motomura, S., Suda, T. (2003) Vasodilative effects of urocortin II via protein kinase A and a mitogen-activated protein kinase in rat thoracic aorta. *J Cardiovasc Pharmacol* **42**(4), 561–5.
92. Chu, C.P., Qiu, D.L., Kato, K., Kunitake, T., Watanabe, S., Yu, N.S., et al. (2004) Central stresscopin modulates cardiovascular function through the adrenal medulla in conscious rats. *Regul Pept* **119**(1–2), 53–9.
93. Abdelrahman, A.M., Lin, L.S., Pang, C.C. (2005) Influence of urocortin and corticotropin releasing factor on venous tone in conscious rats. *Eur J Pharmacol* **510**(1–2), 107–11.
94. Terui, K., Higashiyama, A., Horiba, N., Furukawa, K.I., Motomura, S., Suda, T. (2001) Coronary vasodilation and positive inotropism by urocortin in the isolated rat heart. *J Endocrinol* **169**(1), 177–83.
95. Bradford, M. (1976) A Rapid and Sensitive Method for the Quantitation of Microgram Quantities of Protein Utilizing the Principle of Protein-Dye Binding. *Anal Biochem* **72**, 248–54

Chapter 14

Immune Response Biomarker Profiling Application on ProtoArray® Protein Microarrays

Barry Schweitzer, Lihao Meng, Dawn Mattoon, and Alex J. Rai

Abstract

The development of autoantibodies is observed in autoimmune disorders and numerous cancers. Consequently, autoantibodies form the basis of potential diagnostic and prognostic assays, as well as approaches for monitoring disease progression and treatment response. The effective use of autoantigen biomarkers for these applications, however, is contingent upon the identification of not one but multiple biomarkers. This is a consequence of the observation that the development of autoantibodies to any given protein is typically seen only in a fraction of patients. We have previously demonstrated the utility of functional protein microarrays containing thousands of different human proteins (ProtoArrays®) for discovering novel autoimmune biomarkers in serum and plasma. Here, we describe a protocol for detecting autoantibodies in urine.

Key words: Autoimmune, Autoantibody, Autoantigen, Urine, Protein microarray, Biomarker

1. Introduction

Autoantibody production is the immune response directed against self-antigens. Several diseases have been characterized by the presence of autoantibodies such as lupus and rheumatoid arthritis even prior to the onset of clinical disease (1). Autoantibodies against proteins involved in cancer have also been described for nearly all human tumor types (2). Antibodies directed against tumor proteins arise early in the tumor development process, and are potential biomarkers for early disease detection. Patients could be monitored for the presence and titer of antitumor antibodies in order to facilitate early detection of disease, and to aid in disease prognosis.

Autoantigen microarrays have been developed for carrying out the characterization of autoantibodies from a number of

Alex J. Rai (ed.), *The Urinary Proteome: Methods and Protocols*, Methods in Molecular Biology, vol. 641,
DOI 10.1007/978-1-60761-711-2_14,

human autoimmune diseases (3). Arrays consisting of ~23 different antigens were used to study experimental autoimmune encephalomyelitis (EAE), an animal model for multiple sclerosis. Immune response profiles were used to characterize the evolution of B-cell responses in acute and chronic EAE, and in response to therapy (4). An array of 266 different antigens was used to study a mouse model for type 1 diabetes. While these studies demonstrated the utility of antigen microarrays for autoimmune profiling, the biased antigen set presented on the array makes them unsuitable for the discovery of novel autoimmune biomarkers. In an attempt to address this need, high density microarrays containing >2,400 purified human proteins were generated using *E. coli*-based expression of a human fetal brain cDNA library (5) and used to screen serum from a small number of patients with rheumatoid arthritis and alopecia areata (6). In another study, microarrays made from *E. coli*-expressed proteins derived from a mouse TH1 cDNA expression library were used in a mouse model for SLE, resulting in the identification of novel candidate autoimmune biomarkers (7). Although these results showed the potential of high content human protein microarrays to facilitate discovery of autoimmune biomarkers, they suffered from significant technical drawbacks including the lack of sequence-validation of the expression clones, and the large fraction of clones that were either not full-length open reading frames or were in the wrong reading frame for expression. In addition, the use of an *E. coli*-based expression system and the fact that proteins were purified under denaturing conditions limited the findings to those interactions that did not require native folding or posttranslational processing.

One approach for autoantigen identification in cancer is SEREX: serological analysis of cDNA expression libraries. SEREX involves the generation of tumor-specific λ-GT11 cDNA expression libraries, followed by immunological screening of plaque lifts using patient sera. The SEREX approach was successfully used to identify the cancer autoantigen NY-ESO-1, a protein that is autoantigenic in ~20–50% of patients overexpressing NY-ESO-1 (8). While clearly useful, the SEREX approach is slow, technically challenging, and typically has a high false positive rate. Furthermore, because SEREX relies on bacterial protein expression, it cannot identify autoantigens requiring posttranslational modifications (9). More recently, reverse phase protein microarrays have been used to identify colon cancer and lung cancer autoantigens (10, 11). These arrays are made by fractionating cancer cell homogenates, arraying them in spots on a microarray, probing them with patient sera and detecting antibody binding. Afterward, mass-spectrometry-based techniques are used to identify the actual autoantigen – a process that can be both time-consuming and expensive.

Many of the issues mentioned above can be addressed via newly developed technologies in which full length proteins are purified under native conditions from insect cells. Recent advances in high-throughput cloning, expression, and purification have led to the production of high-density functional protein microarrays that are suitable for discovery-based proteomic analysis (12). We have recently developed high-density human protein microarrays comprised of ~8,000 full-length proteins affinity purified from insect cells to maximize native folding and posttranslational modifications (Fig. 1). The proteins on this array have been shown to be functional in a variety of assays including profiling antibody specificity, autoantibody repertoire, protein–protein interactions, small molecule interactions, and assays to identify the substrates of protein kinases (13). We are currently engaged in several collaborations in which we have identified putative autoimmune biomarkers in autoimmune diseases such as lupus and rheumatoid arthritis as well as in cancer (unpublished observations, B. Schweitzer).

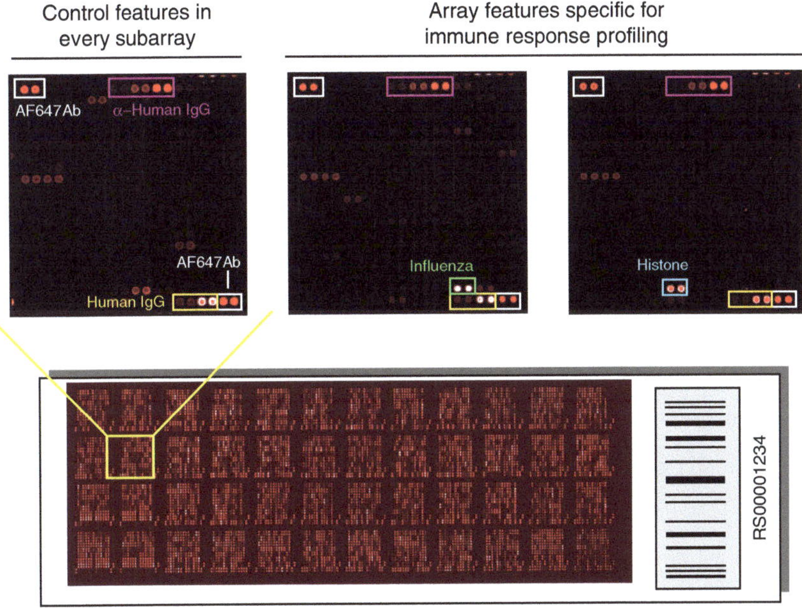

Fig. 1. Immune response profiling on human protein microarrays. An array containing ~8,000 GST tagged full length recombinant human proteins spotted in duplicate is shown in the bottom panel. The left upper panel shows one of the 48 subarrays that comprise the full array. Control features contained in every subarray are highlighted in boxes. Spots of AlexaFluor labeled antibody (boxed in white) are used to align the spotting grid for data acquisition. The human IgG gradient (boxed in yellow) serves as a control for proper performance of the antihuman detection reagent. A gradient of antihuman IgG (boxed in purple) binds IgG present in the sample and provides a control for proper performance of the assay. The two upper right panels show subarrays containing control proteins specific to the immune response profiling. Influenza antigen (upper middle panel, boxed in green) is bound by circulating antiinfluenza antibody in exposed individuals (representing >90% of serum samples tested thus far) and provides an additional assay performance control. Histone protein (upper right panel, boxed in blue) is one of a number of known antigens observed in several autoimmune diseases. Copyright © 2007 Invitrogen Corp.

Essentially all of the autoantibody-profiling studies described above have used serum or plasma as the biological specimen. Autoantibodies have also been shown to be present in other bodily fluids, including cerebral spinal fluid, saliva, and synovial fluid. The presence of antibodies in urine has been known for decades (14), including a correlation of antinuclear antibodies in the urine and serum of patients with lupus (15). More recently, the levels of various immunoglobulin subclasses have been shown to be useful parameters for characterizing patients with diabetic nephropathy (16), ANCA-associated renal vasculitis (17), and bladder cancer (18). To the best of our knowledge, however, no

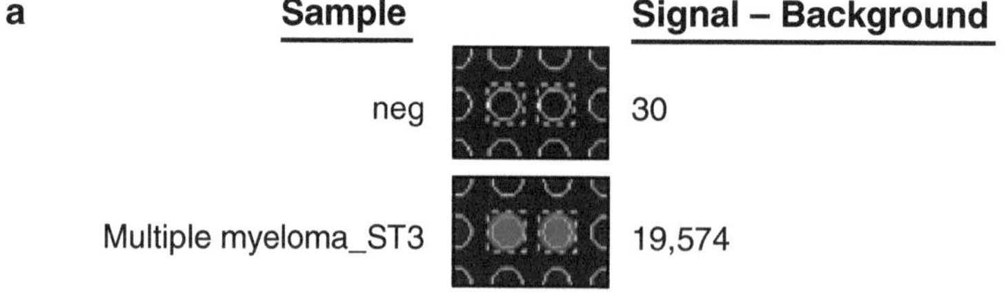

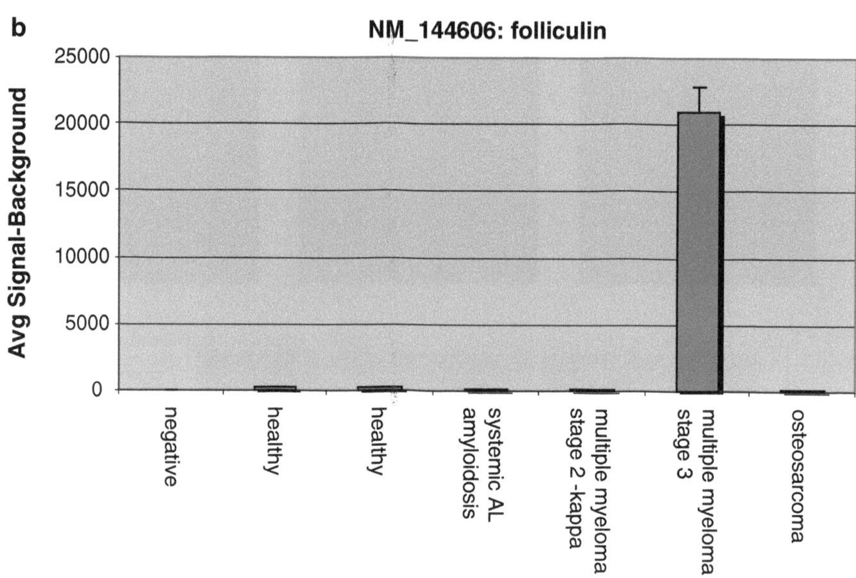

Fig. 2. Immune response profiling of urine on human protein microarrays. Random, clean-catch, urine samples were collected from patients (with disease) or normal, healthy volunteers. Diseased patients include two myeloma cases (one with ISS-stage 2 and one with stage 3 disease) and two "other disease" (one systemic AL amyloidosis and one osteosarcoma) patients. Healthy volunteers included two individuals with no evidence of disease. Briefly, patient was asked to clean genitalia with alcohol wipe prior to voiding to minimize the transfer of surface bacteria to urine. The first >10 mL was voided and discarded. Subsequent urine specimen was collected in sterile container and submitted to the laboratory within 3 h for analysis. Samples were processed within 12 h of collection, were aliquoted, and stored at 4°C. For analysis on protein arrays, specimens were shipped on dry ice to Invitrogen. They were then thawed and processed as described in the text. (**a**) Image and quantitation of replicate spots for a protein on the array that binds IgG in a urine sample collected from a stage 3 myeloma patient. The corresponding spots for the negative control array (processed with detection reagent alone) are also shown. (**b**) Signal intensity of the protein shown in Fig. 2a for all samples.

specific autoantigens have been identified for antibodies present in urine. We show here that high density functional protein microarrays are useful tools for the characterization of the antibody profile in human urine specimens (Fig. 2).

2. Materials

1. ProtoArray Human Protein Microarray (Invitrogen, # PAH052402, Carlsbad, CA).
2. 4-chamber Incubation Tray (Greiner, #96077307, Monroe, NC).
3. Alexa Fluor® 647 goat antihuman IgG (H+L) "2 mg/mL" (Molecular Probes, # A21445, Eugene, OR).
4. Gene Pix Pro Software (Molecular Devices, MDS Analytical Technologies, Sunnyvale, CA).
5. GenePix 4000B Microarray Scanner (Molecular Devices).
6. Eppendorf Centrifuge (5810) (Fisher Scientific, # 05-400-60, Pittsburgh, PA).
7. Lab Rotator (Lab-Line Instruments, # 1314, Melrose Park, IL).
8. Polyacetal Slide Rack (RA Lamb, # E99, Fisher Scientific, Pittsburgh, PA).
9. Blocking buffer (per Liter): 50 mM HEPES pH 7.5, 200 mM NaCl, 0.08% Triton X-100, 25% glycerol, 20 mM reduced glutathione (Sigma, St. Louis, MO), 1.0 mM DTT, 1% BSA (see Note 1).
10. PBST buffer (per liter): 1× PBS (from 10× PBS, pH 7.4, Gibco, Invitrogen, Carlsbad, CA), 1% BSA, 0.1% Tween 20 (American Bioanalytical, Natick, MA) (see Note 2).
11. Antihuman antibody-AlexaFluor conjugate solution: Dilute antihuman IgG antibody-Alexa Fluor® 647 conjugate (2 mg/mL, Molecular Probes, Eugene, OR) to 1.0 μg/mL in PBST buffer.

3. Methods

The following protocol is for probing ProtoArray® Protein Microarrays printed on thin nitrocellulose slides with urine samples to identify antibodies that differ in reactivity and/or abundance in healthy/controls vs. diseased/treated populations. All steps should be carried out at 4°C. Take care to never touch the surface of ProtoArray® Protein Microarrays.

Fig. 3. (**a**) Four chamber incubation tray used for handling protein microarrays. (**b**) Indent in the tray bottom used as the site for buffer removal. Copyright © 2007 Invitrogen Corp.

3.1. Blocking

1. Immediately place the mailer containing the ProtoArray® Human Protein Microarray v4.0 at 4°C upon removal from storage and equilibrate the mailer at 4°C for at least 15 min prior to use.
2. Place one ProtoArray® Human Protein Microarray with the barcode facing up in the bottom of each well of a 4-chamber incubation tray such that the barcode end of the microarray is near the tray end containing an indented numeral (see Fig. 3a). The indent in the tray bottom is used as the site for buffer removal (see Fig. 3b, arrow).
3. Using a sterile pipette, add 5 mL blocking buffer into each chamber. Avoid pipetting buffer directly onto the array surface.
4. Incubate the tray for 1 h at 4°C on a shaker set at 50 rpm (circular shaking).
5. At the end of the incubation, aspirate the blocking buffer using vacuum or a pipette. Position the tip of the aspirator or pipette into the indented numeral and aspirate most of the buffer from each well (see Fig. 4). Tilt the indented numeral end of tray (opposite to the barcode) and aspirate any remaining buffer that accumulates at the base of the well (see Note 5). Proceed immediately to the *Probing Procedure*.

3.2. Probing

1. Add 5 mL urine diluted in PBST buffer (see Note 6).
2. Incubate for 90 min at 4°C with gentle circular shaking (~50 rpm).

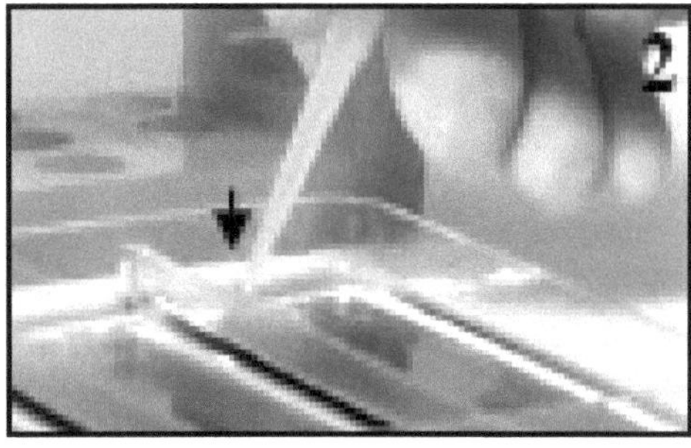

Fig. 4. Positioning of the pipette tip for buffer aspiration. Copyright © 2007 Invitrogen Corp.

3. Remove diluted specimen by aspiration (see Note 5 for details).

3.3. Detection

1. Wash with 5 mL fresh PBST buffer, 5 min incubations per wash with gentle agitation. Remove PBST buffer by aspiration.
2. Repeat wash step 4 more times.
3. Add 5 mL of secondary antibody diluted in PBST buffer (see Note 7)
4. Incubate for 90 min at 4°C with gentle circular shaking (~50 rpm).
5. Remove secondary antibody by aspiration.
6. Wash with 5 mL fresh PBST buffer, 5 min incubations per wash with gentle agitation. Remove PBST buffer by aspiration.
7. Repeat wash step 4 more times.
8. Remove the slide from the 4-well tray using forceps. Insert the tip of the forceps into the indented numeral and gently pry the edge of the slide upward.
9. Pick up the slide with a gloved hand taking care to only touch the slide by its edges.
10. Insert the slide into a slide box and centrifuge at low speed ($200 \times g$ in a centrifuge equipped with a plate rotor) for 1 min.
11. Alternatively, each slide can be inserted into a separate 50 mL conical tube and centrifuged at $200 \times g$.
12. Once slides are completely dry, store the slides in a light-tight slide box (see Note 8).
13. Scan each slide with a fluorescent microarray scanner (see Note 9).

3.4. Analysis

1. Obtain the .gal file (text file describing layout of the ProtoArray®) by submitting the barcode of the ProtoArray® to the ProtoArray Central portal on the Invitrogen web site.
2. Using the appropriate .gal file, acquire image data using microarray image acquisition software (see Note 10).
3. Analyze data using appropriate analysis software (see Note 11).

4. Notes

1. We recommend using protease-free 30% BSA solution from Sigma. Add BSA and DTT to solution prior to use. Freshly prepared blocking buffer is best for blocking slides! Do not store blocking buffer containing BSA for more than 24 h.
2. Add 1% BSA to solution prior to use. Freshly prepared probe buffer is best for probing slides! Do not store probe buffer containing BSA for more than 24 h.
3. Ensure that the barcode is readable and that the barcode end of the slide is near the end of the tray containing an indented numeral. The indent in the bottom of the tray will be used as the site from which buffer is removed. Gently rock the 4-well tray to ensure that each slide is completely immersed in the blocking buffer.
4. Use a shaker that keeps the arrays in one plane during rotation. Nutating or rocking shakers are not to be used because of increased risk of cross-well contamination.
5. Aspiration by vacuum or by the use of a pipettor: Position the tip of the aspirator/pipettor into the indented numeral and remove as much of the liquid as possible. When each well is dry, lift the end of the tray opposite the barcode (indented numeral end) and remove the liquid that pools at the base of the well. Do not aspirate from the surface of the slide since this can cause scratches. Do not allow any part of the array surface to dry before adding the next solution – this can cause high and/or uneven background.
6. We recommend a 1:2 or 1:3 dilution of urine for use on ProtoArray® Protein Microarrays. Pipet the diluted specimen at the indented numeral end of the 4-well tray and allow the solution to flow across the slide surface. To prevent local variations in fluorescence intensity and background, AVOID DIRECT CONTACT with the slide and avoid pouring the diluted specimen directly on to the slide.
7. We recommend using Alexa Fluor® 647 goat antihuman IgG (H+L) diluted to 1 μg/mL (1:2,000) for immune-profiling

applications. Add the diluted secondary antibody at the indented numeral end of the 4-well tray and allow the liquid to flow across the slide surface. To prevent local variations in fluorescence intensity and background, AVOID DIRECT CONTACT with the slide and if at all possible, avoid pouring the specimen directly on to the slide.

8. Prolonged exposure to light will diminish signal intensities. After the slides have been probed and dried, they can be stored either vertically or horizontally.
9. For best results, scan the slides within 24 h of probing. We recommend using the GenePix 4000B scanner (Molecular Devices) at 635 nm with a PMT gain of 600, a laser power of 100%, and a focus point of 0 μm. Laser power and PMT should be adjusted such that the control feature signals (Human IgG and/or antiHuman IgG) on the array are not saturated. This is done to keep signals within the linear range.
10. We recommend GenePix Pro (Molecular Devices).
11. We recommend ProtoArray® Prospector, which is available for free download at http://www.invitrogen.com/protoarray.

References

1. Arbuckle, M.R, McClain, M.T., Rubertone, M.V. et al. (2003) Development of autoantibodies before the clinical onset of systemic lupus erythematosus. *N. Engl. J. Med.*, **349**, 1526–1533.
2. Disis, M.L., Montgomery, R.B., Goodell, V., delaRosa, C., and Salazar, L.G. (2005) Antibody Immunity to Cancer-Associated Proteins. AACR 96th Annual Meeting, April 16–20, 166–169.
3. Robinson, W.H., DiGennaro, C., Hueber, W. et al. (2002) Autoantigen microarrays for multiplex characterization of autoantibody responses. *Nat. Med.*, **8**, 295–301.
4. Robinson, W.H., Fontoura, P., Lee, B.J. et al. (2003) Protein microarrays guide tolerizing DNA vaccine treatment of autoimmune encephalomyelitis. *Nat. Biotechnol.*, **21**, 1033–1039.
5. Braun, P., Hu, Y., Shen, B. et al. (2002) Proteome-scale purification of human proteins from bacteria. *Proc. Natl. Acad. Sci. U.S.A.*, **99**, 2654–2659.
6. Lueking, A., Huber, O., Wirths, C. et al. (2005) Profiling of alopecia areata autoantigens based on protein microarray technology. *Mol. Cell Proteomics*, **4**, 1382–1390.
7. Gutjahr, C., Murphy, D., Lueking, A. et al. (2005) Mouse protein arrays from a TH1 cell cDNA library for antibody screening and serum profiling. *Genomics*, **85**, 285–296.
8. Chen Y.T., Scanlan, M.J., Sahin, U., Tureci, O., Gure, A.O., Tsang, S. et al. (1997) A testicular antigen aberrantly expressed in human cancers detected by autologous antibody screening. *Proc. Natl. Acad. Sci. U.S.A.*, **94**, 1914–1918.
9. Sahin, U., Tureci, O., Schmitt, H., Cochlovius, B., Johannes, T., Schmits R. et al. (1995) Human neoplasms elicit multiple specific immune responses in the autologous host. *Proc. Natl. Acad. Sci. U.S.A.*, **92**, 11810–11813.
10. Nam, M.J., Madoz-Gurpide, J., Wang, H., Lescure, P., Schmalbach, C.E., Zhao, R. et al. (2003) Molecular profiling of the immune response in colon cancer using protein microarrays: occurrence of autoantibodies to ubiquitin C-terminal hydrolase L3. *Proteomics*, **3**, 2108–2115.
11. Brichory, F.M., Misek, D.E., Yim, A.M., Krause, M.C., Giordano, T.J., Beer, D.G., and Hanash, S.M. (2001) An immune response manifested by the common occurrence of

annexins I and II autoantibodies and high circulating levels of IL-6 in lung cancer. *Proc. Natl. Acad. Sci. U.S.A.* **98**, 9824–9829.

12. Zhu, H., Bilgin, M., Bangham, R. et al. (2001) Global analysis of protein activities using proteome chips. *Science*, **293**, 2101–2105.
13. Predki, P.F., Mattoon, D., Bangham, R., Schweitzer, B., and Michaud, G. (2005) Protein microarrays: A new tool for profiling antibody cross-reactivity. *Hum. Antibodies*, **14**, 7–15.
14. Hanson, L.A. and Tan, E.M. (1965) Characterization of antibodies in urine. *J. Clin. Invest.*, **44**, 703–715.
15. Stevens, M.B. and Knowles, B. (1962) Significance of urinary gamma globulin in lupus nephritis. *N. Engl. J. Med.*, **267**, 1159–1164.
16. Gambardella, S., Morano, S., Cancelli, A., Pietravalle, P., Frontoni, S., Bacci, S., Di Mario, U., and Andreani, D. (1989) Urinary IgG4: an additional parameter in characterizing patients with incipient diabetic nephropathy. *Diabetes Res.*, **10**, 153–157.
17. Bakoush, O., Segelmark, M., Torffvit, O., Ohlsoon, S., and Tencer, J. (2006) Urine IgM excretion predicts outcome in ANCA-associated renal vasculitis. *Nephrol. Dial. Transplant.*, **21**, 1263–1269.
18. Huland, H., Otto, U., and Droese, M. (1983) The value of urinary cytology, serum and urinary carcinoembryonic antigen, rheumatoid factors, and urinary immunoglobulin concentration as tumor markers or prognostic factors in predicting progression of superficial bladder cancer. *Eur. Urol.*, **9**, 346–349.

Chapter 15

Design and Validation of an Immunoaffinity LC–MS/MS Assay for the Quantification of a Collagen Type II Neoepitope Peptide in Human Urine: Application as a Biomarker of Osteoarthritis

Olga Nemirovskiy, Wenlin Wendy Li, and Gabriella Szekely-Klepser

Abstract

Biomarkers play an increasingly important role for drug efficacy and safety evaluation in all stages of drug development. It is especially important to develop and validate sensitive and selective biomarkers for diseases where the onset of the disease is very slow and/or the disease progression is hard to follow, i.e., osteoarthritis (OA). The degradation of Type II collagen has been associated with the disease state of OA. Matrix metalloproteinases (MMPs) are enzymes that catalyze the degradation of collagen and therefore pursued as potential targets for the treatment of OA. Peptide biomarkers of MMP activity related to type II collagen degradation were identified and the presence of these peptides in MMP digests of human articular cartilage (HAC) explants and human urine were confirmed. An immunoaffinity LC/MS/MS assay for the quantification of the most abundant urinary type II collagen neoepitope (uTIINE) peptide, a 45-mer with 5 HO-proline residues was developed and clinically validated. The assay has subsequently been applied to analyze human urine samples from clinical studies. We have shown that the assay is able to differentiate between symptomatic OA and normal subjects, indicating that uTIINE can be used as potential biomarker for OA. This chapter discusses the assay procedure and provides information on the validation experiments used to evaluate the accuracy, precision, and selectivity data with attention to the specific challenges related to the quantification of endogenous protein/peptide biomarker analytes. The generalized approach can be used as a follow-up to studies whereby proteomics-based urinary biomarkers are identified and an assay needs to be developed. Considerations for the validation of such an assay are described.

Key words: Osteoarthritis, Matrix metalloproteinase, MMP, Immunoaffinity, Antibody-capture, LC–MS/MS, Peptide quantification, Biomarker assay validation

1. Introduction

Osteoarthritis (OA) is a chronic musculoskeletal disease characterized by the degradation and erosion of the articular cartilage in the synovial joints. In OA, an imbalance in matrix synthesis and

Alex J. Rai (ed.), *The Urinary Proteome: Methods and Protocols*, Methods in Molecular Biology, vol. 641,
DOI 10.1007/978-1-60761-711-2_15, © Springer Science+Business Media, LLC 2010

breakdown leads to the destruction and eventual loss of articular cartilage, which results in restricted joint movement, joint instability and pain. Articular cartilage is composed of two major macromolecules, the aggrecan and type II collagen. In OA, collagen type II is degraded by proteolytic enzymes secreted by the chondrocytes and synoviocytes. The secreted collagenases, a sub-family of the matrix metalloproteinase (MMP) family, collagenase-1 (MMP-1), collagenase-2 (MMP-8), and collagenase-3 (MMP-13) are uniquely capable of cleaving the three chains of collagen fibrils at a preferred intra-helical site, which results in the formation of a large (3/4-length) amino-terminal fragment and a short (1/4-length) carboxy-terminal fragment. This cleavage is considered to be the rate-limiting step in the degradation of collagen fibrils. Initial cleavage of type II collagen by collagenases unwinds the triple helical structure rendering it susceptible to further degradation by collagenases, gelatinases, and/or other proteases. MMP-13 is the collagenase with higher affinity for type II collagen and has been implicated in the pathogenesis of OA (1–3).

The discovery and development of biochemical markers that reflect MMP activity in vivo may aid in the development of selective MMP inhibitors as novel therapeutics for OA. In other words, these biomarkers can be used for proof of pharmacology for MMP inhibitors and for use in early decision making during the drug development process. These target/mechanism specific biochemical markers may also be outcome/efficacy markers if shown to correlate with joint structural and functional changes. Moreover, the same collagen degradation markers may predict the likelihood that an individual will exhibit a rapid rate of disease progression and may be useful as diagnostics for early OA. Capitalizing on the knowledge of OA pathology and the involvement of collagenase activity in OA, a targeted approach has been applied to the discovery and development of OA biomarkers. Specifically, the production of neoepitope peptides by the key collagenase MMP-13 was studied in a complex biological system such as urine (4). Urine is the easiest and most practical body fluid to analyze since it does not involve an invasive procedure. Even though urinary levels of biomarkers reflect systemic metabolism and contribution of the biomarker from a single joint might represent a small portion of the measured signal, its use as a matrix for biomarkers quantification for oral therapies is widely accepted. A large variation in the concentration of biomarkers in urine because of individual variations in the urinary volume output creates an additional challenge for robust assay development. Therefore, urinary levels of biomarkers should always be normalized for creatinine levels. After identification of the most abundant collagen type II neoepitope (TIINE) peptide, a 45-mer with five hydroxy-proline residues, an immunoaffinity LC–MS/MS assay was developed (4). The 45-mer uTIINE peptide is concentrated from human urine samples by using an online immunoaffinity-based sample preparation methodology. A d_5- labeled isomer of the 45-mer uTIINE peptide is used as the internal standard. Prior to analysis the pH of the urine is buffered to neutral pH to facilitate the capture of the peptides on the immunoaffinity column. After concentration and washing, the peptides are released from the column using an acidic eluent at pH ~2.5 and undergo further separation using reverse phase HPLC before detected using electrospray mass spectrometry operated in the selected reaction monitoring mode.

Validation of a bioanalytical assay for its intended purpose is critical to assure data quality and appropriate application of the assay in decision making. While validation guidelines from the regulatory agencies specific to biomarker assays are not available at this time, existing guidelines applied for LC–MS/MS based chromatographic assays can be adopted (5–10). The method described here was validated for measuring the uTIINE biomarker concentration in human urine (11). The clinical validation of uTIINE assay included both the technical validation of the immunoaffinity LC–MS/MS measurement and biologic validation of the uTIINE peptide as a biomarker relevant to OA disease state. The technical validation comprised the determination of the linear range of quantification, intra- and inter-assay accuracy and precision, selectivity, recovery, and evaluation of stability for storage and process conditions. Assay selectivity and consistent matrix and concentration independent recovery was also demonstrated. The biologic validation of the assay included the understanding of the uTIINE peptide biomarker's intra- and inter-subject variability in female and male subjects and in normal and OA populations (11). Relatively large inter-subject variability (>60% CV) was observed, while the intra-subject variability was significantly lower (<30% CV), indicating that changes in the level of this biomarker may be a better indication of subject disease status or response to drug treatment than absolute concentrations. The assay was subsequently used in a clinical study to understand the differentiation capability of the uTIINE biomarker between age/sex matched normal, patients with confirmed radiographic OA in the hip(s)/knee(s) or spine and subjects with symptoms of OA that cannot be confirmed by radiographic measurements. Creatinine normalized uTIINE 45-mer peptide levels in the symptomatic OA group were 100% greater compared to the mean normalized uTIINE values in the non OA group (P=0.0012). Mean uTIINE levels were similar in the hip(s)/knee(s) radiographic OA (ROA), hand(s)/spine ROA and non OA groups (12). Finally, the correlation of uTIINE with joint space narrowing (JSN), the currently accepted surrogate biomarker for OA, has also been studied in patients with knee OA undergoing doxycycline treatment (13). This study found that although uTIINE was not a consistent predictor of JSN in subjects with knee OA, serial measurements of uTIINE levels in longitudinal studies reflected concurrent JSN.

In this chapter, the procedures to run the clinically validated immunoaffinity LC–MS/MS assay for a collagen type II neoepitope (TIINE) peptide are described. This assay provides a specific example for the development and validation of an assay for protein/peptide-based biomarkers in urine. Details and considerations for technical, as well as, biological validation are described. Such an approach can be generalized and used for other protein/peptide-based markers that are identified from large scale biomarker discovery efforts.

2. Materials

2.1. Antibody Column Preparation

1. Monoclonal antibodies raised against the GEPGDDGPS sequence portion of human collagen type II, purified using protein G (mAB 5109 LN15828) purchased from American Type Culture Collection (ATCC) (Manassas, VA).
2. Poros™ Epoxide immunodetection cartridge, 50 mm (Applied Biosystems, Framingham, MA).
3. Mixture of 0.5 M sodium sulfate and 0.1 M sodium phosphate (Sigma, St. Louis, MO) (pH = 8.5–9.0) in HPLC water.
4. 0.2 M Tris–HCl (pH = 10.5) prepared in HPLC water.
5. Standard HPLC equipment with a UV detector at 280 nm.

2.2. Calibration Standards, Internal Standard

1. Stock solution of calibration standard of 45-mer uTIINE peptide with five HO-proline residues (American Peptide Company, Sunnyvale, CA) prepared in acetonitrile:water (1:1, v/v) at a concentration of 0.5 mg/mL. This stock solution can be stored at –20°C for 2 months (see Fig. 1 for the structure of the standard peptide and its stable isotope labeled internal standard).
2. Stock solution of the internal standard (IS): d5-labeled standard of 45-mer uTIINE peptide with 5-HO-proline residues (American Peptide Company, Sunnyvale, CA), prepared in acetonitrile:water (1:1 v/v) at 0.5 mg/mL. This stock solution can be stored at –20°C for 3 months (see Note 1).

2.3. Quality Control

1. Urine samples (100–200 mL of 24 h collection) from 20 subjects, representative of the population (age, sex, body mass index (BMI)) were used to measure the uTIINE levels as well as to prepare a pooled urine matrix. Stability of the uTIINE peptide in human urine has been confirmed at –70°C for up to 1 year.

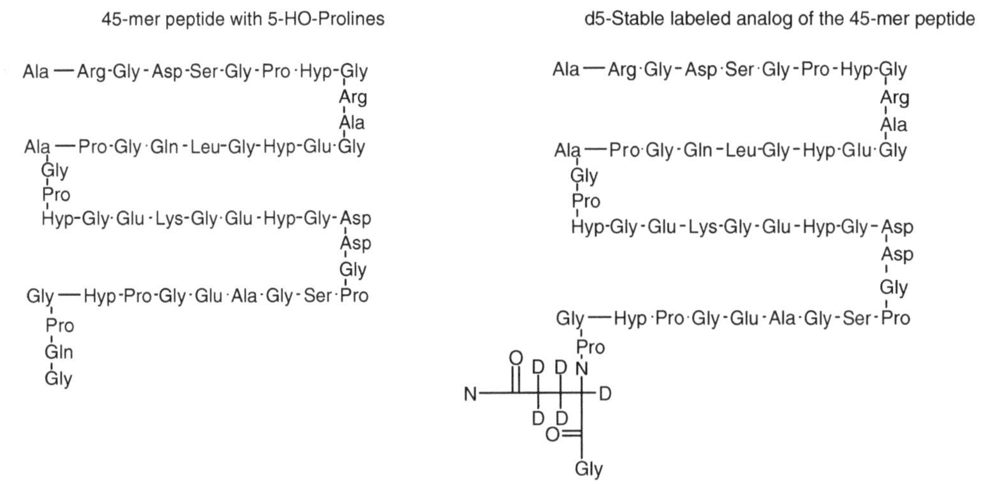

45-mer peptide with 5-HO-Prolines

Ala—Arg-Gly-Asp-Ser-Gly-Pro-Hyp-Gly
Arg
Ala
Ala—Pro-Gly-Gln-Leu-Gly-Hyp-Glu-Gly
Gly
Pro
Hyp-Gly-Glu-Lys-Gly-Glu-Hyp-Gly-Asp
Asp
Gly
Gly—Hyp-Pro-Gly-Glu-Ala-Gly-Ser-Pro
Pro
Gln
Gly

d5-Stable labeled analog of the 45-mer peptide

Ala—Arg-Gly-Asp-Ser-Gly-Pro-Hyp-Gly
Arg
Ala
Ala—Pro-Gly-Gln-Leu-Gly-Hyp-Glu-Gly
Gly
Pro
Hyp-Gly-Glu-Lys-Gly-Glu-Hyp-Gly-Asp
Asp
Gly
Gly—Hyp-Pro-Gly-Glu-Ala-Gly-Ser-Pro
Pro
O D D N
N— —D
D D
O=
Gly

Hyp = HO-Proline

Fig. 1. Structures of the 45-mer uTIINE peptide with 5 HO-Proline (Hyp) residues and its stable isotope labeled analog used as internal standard (IS).

2.4. Sample Preparation Buffers

1. Ammonium acetate (EM Science, Gibbston, NJ, purity: 99%) 25 mM, pH = 6.6–7.4), store at room temperature (RT) for up to a month.
2. Bovine serum albumin (BSA, Fatty Acid Ultra Free, Roche, Indianapolis, IN), 10 μg/mL, store at 4°C for 3 month.

2.5. HPLC Separation Buffers

1. HPLC loading pump eluent A: 0.5% formic acid (Mallinkrodt, Paris, KY) in HPLC water, store at RT for a maximum of 1 month.
2. HPLC loading pump eluent B: 25 mM ammonium acetate (EM Science, Gibbston, NJ) in HPLC water, store at RT for a maximum of 1 month.
3. HPLC analytical pump eluent A: 0.2% formic acid (Mallinkrodt, Paris, KY) in HPLC water, stored at RT for a maximum of 1 month.
4. HPLC analytical pump eluent B: acetonitrile (Sigma, St. Louis, MO) with 0.2% formic acid (Mallinkrodt, Paris, KY), stored at RT for a maximum of 1 month.
5. Autosampler needle wash solvent 1: methanol: water: formic acid (600:400:2 v: v: v), may be stored at RT up to 1 month.
6. Autosampler needle wash solvent 2: acetonitrile with 0.2% formic acid may be stored at RT for up to 1 month.

2.6. Creatinine Determination

1. Colorimetric assay kit for the quantification of creatinine in human urine (Quidel Corporation, San Diego, CA).

2.7. Other Laboratory Equipment

1. 2 mL 96-well, deep well polypropylene assay plate (VWR Scientific, Costar).
2. 96-well plate sealer (Velocity 11, Plateloc).
3. Eppendorf adjustable pipettes (2–20, 10–100, 100–1,000 μL) with disposable tips (VWR Scientific).
4. MetaSil AQ, C-18, 100 × 2.0 mm, 5 μm HPLC column (Varian).
5. 0.5 μm pre-column filter (Phenomenex 0.5 μ).
6. 3 × 8 mm peptide trap cartridge and holder (Michrom).
7. 10-port switching valve (Valco).
8. Standard multitude vortexer (VWR).
9. Safe-lock 1.5 mL polypropylene microcentrifuge tubes (snap-cap type) (Eppendorf, VWR).
10. 5 mL glass vials (screw-cap) (VWR).
11. CTC PAL autosampler 5000 IL (Shimadzu) or equivalent.
12. Four LC-10ADvp HPLC pumps with two SCL-10Avp controllers (Shimadzu) or equivalent.
13. Sciex API 4000 triple quadrupole mass spectrometer (Applied Biosystems Inc.) or equivalent.
14. PEEK tubing, 1/16 OD × 0.005, 0.01, and/or 0.020 inches ID (Upchurch Scientific Inc.).
15. 2.5 mL syringe (Hamilton).
16. Plate reader capable of optical density readings between 0.0 and 2.0 for determination of creatinine using a standard assay kit.

3. Methods

3.1. Peptide Identification by LC/MS/MS

Samples were analyzed on a Q-TOF (quadrupole time-of-flight) mass spectrometer (Waters, Beverly, MA) coupled to a CapLC (Waters, Milford, MA) system equipped with an autosampler, gradient and auxiliary pump. Five microliters of sample was injected via "microliter pickup" mode and desalted and concentrated online through a peptide CapTrap trapping cartridge (Michrom BioResources, Auburn, CA). The samples were desalted at high flow (20 μL/min) for 3 min. The peptides were separated on a Pepmap column (75 μm × 150 mm, 3 μm particle) (LC Packings, San Francisco, CA) prior to introduction into the mass spectrometer. Separation was achieved using a 50 min linear gradient of 95% H_2O, 0.1% HCOOH to 95% MeCN, and 0.1% HCOOH. The flow rate was also increased linearly with organic mobile phase from 250 to 400 nL/min.

The data was processed by ProteinLynx version 3.5 (Waters, Beverly, MA) to generate searchable .pkl files. These files were searched using the search engine MASCOT (Matrix Science, Ltd., London, UK). The data was searched against NCBI nonredundant or a custom database containing relevant cartilage matrix proteins with no enzyme specificity to identify endogenously produced peptides. Several variable modifications, including methionine oxidation and proline or lysine hydroxylation were added to the search parameters. All identifications of peptides exceeded the significant homology threshold. The sequences of the peptides identified can be found in ref. (4).

3.2. Creatinine Analysis

The creatinine concentration in the urine samples was determined using a colorimetric assay procedure with a standard enzyme immunoassay kit (Metra) from Quidel Corporation (San Diego, CA). The assay kit was validated in house for this application and the measurements were performed using 20 μL urine following the procedures described in the kit insert without any modifications.

3.3. Preparation of Antibody Column for Immunoaffinity LC–MS

Online immobilized antibody (immunoaffinity) columns are prepared as follows:

1. Concentrate antibodies to greater than 2 mg/mL and dialyze into 0.5 M Na_2SO_4, 0.1 M sodium phosphate (pH = 8.5–9.06) using a 10 kDa cutoff membrane (Pierce, Rockford, IL). The antibody solutions can be stored at −70°C for at least 6 months.
2. The antibody is immobilized onto a Poros™ Epoxide immunodetection cartridge (Applied Biosystems, Framingham, MA) using a standard HPLC system. The cartridge is equilibrated with 0.5 M Na_2SO_4, 0.1 M sodium phosphate (pH ~8.8) at 1 mL/min. The concentrated antibody solution (0.5–1 mL) is recirculated through the cartridge at 1 mL/min and the A280 absorbance is monitored.
3. After the A280 stabilized, a volume of 1.5 M Na_2SO_4, 0.1 M sodium phosphate (pH ~8.8) equal to the initial antibody volume is added and recirculated until the A280 stabilizes again.
4. The 1.5 M Na_2SO_4, 0.1 M sodium phosphate solution is added until the A280 is less than 10% of the maximum A280 and recirculated for an additional 30 min.

5. The column is then removed and incubated at RT for 24–72 h. After incubation, the remaining epoxide sites are blocked with 0.2 M Tris base (pH ~10.5) at RT for 2 h.
6. Test the performance of the antibody columns before routine assay use by running a full calibration curve and testing for the linearity of the concentration response as well as for minimum carry over (see Notes 2 and 3).

3.4. Preparation of Calibration Standards and Internal Standard Spiking Solution

1. Working solutions of calibration standards are prepared by further diluting the 0.5 mg/mL stock solution to concentrations of 750, 500, 250, 125, 2.50, 31.3, and 15.6 ng/mL in methanol:water (1:9, v/v) using serial dilutions. These solutions are prepared fresh prior to each assay run.
2. Final concentrations of the calibration standards at concentrations of 7.50, 5.00, 2.50, 1.25, 0.625, 0.313, and 0.156 ng/mL are prepared by diluting the working solutions of the calibrating standards 1:100 in 10 μg/mL BSA solutions (see Note 4). Figure 2 shows representative calibration curves obtained on three separate days demonstrating the linearity of the dynamic range and the reproducibility of the concentration response (see Note 5).
3. To prepare the final spiking solution of the IS, the stock solution is diluted in 10 μg/mL BSA solution to final concentration of 40 ng/mL. This solution is prepared fresh prior to each assay run.

3.5. Preparation of Quality Control Samples

1. Quality control (QC) samples are prepared by spiking known amounts of the 45-mer uTIINE peptide into a pooled urine matrix that is collected from subjects, representative of the population in which the assay is applied.

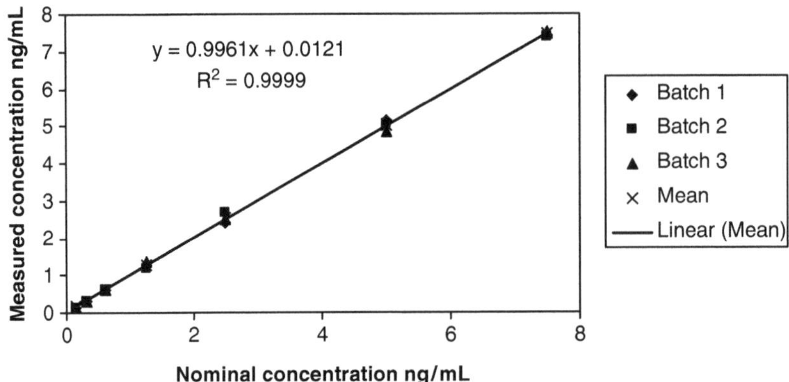

Fig. 2. Representative calibration curves showing the nominal and back calculated standard concentrations of the 45-mer uTIINE peptide in three validation batch runs. The equation of the linear regression fit and the regression coefficient is shown for the mean of the three calibration curves.

2. The pooled urine matrix is prepared by collecting 24 h urine samples from at least 20 healthy subjects of the same age as the studied population (see Note 6).
3. Basal concentrations of the 45-mer uTIINE peptide in the individual urine samples are determined using the immunoaffinity LC–MS/MS method.
4. Urine samples from at least 7–10 subjects that have the lowest level of 45-mer peptide basal concentration are pooled by mixing equal amounts of urine from each of these subjects. The mixture needs to be mixed well using a stirrer or vortexer. The endogenous baseline concentration in this pooled matrix is then measured again with the immunoaffinity LC–MS/MS method.
5. Prepare a 1,000 ng/mL 45-mer uTIINE peptide working standard by diluting the standard peptide stock solution in methanol:water 1:9 v/v.
6. Quality control samples at five concentration levels (denoted as LLQC1, LLQC2, LQC, MQC and HQC) covering the full range of the intended calibration curve are prepared using the pooled urine matrix as follows. Into a 250 mL volumetric flask, add about 100 mL pooled urine and an adequate amount of the 1,000 ng/mL QC working standard to prepare final concentrations of LLQC1 = endogenous baseline (EB) diluted twofold, LLQC2 = EB + 0 ng/mL, LQC = EB + 0.3 ng/mL, MQC = EB + 3 ng/mL, and HQC = EB + 6 ng/mL. Add pooled urine to 250 mL and mix well. Aliquot 7 mL into a 10 mL labeled plastic tube, to be used in validation experiments and study samples assays. LLQC1 QC samples were prepared fresh at each batch run by diluting the LLQC2 with HPLC Grade water 1:1. These QCs are used for validation, sample assay and long-term stability assessment. All the QC samples are stored at −70°C. Freeze/thaw stability of the 45-mer TIINE peptide in human urine matrix for up to three F/T cycles at −70°C had been established. Representative data on the back-calculated concentrations of the QC samples in three separate batch runs relative to their nominal concentrations is shown in Fig. 3 (see Note 7).

3.6. Preparation of Samples for Analysis

1. Before beginning the sample preparation procedure, remove samples and adequate number of standard and QC matrix samples, as well as 45-mer TIINE peptide and IS peptide stock solutions from the freezer. Allow samples and QCs to thaw unassisted at RT. Vortex the samples and QCs for 30 s. All the samples and QCs are centrifuged for 4 min under 4,000 rpm = 3,220 g-force or rcf (relative centrifugal force) at RT.

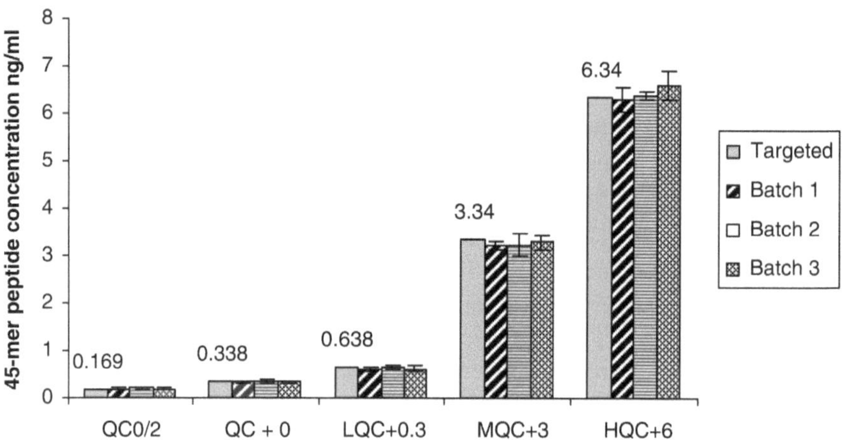

Fig. 3. Back-calculated concentrations of the quality control samples in three validation batch runs. QC samples were prepared at concentrations of 0.169, 0.338, 0.638, 3.34, and 6.34 ng/mL and assayed in replicates of 5 at each concentration level.

2. Aliquoting into 96-well plate

   Blanks: There different blank samples are prepared and run in the assay to assess chromatographic carry over, cross-talk of the internal standard with the standard curve matrix and to assess the cross-talk of the internal standard with the pooled biological matrix. These three blank samples are prepared as follows:

   (a) Standard curve matrix (10 μg/mL BSA) with IS added (carry over check injected after the highest calibrating standard).
   (b) Standard curve matrix 10 μg/mL BSA without IS added.
   (c) Pooled QC matrix without IS added.

   Calibrating standards: To a well of a 2 mL-96-plate, add 990 μL of 10 μg/mL BSA followed by 10 μL of the corresponding working standard to prepare the final concentration as described in Subheading 3.4, step 2. Calibrating standards are prepared fresh prior to each assay run.

   Quality control samples: To a well of a 2 mL-96-plate, add 1 mL of the corresponding quality control sample.

   Samples: To a well of a 2 mL-96-plate, add 1 mL of the urine sample.

3. Sample dilution, if required:

   The twofold diluted sample is prepared by adding 500 μL urine sample and 500 μL HPLC Grade water. The fourfold diluted sample is prepared by adding 250 μL urine sample and 750 μL HPLC Grade water (see Note 8).

4. Sample processing prior to injection for analysis

   Samples are processed as follows: Add 50 μL IS solution (40 ng/mL d5-labeled 45-mer uTIINE peptide) to all sample-wells, except Blank 1 and Blank 3. Add 1 mL 25 mM ammonium acetate to all the wells and mix well. A Tomtec Quadra 96 liquid handling robot can be used for mixing. Inject 900 μL from each vial into the LC/MS/MS system for analysis. The bench top stability of unprocessed urine samples thawed at RT was confirmed up to 24 h. The stability of processed samples has been established for up to 48 h at RT in the autosampler 96-well plate.

3.7. Immunoaffinity LC–MS/MS Analysis

The immunoaffinity LC–MS/MS quantification method was validated on a system utilizing four LC-10ADvp HPLC pumps (Shimadzu) operated by two SCL-10Avp controllers (Shimadzu), a CTC- PAL autosampler (Shimadzu), and a 10-port switching valve (Valco). The chromatographic system was interfaced to an API 4000 mass spectrometer (MDS-Sciex, Toronto, Canada) operated in the positive ion electrospray selected reaction monitoring (SRM) mode. Any equivalent quaternary HPLC system capable of injecting 1–2 mL sample and interfaced with a triple quadrupole mass spectrometer can be adapted for this assay. Typically, 900 μL of urine spiked with internal standard peptide is injected onto an immunoaffinity column. The switching valve configurations are shown in Fig. 4.

1. Chromatographic separation: Prior to starting sample analysis the antibody column is washed at 1.2 mL/min with 25 mM NH_4OAc, at pH ~7. After 3.5 min of washing, the valve is

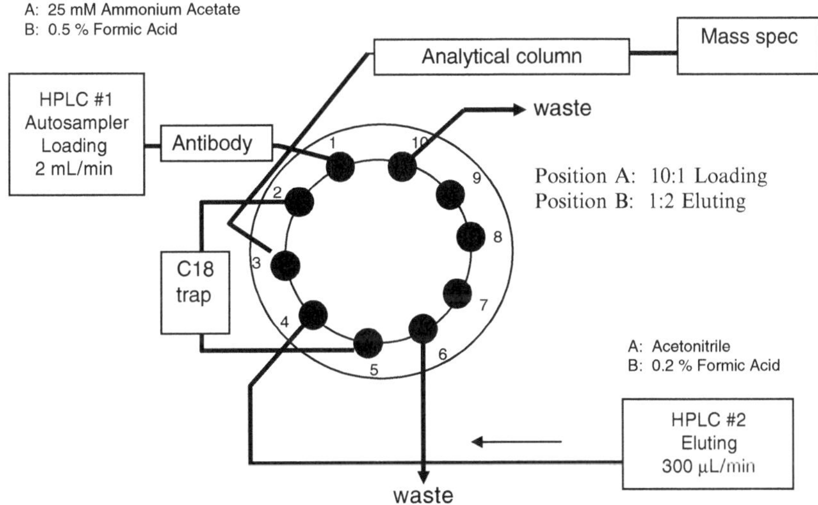

Fig. 4. Schematics of the switching valve set-up.

switched and the captured peptides are eluted off the immunoaffinity column with a 0.5% formic acid solution at a flow rate of 2.0 mL/min onto a C-18 peptide 3×8 mm peptide trapping column (Michrom BioResources, Auburn, CA) for an additional 3 min. The valve is switched back to the original position and peptides are then eluted off the trapping column onto a 2.0×100 mm, 5 μm, C-18 analytical column (Metasil AQ) with the following gradient flow at 0.3 mL/min: 95% H_2O with 5% of 0.5% HCOOH to 15% MeCN, 0.5% HCOOH in 2 min and 55% in an additional 3 min. At the end of the program, the analytical column is re-equilibrated with 95% water and 5%, 0.5% Formic acid for 2 min. The gradient programs are summarized in Tables 1 and 2 for the immunoaffinity capture and analytical separation.

Table 1
HPLC Program for the loading pumps for the immunoaffinity capture

Time [min]	Flow rate [mL/min]	Mobile phase A (0.25 mM NH4OAc)	Mobile phase B (0.5% HCOOH)
0.0	1.2	100	0
3.5	1.2	100	0
3.6	2.0	0	100
9.7	2.0	0	100
9.8–13.0	2.0	100	0

Table 2
HPLC gradient program for the analytical separation

Time [min]	Flow rate [mL/min]	Mobile phase A (water with 0.2% HCOOH)	Mobile phase B (acetonitrile with 0.2% HCOOH)
0	0.3	95	5
6.0	0.3	95	5
7.0	0.3	85	15
8.0	0.3	85	15
8.1	0.3	45	55
11.0	0.3	45	55
11.1	0.3	95	5
13.0	0.3	95	5

Table 3
Selected reaction monitoring (positive ion mode) parameters

Analyte	Q1 m/z (M + 4H)	Q3 m/z	Dwell time (ms)	Declustering potential	Collision energy	CXP
uTIINE 45-mer	1,039.4	568.4	200	80	43	12
Labeled IS	1,040.4	573.4	150	80	43	12

2. Mass spectrometry detection: The 45-mer uTIINE peptide and its internal standard are specifically detected by monitoring HPLC elution times and ion pairs corresponding to the precursor and specific product ion mass-to-charge ratios (m/z). Peptide abundances are determined by comparing the LC–MS/MS peak areas of the analyte at m/z 1039.4/568.4 to that of the deuterated internal standard at m/z 1040.4/573.4. These mass transitions were determined on a tuned and calibrated mass spectrometer and need to be verified on the actual system being used. Typical tuning conditions for the mass spectrometer are summarized in Table 3. The selectivity of the method is inherently derived from the combination of the immunoaffinity capture, chromatographic separation coupled with precursor/product ion specific mass selective detection. During the method validation experiments, the selectivity of the method has been confirmed (see Note 9).

3.8. Data Acquisition and Reduction

Data acquisition and integration of the peak areas of the SRM signal for the 45-mer uTIINE peptide and its stable labeled internal standard are performed using the standard quantitative analysis package of the Analyst 1.2 software (Applied Biosystems, MDS Sciex, Toronto, CA). All regression analysis and sample quantification can be performed using the Analyst 1.2 software or the regression utility of the Watson LIMS database software. The calibration curve is constructed based on the response ratio of peak area ($P_{analyte}/P_{IS}$) versus nominal standard concentrations by least squares linear regression using a weighting factor of 1/concentration. Concentrations of 45-mer peptide are determined using response ratio from unknown samples and the linear regression curve. Example of a typical chromatogram is shown in Fig. 5 for the lowest calibration standard. Figure 6 shows representative creatinine normalized uTIINE data from a clinical study comparing subjects diagnosed with symptomatic OA, knee/hip and hand/spine radiographically confirmed OA, and non OA subjects.

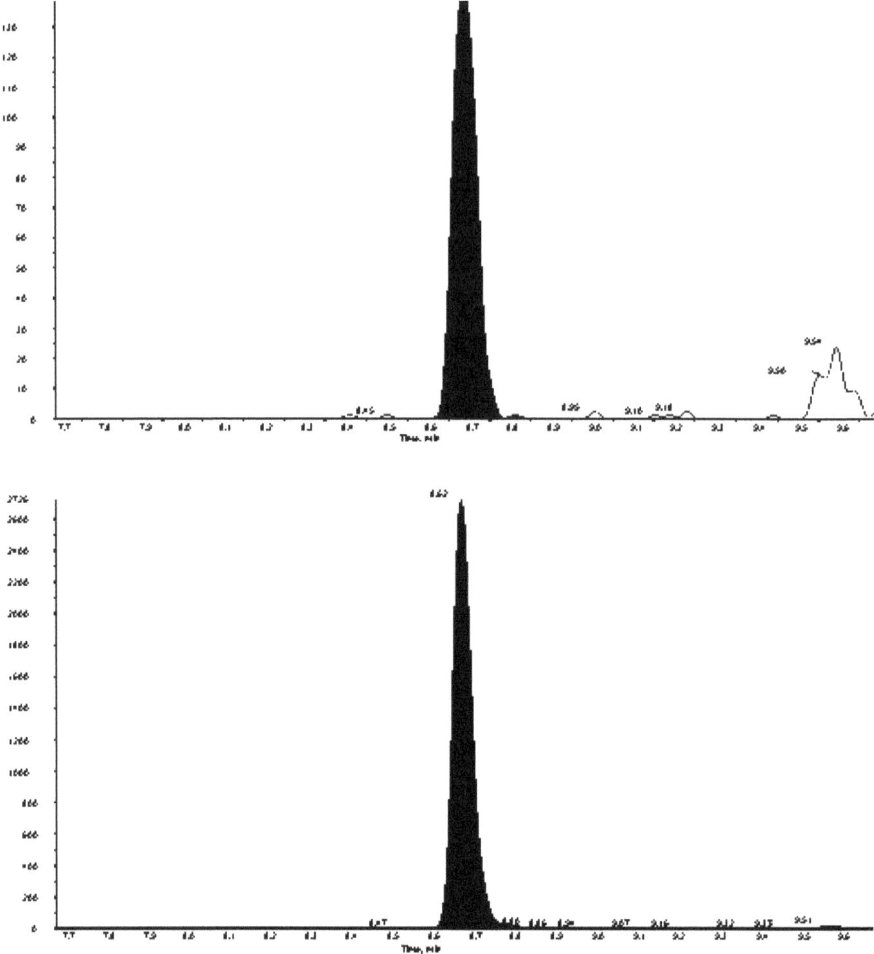

Fig. 5. Representative chromatogram at the LLOQ standard concentration (0.156 ng/mL) (45-mer uTIINE peptide standard, *top trace*) and IS (*bottom trace*).

4. Notes

1. Peptide and IS stock solutions are stored in glass vials to avoid loss due to adhesion to plastic surfaces.
2. The immunoaffinity capture using 5109 monoclonal antibody column is essential to enrich the peptide from urine to reach quantifiable levels. The antibody columns are custom prepared in-house and the reproducibility and robustness of these columns are evaluated prior to use by testing the curve linearity and carry over by running a full calibration curve and a set of urine-based QCs on each column. The performance of the columns was found to be very robust. Over 1,000 sample injections can be made on a single column without loss of performance. When not in use the columns need to be washed

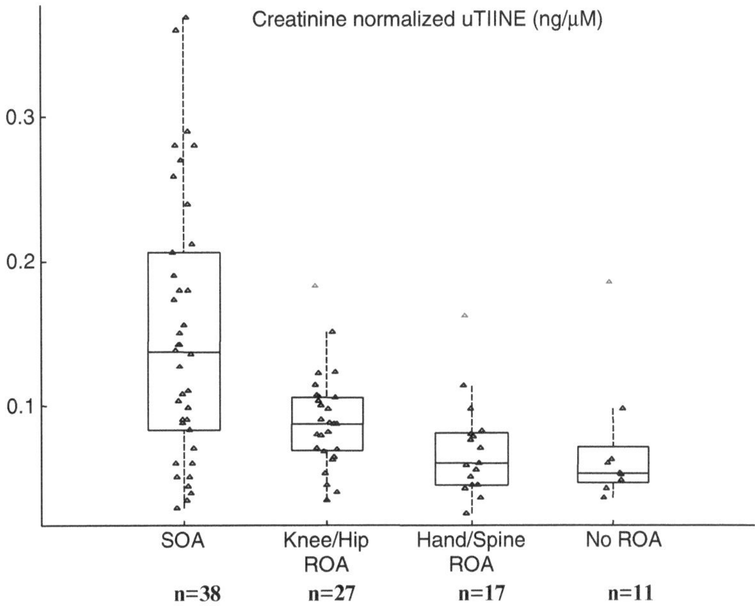

Group Comparisons	P-value* *P-value based on one-way ANOVA P< 0.004 is significant
Symp. vs Knee/Hip	0.0068
Symp. vs Hand/Spine	0.0001
Symp. vs No ROA	0.0012
Knee/Hip vs Hand/Spine	0.0965
Knee/Hip vs. No ROA	0.1623
Hand/spine vs. No ROA	0.9572

Fig. 6. Representative data from a clinical study on the differentiation capability of uTIINE 45-mer peptide between patients with symptomatic OA (SOA), knee/hip or hand/spine Radiographic OA (ROA) and normal subjects.

and equilibrated with ammonium acetate buffer and stored in the refrigerator at 4°C.

3. Carry over needs to be evaluated for all chromatographic assays. During method development, it was found that the antibody column can be a major source of carry over. By increasing the strength and timing of the acidic eluent, the carry over was reduced to 1%. The impact of the carry over on the accuracy of the results was investigated and was determined to be negligible for most representative study samples in the linear range of the assay. However, to minimize any potential carry over contribution to sample concentrations near the Lower Limit of Quantification (LLOQ), samples with concentrations that are 20-fold lower than the previously

injected sample concentration should be re-analyzed to ensure that the carry over is not significantly contributing to their reported concentration values.

4. Selection of Standard Curve Matrix

   In bioanalytical assays, the standard curve is usually prepared in a biological matrix that is representative of the study samples. However, in the case of endogenous biomarker assays, analyte free matrix is not available. Therefore, assay accuracy measurements must take into account the endogenous basal analyte concentration or a substitute matrix free or stripped of endogenous analytes that should be used to prepare the calibration curve. In any case, matrix-based QC samples should be used to test the accuracy and precision of the quantification. During method optimization, several substitute matrices such as ammonium acetate buffer, 20-fold diluted urine and aqueous solutions of BSA protein were investigated. Adequate peptide recovery, ionization response comparable to urine samples and linearity for accurate quantification was found using the BSA protein solutions with concentrations ranging from 5 to 100 μg/mL. To avoid introducing excess protein into the system, 10 μg/mL BSA was selected as standard curve matrix.

5. During the method validation experiments, the LLOQ for the 45-mer uTIINE peptide as determined by suitable accuracy and precision was defined as 0.156 ng/mL based on an injection volume of 900 μL buffered urine sample. The assay range of 0.156–7.50 ng/mL was set based on the anticipated study sample concentrations and the acceptable signal-to-noise ratio and accuracy and precision at the LLOQ.

6. Since the collagen turnover in the body is dependent on age and sex, matching the age and sex of the subjects is important.

7. Quality control (QC) samples prepared in a representative pooled biological matrix were used to determine the intra- and inter-assay accuracy and precision of the assay in three separate batch runs. The intra-run precision in three batch runs with $n = 5$ QC replicates at each QC level was 3.2–12.4% CV and 14.7% CV at LLQC of 0.169 ng/mL, while the intra-run accuracy was 0–7.6% RE and 19.5% RE at LLQC of 0.169 ng/mL. The inter-run precision was in the range of 4.0–8.7% CV and 12.3% CV at LLQC of 0.169 ng/mL, and inter-run accuracy in the range of 0.6–2.7% RE and 14.2% RE at LLQC of 0.169 ng/mL.

8. Dilution linearity of the assay was independently tested prior to the validation using two distinct methods. For the first test, 20.72 ng/mL of 45-mer uTIINE peptide was spiked

into pooled human urine that had a basal concentration of 0.886 ng/mL, then diluted to 2-, 4-, 10-, and 20-fold, with three replicates at each concentration level. For the second test, four urine samples were selected from a clinical study, then, each sample was diluted two and fourfold separately with three replicates at each concentration level. Results demonstrated that the dilution of human urine samples was concentration and matrix independent and the accuracy of the back calculated concentration was <15% RE with better than 15% precision.

9. The intrinsic selectivity of the assay was provided by a combination of antibody affinity capture, HPLC separation and mass selective detection. Triple quadrupole mass spectrometry "cross-talk" between the 45-mer uTIINE peptide and its d_5- labeled isomer (IS) was found negligible. The Q1 full MS Scan from 800 to 1,300 m/z was performed for selected blank, standard and human urine samples, and the precursor ion of the 45-mer peptide was confirmed in the human urine sample.

References

1. Huebner, J.L., Otterness, I.G., Freund, E.M., Caterson, B. and Kraus V.B. (1998) Collagenase 1 and collagenase 3 expression in a guinea pig model of osteoarthritis, *Arthritis Rheum* **41**, 877–890.
2. Kraus, V.B., Stabler, T.E., Le, T., Saltarelli, M. and Allen, N.B. (2003) Urinary type II collagen neoepitope as an outcome measure for relapsing polychondritis, *Arthritis Rheum* **48**, 2942–2948.
3. Poole, A.R. (2003) Biochemical/immunochemical biomarkers of osteoarthritis: Utility for prediction of incident or progressive osteoarthritis, *Rheum Dis Clin North Am* **29**(4), 803–818.
4. Nemirovskiy, O.V., Dufield, D.R., Sunyer, T., Aggarwal P., Welsch, D.J. and Mathews, W. R. (2007) Discovery and development of a type II collagen neoepitope (TIINE) biomarker for MMP activity: From in-vitro to in-vivo. *Anal Biochem* **391**, 93–101.
5. Kindt, E., Shum, Y., Badura, L., Snyder, P.J., Brant, A., Fountain, S. and Szekely-Klepser, G. (2004) Development and validation of an LC/MS/MS procedure for the quantification of endogenous *myo*-inositol concentrations in rat brain tissue homogenates, *Anal Chem* **76**(16), 4901–4908.
6. Li, W. and Cohen, L. (2003) Quantitation of endogenous analytes in biofluid without a true blank matrix, *Anal Chem* **75**, 5854.
7. Jemal, M., Schuster, A. and Whigan, D.B. (2003) Liquid chromatography/tandem mass spectrometry methods for quantitation of mevalonic acid in human plasma and urine: Method validation, demonstration of using a surrogate analyte, and demonstration of unacceptable matrix effect in spite of use of a stable isotope analog internal standard, *Rapid Commun Mass Spectrom* **17**(15), 1723–1734.
8. Szekely-Klepser, G., Wade, K., Woolson, D., Brown, R., Fountain, S. and Kindt, E. (2005) A validated LC/MS/MS method for the quantification of pyrrole-2,3,5-tricarboxylic acid (PTCA), a eumelanin specific biomarker, in human skin biopsies, *J Chromatogr B* **826**, 31–40.
9. Conference on "Analytical methods validation: Bioavailability, bioequivalence, and pharmacokinetic studies" held in December 1990 in Crystal City, VA, and sponsored by the FDA, AAPS, HPB of Canada, AOAC, and Federation International Pharmaceutique.
10. "AAPS Workshop on Bioanalytical Methods Validation- A Revisit with a decade of progress" held in January 2000 in Crystal City, VA cosponsored with the FDA. Guidance for Industry, Bioanalytical method validation", issued by U.S. Department of Health and Human Services, Food and Drug Administration, Center for Drug Evaluation and Research (CDER) Center for Veterinary Medicine (CVM) in May 2001.

11. Li, W., Nemirovskiy, O.V., Fountain, S., Mathews, W.R. and Szekely-Klepser, G. (2007) Clinical validation of an immunoaffinity LC-MS/MS assay for the quantification of a collagen type II neoepitope peptide: A biomarker of matrix metalloproteinase activity and osteoarthritis in human urine, *Anal Biochem* **369**, 41–53.
12. Pickering E., Szekely-Klepser G., Nemirovskiy O., Li W., Brown R., Sunyer T. and Hellio Le Graverand M.-P. (2004) Urinary type II collagen neo-epitope (uTIINE): A marker of osteoarthritis activity, *Osteoarthr Cartil* **12**(Suppl. B) S93.
13. Hellio Le Graverand, M-P., Brandt, K.D., Mazzuza, S.A., Katz, B.P., Buck, R., Lane, K.A., Pickering, E., Nemirovskiy, O.V., Sunyer, T. and Welsch, D.J. (2006) Association between concentrations of urinary type II collagen neoepitope (uTIINE) and joint space narrowing in patients with knee osteoarthritis, *Osteoarthr Cartil* **14**(11), 1189–95.

Chapter 16

Cell-Specific Biomarkers in Renal Medicine and Research

Martin Shaw

Abstract

Histopathology is the gold standard for defining renal injury, but it is invasive, time-consuming and expensive, plus it is seldom used in subjects with mild renal injury. Using biomarkers linked to distinct, defined cell types and tissues provides a direct link to histopathology without its drawbacks, plus it provides increased sensitivity, and specificity. The nephron consists of several sections, each with its own specific biomarkers; therefore, by the use of a battery of tests injuries can be localised to distinct areas of it. Using urine samples simplifies repeated sampling from the same subject or animal leading to better defined toxicokinetics and disease monitoring.

Serum creatinine is the most widely used renal biomarker in spite of its known shortcomings. Cell-specific biomarkers are more specific and sensitive and have been known for over 40 years, but they are still underused in renal medicine and research. In particular, while many studies have shown cell-specific biomarkers to be valuable in diagnosis, there are few studies where they have been used to guide therapy or linked to quantitative changes in the kidney. Furthermore, the great majority of cell-specific biomarkers are from the proximal tubule, which may have hindered research into the study of conditions where the distal tubules are affected. Recently, the range of biomarkers and their applications has been expanded by the introduction of indicators of cellular regeneration.

This chapter will discuss how using biomarkers with a known cellular origin, renal effects may be found earlier and at lower levels of injury. Their use in both renal medicine and drug research will be presented. Knowledge of these existing markers lays the foundation for evaluation, comparison, and characterisation of new markers that will be identified in the future.

Key words: Renal physiology, Renal disease, Kidney biomarkers, Multiple markers, Cell-specific markers

1. Introduction

Renal diseases are an expanding problem, half a million Americans could be on dialysis in 2010 at a cost of $46 billion dollars a year (1). Furthermore, drug-induced renal injury is the second most common cause of toxicity-related failure in drug development. Renal injury may be an under-recognised problem due to the

Alex J. Rai (ed.), *The Urinary Proteome: Methods and Protocols*, Methods in Molecular Biology, vol. 641,
DOI 10.1007/978-1-60761-711-2_16,

ability of the kidneys to regenerate and the great reserve capacity of the kidneys (2). However, as renal injury continues, the kidney's ability to compensate may be exceeded when a further renal injury may tip the subject into renal failure. The commonest test of renal function is serum creatinine. Creatinine is produced by the muscles at an approximately constant rate and mainly removed from the blood by glomerular filtration; therefore, the serum level is approximately inversely related to the glomerular filtration rate (GFR), the rate at which plasma is filtered by the glomeruli. However, serum creatinine is affected by many non-renal factors (3) and, furthermore, it is a late biomarker as there must be a considerable loss of glomerular function before significant increases occur (4). Finally, most toxins affect the renal tubules and, even in cases of glomerular injury, changes in renal tubular biomarkers are important diagnostic indicators (5). Because of the above, the measurement of biomarkers of renal tubular injury and function are important in the understanding and monitoring of renal effects.

1.1. Urinary Proteins

Urinary proteins can be derived from several sources. In the normal kidney, proteins with a molecular weight below about 40 kDa are freely filtered across the glomerular membrane and then reabsorbed in the proximal tubules. This reabsorption is over 99% efficient (6); therefore, even a tiny decrease in tubular function leads to great increases in urinary levels of low molecular weight proteins (low molecular weight proteinuria, tubular proteinuria). The finding of increased levels of low molecular weight proteins in urine is, therefore, a very sensitive indicator of proximal tubular dysfunction or injury. Injury to the glomerulus leads to increased levels of high molecular weight proteins, which can saturate the tubular reabsorption mechanisms and spill over into the urine (high molecular weight proteinuria, glomerular proteinuria). Pre-renal, or spillover, proteinuria is where the level of a low molecular weight protein in the plasma is so high that its concentration in the glomerular filtrate exceeds the resorptive capacity of the tubular reabsorption mechanism and appears in the urine. Such a case is Bence Jones proteinuria in subjects with gammopathies (6). Finally, nephrogenic proteinuria is when proteins derived from the kidney itself appear in the urine. These could be derived from the cells lining the nephron, or matrix proteins derived from the basement membranes (7). Cell-derived proteins can originate in the cell membrane, lysosomes or the cell cytoplasm and it is upon these that the rest of this chapter will concentrate.

1.2. Detection and Measurement of Renal Proteins

The first measurement of urinary protein was by Richard Bright (1789–1858) who showed the presence of protein (albumin) in the urine of a subject with renal disease by heating it, causing the protein to coagulate. Proteinuria was established as a symptom of renal disease but investigation of the individual proteins began with the development of chromatography (8), electrophoresis (9), and then polyacrylamide gel electrophoresis (PAGE) (10), which enabled more refined investigations of the patterns of urinary proteins. Two main patterns are seen, high molecular weight proteinuria (glomerular proteinuria) and low molecular weight proteinuria (tubular proteinuria). For convenience and to enable more quantitative results, representative proteins from each of these groups may be studied, e.g. albumin for glomerular proteinuria and $\alpha 1$ microglobulin for proximal proteinuria (11). Generally, low molecular weight proteinuria is associated with increased urinary levels of proteins derived from tubular cells, e.g. enzymes and cytosolic proteins. Their presence enables the site of renal injury to be localised to precise cell groups. The presence of renal-specific proteins in pathological urine was first shown by Grant in 1935 (12) and their presence in normal urine was confirmed by Gillman in 1959 (13). The presence of urinary proteins derived from the kidney was termed "Histuria" by Antoine and Neveu in 1967 (14) and "Nephrogenic Proteinuria" by Hardwicke in 1975 (15). The concept was developed further by Scherberich et al. (16). The cell-specific localisation of enzymatic biomarkers was confirmed by microdissection (17) or histological staining (18). Non-enzymatic biomarkers were identified and localised immunohistologically, utilising antisera raised against renal extracts or defined proteins (19). Enzymatic biomarkers have been mostly assayed using colorimetric or fluorometric substrates (20). ELISA methods have been developed for other proteins. Methods for simultaneously measuring panels of cell-specific biomarkers are in development (21).

Recently, proteomic techniques have been applied to the identification of urinary biomarkers. The two main methods are MALDI (22) where the proteins are separated using two-dimensional electrophoresis and then identified using mass spectrometry and SELDI (23) where proteins are initially concentrated from the sample using a strip bearing absorptive agents, e.g. hydrophobic groups, and then identified using mass spectrometry. These have the ability to detect many proteins simultaneously and can identify new biomarkers, but provide little initial information as to their origin. This can be obtained by comparing the properties of the proteins detected with proteins with known origins.

Genomic investigation of renal tissue can aid in the discovery and identification of biomarkers describing the kidneys response to toxic stress (24). Urinary assays can then be developed for the proteins related to the genes discovered.

Table 1
A selection of cell-derived biomarkers with their main locations

	Cytoplasm	Lysosomes	Brush border
Proximal tubule	Lactate dehydrogenase (LDH) isoforms 1 and 2 Bis fructose phosphatase Kidney/liver type fatty acid binding protein α Glutathione S-transferase	*N*-acetyl-β-D-glucosamidase (NAG) β galactosidase Acid phosphatase	Intestinal alkaline phosphatase Malate dehydrogenase Leucine amino peptidase γ glutamyltransferase Dipeptidyl-aminopeptidase IV
Loop of Henle	Renal papillary antigen 2 (in rats)		
Distal tubules	Pi glutathione S-transferase (πGST, in humans), glutathione S-transferase Yb1 (in rats) Heart fatty acid binding protein		
Collecting duct/ renal papilla	LDH isoforms 4 and 5 Renal papillary antigen 1 (in rats)		

1.3. Cell-Specific Biomarkers

By measuring a selection of biomarkers with known origins, injury can be localised to distinct cell types. Furthermore, since different biomarkers come from different compartments of the cell, it is also potentially possible to localise injury to distinct sub-cellular compartments. The absence of a biomarker from the urine also provides valuable information in that it shows where injury is *not* occurring and serves as a negative control. A selection of renal cell-specific biomarkers with their locations is shown in Table 1.

The proximal convoluted tubules make up the bulk of the kidney (see Fig. 1) (25). The predominance of proximal tubular tissue makes it more difficult to find and utilise biomarkers for other parts of the kidney, as greater demands will be placed on their specificity and sensitivity. Responses for biomarkers that are found mainly in other parts of the kidney will be obscured if those biomarkers are also found in the proximal tubules, albeit at much lower levels.

1.4. Urinary Enzymes as Indicators of Renal Injury

Over 40 enzymes have been identified in the urine (26) but only a handful have been broadly used as renal biomarkers (20). Urinary enzymes have the advantages of being rapidly and easily measurable using standard colorimetric or fluorometric techniques.

1.4.1. N-Acetyl-β-D-Glucosamidase

N-acetyl-β-D-glucosamidase (NAG) is predominantly found in the lysosomes of proximal tubular cells, but it is also found at

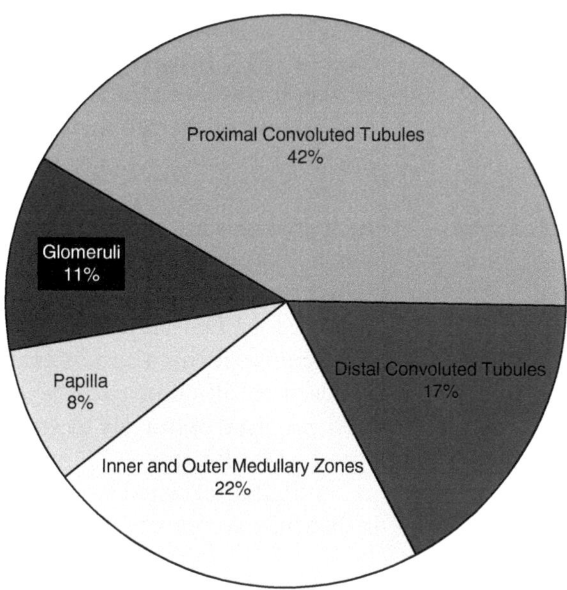

Fig. 1. Kidney composition. Mattenheimer et al. 1968. Based on reference 25 with permission.

lower levels elsewhere in the nephron (27). It has been widely studied as a biomarker of proximal tubular injury in many conditions. Its drawbacks are that it is upregulated in response to proteinuria (28) and that it is not totally cell specific. There are three isoforms of NAG, of which the A and B forms are most important. The A form is found in solution in the lysosome and may be released during the normal process of exocytosis. The B form is found on the membrane of the lysosome and may only be released when there is injury to the lysosome (29). While NAG is found along the whole nephron, the B form forms a relatively larger proportion of the NAG present in the proximal tubule, potentially making it a more specific biomarker for injury to it. Large increases in the B form can be masked by only modest increases in the total urinary NAG activity (29).

1.4.2. Brush Border Enzymes

There are a number of enzymes located on the brush border villi of the proximal tubular cells, e.g. alanine aminopeptidase (AAP), alkaline phosphatase (ALP), and gamma glutamyltransferase (GGT). Alkaline phosphatase is found in two forms, tissue non-specific alkaline phosphatase (T-NAP) and intestinal alkaline phosphatase (I-ALP). I-ALP is restricted to the S3 segment of the proximal tubule, their simultaneous assay allowing renal injury to be localised to S3, S1/2 or the whole proximal convoluted tubule (30). The main problems in measuring enzymes in general are their lack of stability, interference from factors in the urine, and the fact that their release varies according to dieresis (31). Brush border enzymes may be found in the urine in two forms: soluble or bound to cell membrane fragments. The finding of the cell membrane found form may indicate more severe injury (18).

1.4.3. Cytosolic Enzymes

Such enzymes are, for example, fructose bis phosphatase (FBP), which is found in the proximal tubular cells (32), and lactate dehydrogenase (LDH), which has a broader distribution. Different isoforms of LDH are found in the cortex (forms 1 and 2) and in the medulla (forms 4 and 5) enabling renal injury to be potentially localised to different parts of the kidney (33, 34). The need to perform electrophoresis to separate the isoforms has limited the use of LDH for this.

A limitation when using renal enzymes to study renal injury is that they are mostly from the proximal tubule.

1.4.4. Cell-Specific Proteins

There are a number of proteins specific for distinct parts of the nephron (Table 1) but the most studied are the fatty acid binding proteins (FABPs) and the glutathione S-transferases (GSTs).

1.4.5. Fatty Acid Binding Proteins

Two forms of FABP are found in the kidney. A renal form, very similar to liver FABP, is mostly found in the proximal tubules and a heart form, which is mostly found in the distal tubules (35). In the adult female rat only the heart form is expressed and it is found in both the proximal and distal tubules (36).

1.4.6. Glutathione S-Transferases

The glutathione S-transferases are a family of phase II enzymes with distinct distributions along the nephron (19). They are found in high concentrations (4% of the soluble protein) in the cytosol and they are rapidly released into the urine in the event of injury to the tubular cells. The alpha form of glutathione S-transferase (αGST) is localised to the proximal convoluted tubule and other forms are distributed in the distal tubules (Fig. 2) (37). In human, the pi form (πGST) is localised to the distal tubules while in the rat a mu form of GST (GSTYb1) has a similar distribution (38). Early studies on urinary GSTs utilised enzyme activity assays but these are not sensitive enough to detect GST activity in normal urine samples (39) and GSTs non-competitively bind toxins, potentially leading to falsely low values (40). More importantly, enzyme methods are not specific enough to enable the separate GST isoforms to be individually measured, losing the advantage of the cell-specific distribution of GST isoforms. The GSTs have been widely studied in renal medicine and research with over 100 published studies.

1.4.7. Other Cell-Specific Biomarkers

Other cell-specific biomarkers have been identified but have not been used widely in renal research. Scherberich (11) and

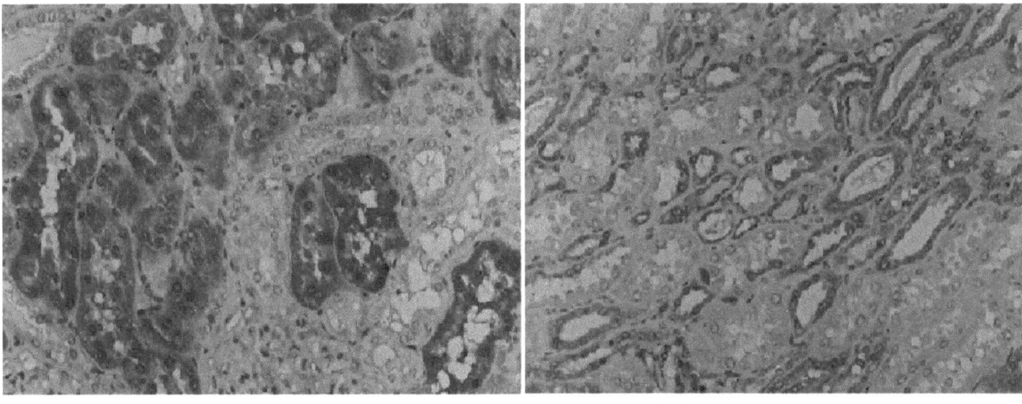

Fig. 2. Immunohistochemical localisation of GST isoforms in the human kidney. Shaw 2005 Reproduced from reference 37 with permission.

Falkenberg et al. (40), for example, discovered segment-specific biomarkers by developing antisera against extracts of renal extracts, but they have not been widely employed. They will be presented below as appropriate.

2. Cell-Specific Biomarkers in Renal Disease

2.1. Introduction

The kidney has great powers of recovery and a large reserve capacity; therefore, apparently mild renal injury may be accepted since one can expect full recovery, as judged by near normalisation of serum creatinine. However, this normalisation may be achieved at the expense of future reserve capacity (42). Using cell-specific biomarkers enables renal injury to be detected and monitored more closely with potential advantages for the subject's long-term health.

2.2. Chronic Renal Disease

Chronic renal diseases, e.g. SLE, rheumatoid arthritis, diabetes, and IgA nephropathy, are followed by measuring changes in serum creatinine and proteinuria. This provides information as to long-term changes in renal function, but little information as to *how much destruction of the kidneys is occurring at that moment.* Cell-specific biomarkers provide complimentary valuable information here. They are markers of disease activity.

In these diseases, proteinuria is initially of the glomerular type, but from a very early stage, increases in low molecular weight proteinuria occur, associated with increases in urinary enzymes and cell-specific proteins (43). Plasma proteins cause inflammation in the proximal tubules with the release of proximal tubular enzymes (44).

For example, proximal enzymuria was associated with a more serious outcome in subjects with glomerulonephritis (45). The most widely used enzyme biomarker is NAG (46), but it is upregulated in response to protein load and, therefore, in cases of mild renal injury, it may not be showing renal injury, rather renal adaptation (28).

Few studies have been published on cell-specific cytosolic proteins in chronic renal injury. Increased levels of liver type FABP (proximal tubular injury) are found in subjects with declining glomerular filtration rate (47) but no studies have been published with heart type FABP (distal tubular injury). Studies with the GSTs show that the early stages of chronic renal injury are associated with increased levels of αGST indicating mainly proximal tubular injury, but as the disease progressed, πGST increases relatively faster, indicating increasing involvement of the distal renal tubules (48).

The kidney has great powers of reserve capacity and regeneration, but beyond a certain point regenerative powers are reduced and vicious circle of progressive renal failure can ensue. Venov et al. presented the concept that chronic renal injury may result from "multiple hits" of acute injury that progressively decrease the number of nephrons and increase the load on those remaining (49). Furthermore, even very limited renal capacity can be important for the well-being of subjects with renal failure (50). Therefore, it is important to reduce the exposure of subjects to potentially nephrotoxic substances. However, 25% of subjects with compromised renal function were still given drugs that are recommended to be avoided by subjects with renal insufficiency (51). The early assay of cell-specific biomarkers may offer the potential to detect adverse renal effects of drugs or procedures early, facilitate the choice of alternative therapies, and preserve remaining renal function.

2.3. Diabetes

In diabetes, elevations in urinary enzymes are found from a very early stage and improved glycaemic control leads to decreases in them (52). This has been proposed as showing a direct effect of hyperglycaemia on the proximal tubules, but it could be indirect, following on from the effects of proteinuria on the proximal tubule. In particular, increases in urinary NAG could be the result of adaptation of the renal tubules to increased protein load.

There has been little published on cell-specific proteins in diabetes but Maxwell et al. showed an early increase in urinary πGST in diabetics indicating early distal tubular injury (53). The finding of early elevations in distal tubular biomarkers could be expected from the pathology of diabetic nephropathy. Proteinuria leads to the secretion of pro-inflammatory cytokines by the proximal tubules into the inter-tubular space (43) and injury to the distal tubules and glomeruli. The development of biomarkers for the distal tubule could provide a new insight into the development of diabetic nephropathy. In this study αGST was not elevated, while NAG correlated with urinary albumin, indicating again that NAG is influenced by protein load. That urinary αGST is little affected by diabetic nephropathy provides the possibility of detecting renal effects other than diabetic nephropathy, e.g. nephrotoxicity, in diabetics.

2.4. Environmental Medicine

Many studies have been performed on the effects of environmental toxins on the kidneys. The majority of the proximal tubular biomarkers are elevated in subjects exposed to, for example, high levels of cadmium (54, 55), lead (55), and industrial solvents (56) even where the majority of subjects have normal serum creatinine. Few studies have been performed where distal tubular biomarkers have been assayed. There is evidence that changes even within the normal range of these biomarkers may be significant. For example, in subjects removed from exposure to cadmium, urinary B-NAG (54) correlated with the body burden of cadmium even within the normal range, as did urinary αGST for subjects removed from exposure to trichloroethylene (56).

2.5. Renal Transplantation

While 1-year survival of renal transplants is approaching 95%, the long-term half-life of grafts is only about 8 years (57). Better biomarkers, enabling earlier and better diagnosis and improved monitoring of therapy, could improve graft survival. In transplantation, immunosuppressive drugs, e.g. calcineurin inhibitors, must be given to prevent and control rejection, leading to the risk of nephrotoxicity while too little immunosuppression leads to the risk of rejection. In both instances serum creatinine increases; therefore, clinical expertise and other biochemical tests are required to make the distinction. Cell-specific biomarkers have the potential for early detection and discrimination between these conditions. Earlier treatment could reduce the severity of acute rejection, which, in turn, could lead to a reduced risk of chronic rejection, while the early detection of nephrotoxicity will also help to maintain renal capacity.

Immediately post-transplantation, recipients show both glomerular and tubular proteinuria associated with high urinary levels of renal enzymes (58, 59) and cell-specific proteins (60). With a well-functioning graft, the proximal proteinuria, enzymuria, and cell-specific proteins decline rapidly (61). Rejection episodes and nephrotoxicity are associated with the reappearance of enzymuria days in advance of an increase in serum creatinine (62). However, increases in urinary enzymes provide little information as to the causes of decreased renal function in transplant recipients. Cell-specific proteins offer the potential to offer further diagnostic assistance.

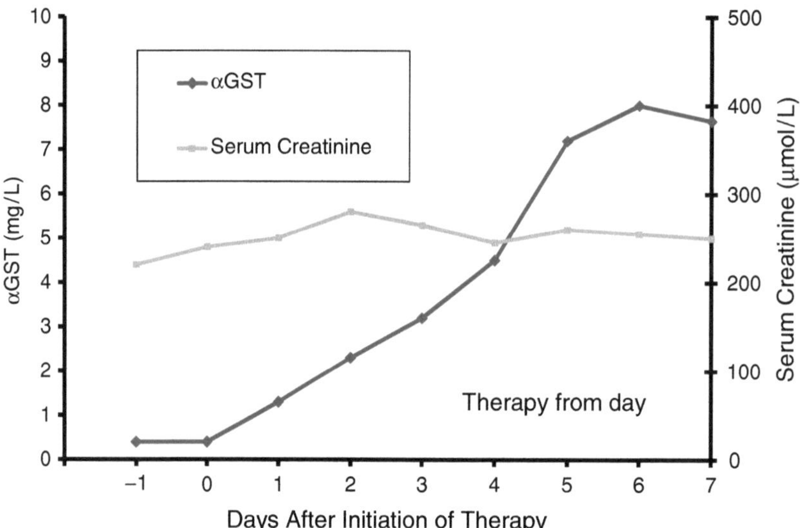

Fig. 3. Proximal tubular injury: aminoglycoside therapy. Bäckman et al. 1988. Based on reference 64 with permission.

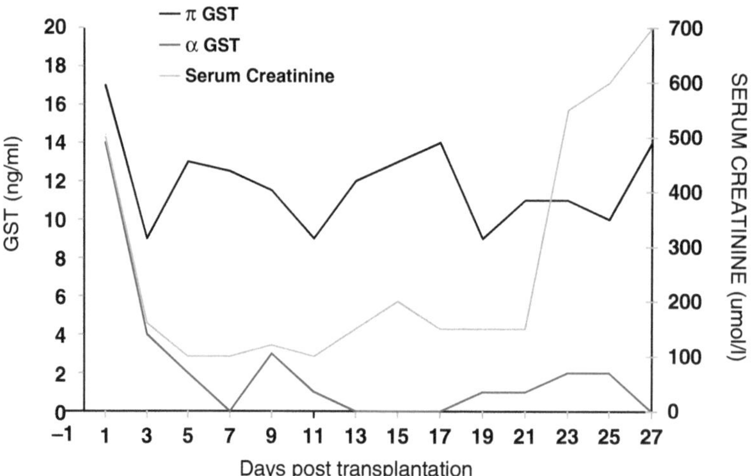

Fig. 4. Renal biomarkers in acute rejection. Sundberg et al. 1994. Based on reference 60 with permission.

Most nephrotoxic drugs affect the proximal tubules; therefore, specific biomarkers for the proximal tubule are early indicators of it (62, 63) (Fig. 3). Rejection starts initially in the distal tubules (64); therefore, distal tubular biomarkers, e.g. πGSTs are potentially sensitive biomarkers for it (60) (Fig. 4). This supported by a subject studied by Falkenberg et al. (65) where rejection episodes were accompanied by increases in the distal tubular biomarkers. Simultaneous assay of proximal and distal tubular biomarkers offer the potential to diagnose the cause of declining graft function earlier than using serum creatinine and reduce the need for biopsies. For example, considering the commonest causes of transplant injury, urinary biomarkers could provide the following assistance:

- Rejection > Relative elevation of distal tubular biomarkers.
- Nephrotoxicity > Relative elevation of proximal tubular biomarkers.
- Ischaemia > Elevation of both types of biomarker.

Furthermore, cell-specific biomarkers fall much faster than serum creatinine as the rejection episode or nephrotoxic episode is resolved, potentially enabling the physician to be surer of the diagnosis and shorten courses of therapy. Shorter, better-timed, and more appropriate therapy could improve the long-term survival of renal grafts. The discriminatory power of cell-specific biomarkers is likely to be greater when injury is mild, e.g. at the beginning of a toxic or rejection episode, since severe damage to one part of nephron will cause injury elsewhere.

2.6. Renal Graft Selection

Renal transplantation is the treatment of choice for end stage renal failure, but there is a severe shortage of organs. Potential sources are non-heart beating donors (NHBD) and grafts that have been exposed to significant warm ischaemia time (66). These grafts have an increased risk of non-function due to ischaemia-reperfusion injury and it is important to be able to *select-out* those grafts that are non-viable. The proximal tubules have a high oxygen requirement and are especially susceptible to ischaemia–reperfusion injury; therefore, biomarkers of injury to the proximal tubule in the graft perfusate have the potential to detect non-viable grafts (67). For example, Daeman et al., using a GST EIA, showed that a low αGST in the graft perfusate was 98% predictive of a functional graft while LDH was of little value. They suggested that this was due to the greater specificity of αGST for the proximal convoluted tubules, the most sensitive part of the nephron to ischaemic injury.

2.7. Renal Infection

Upper urinary tract infection leads to proximal proteinuria and the release of renal enzymes and cellular proteins (68, 69) and their presence can be a diagnostic aid to differentiate between upper and lower urinary tract infections. However, while NAG and most other urinary enzymes originate in the proximal tubule, bacteria are initially localised around the distal tubules (70) suggesting that distal tubular biomarkers could be especially sensitive indicators of it. In support of this, πGST has been shown to be a sensitive indicator of renal infections (71, 72). A problem in monitoring of renal infections is that the drugs used, e.g. aminoglycosides, could be nephrotoxic. Most nephrotoxins affect the proximal tubules; therefore, using a combination of proximal tubular biomarkers (nephrotoxicity) and distal tubular biomarkers (renal infection) offers the theoretical possibility to independently monitor the effects of the infection and the therapy on the kidney.

2.8. Acute Renal Failure

Acute renal failure (ARF) is a condition that occurs in about 10–30% of subjects in intensive care and has a high mortality (73). Risk factors for it are well known. Elevated serum creatinine is a risk factor as it indicates subjects with already compromised renal function and who, therefore, have less renal reserve. Early detection of further renal injury can aid in preserving this renal function and preventing further renal insult, e.g. by avoiding the use of nephrotoxic drugs.

Most studies on biomarkers in ARF are on subjects already in intensive care or with pathological conditions when most biomarkers are already elevated. To see the true power of specific biomarkers in this condition, a study would need to be performed where subjects at risk of ARF are tested for cell-specific biomarkers to see if these can be used to predict those who will develop ARF. A study by Cressey et al. (74) on subjects undergoing major vascular surgery indicated that this could be possible. They showed that an elevated urinary αGST during surgery was predictive of an elevation in serum creatinine 2 days later and that there was a close correlation between urinary αGST levels and the later increase in serum creatinine (Fig. 5).

2.9. Nephrotoxicity

Nephrotoxicity is a significant problem as many drugs in use, e.g. antibiotics, chemotherapeutics, immunosuppressive drugs, and contrast media are potentially nephrotoxic (75). Nephrotoxicity can be common with certain drug classes, 36% of course of aminoglycosides and 25% of courses with cisplatin. The proximal tubules are most frequently affected by drugs and these toxicities are associated with the release of enzymes and cell-specific proteins. For example, anti-cancer drugs (76), contrast media (77, 78),

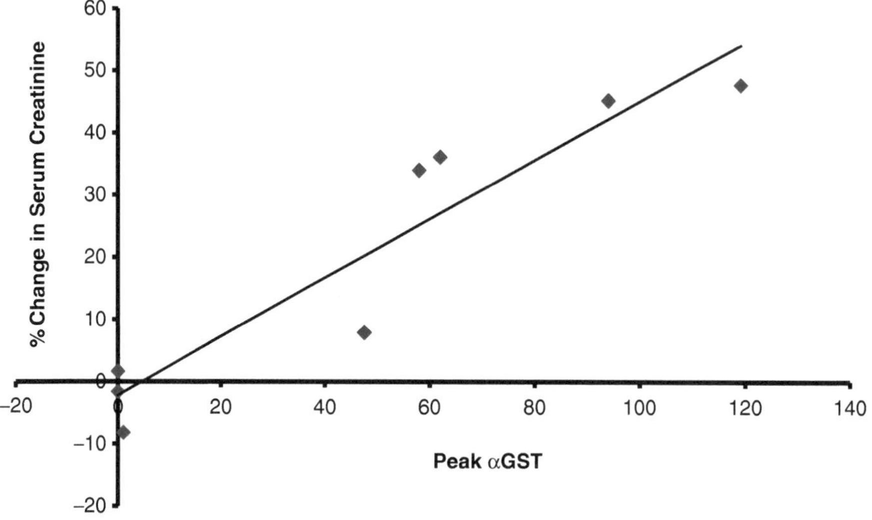

Fig. 5. αGST release during vascular surgery. Cressey et al. 2002. Based on reference 74 with permission.

and antibiotics (33) are associated with the release of NAG, brush border enzymes, and cell-specific proteins (79–81). Urinary levels of these biomarkers are released in advance of increases in serum creatinine. For example, Donta and Lembke found that urinary NAG rose faster and higher following gentamicin than tobramycin and that the subjects who had the most rapid increase in urinary NAG showed the largest later increase in serum creatinine (82). Dupond found that elevated B-NAG is more closely related to the nephrotoxicity of aminoglycoside than the A form (83). Similarly, Scherberich found that urinary AAP was elevated to a greater extent and for a longer period in subjects who later showed the greatest increase in serum creatinine (18). Fewer drugs are recognised as affecting the distal tubules, but this could be due to the paucity of biomarkers for this part of the kidney. However, πGST enables the distal tubules to be studied and it is released in response to drugs known to affect the distal tubules, e.g. Foscarnet (60) and Amphotericin B (79).

2.10. Summary

Urinary cell-specific biomarkers have the potential to reduce renal injury in subjects with renal disease or injury. Injury can be detected earlier and more precisely, offering the potential to terminate or change ineffective or damaging therapies. The release of the distal tubular biomarker πGST seems to be linked to more severe, progressive disease, offering a potential new diagnostic aid. Unfortunately, no controlled studies seem to have been performed where therapy based on urinary biomarkers has been compared with therapy based on clinical judgement and serum creatinine. Furthermore, few studies have been performed in humans where biomarker release is compared with histologically confirmed changes in the kidney. The specificity and sensitivity of renal biomarkers are still compared with serum creatinine, which is known to be a tarnished standard. Furthermore, temporary renal impairment (increased serum creatinine) may be associated with increased long-term risks of renal failure and other morbidities (84). Faster reacting and more sensitive biomarkers have the potential to reduce these risks. Controlled trials where therapy is based on cell-specific biomarkers have the potential to reduce iatrogenic renal injury and improve patient health and prognoses.

3. Cell-Specific Biomarkers in Drug Discovery

3.1. Introduction

There is a great need to reduce the cost of developing new pharmaceuticals. Costs of developing a new drug are now approaching $900 million (85) and, of this, about 75% is due to the costs of failed candidates. A means to reduce this is to use more discriminating tests of function and adverse side effects earlier in development. This will remove compounds early in their development when it is still relatively cheap and free-up resources for the surviving candidates. Dr Crawford of the FDA suggested that the use of better biomarkers could reduce average drug development costs by $100 million (86).

Nephrotoxicity is an important cause of drug failure, but traditional biochemical tests for it were not designed for use in drug development and, therefore, may be inappropriate or lack sufficient sensitivity and specificity. Histopathological techniques are the gold standard, but they are labour intensive, subjective, and may miss localised injury. Creatinine and blood urea nitrogen (BUN) are used although it is known that they are insensitive and affected by other factors, e.g. age and diet. These are used for convenience, history, and because they will detect significant injury, but they need to be interpreted with regard to possible confounding factors.

Increasing sensitivity requires the ability to distinguish a weak positive signal from background (high specificity). Using biomarkers with a known origin reduces the background variation and makes this possible. Having biomarkers with a known origin simplifies cross-species comparisons in that it enables comparisons between cell culture, animal species, and human studies. Knowing the site of origin of a biomarker from animal studies provides information as to the likely site of renal effects in human subjects where renal biopsy may be impractical.

3.2. Nephrotoxicity Testing in Drug Development

The most commonly used test for nephrotoxicity is serum creatinine. However, it is a late biomarker of reduced glomerular function, while most nephrotoxic injury is to the renal tubules and one needs to detect it quickly. Here, the use of biomarkers localised to different parts of the nephron can be very advantageous. Using urine as a medium for assaying renal-specific biomarkers leads to greater sensitivity and specificity, plus it is minimally invasive enabling multiple sampling from animals or volunteers. It is difficult to predict if a compound will be successful based on biomarkers, since it must eventually be evaluated in humans where the clinical condition has its own natural history. However, biomarkers can identify compounds likely to be toxic; therefore, freeing up resources for other projects. Detecting likely toxic compounds early in development requires a sensitive test, but one does not wish to terminate potentially useful compounds due to false positives; hence specificity is also important. Cell-specific biomarkers combining high sensitivity and specificity can play a role here. Urinary biomarkers also have the advantage in that urine is a relatively non-invasive sample. Animals can serve as their own control, reducing inter-animal variation and potentially reducing the number of animals required. Similarly, in humans, using urine simplifies multiple sampling.

Several proximal tubular enzymes have been used to study nephrotoxicity e.g. NAG (87), AAP, and ALP (20, 33, 34). However, convenient enzymatic biomarkers for the distal tubules are lacking although LDH isoenzymes have been used for this (33, 34). Enzymatic biomarkers have been used to detect nephrotoxicity due to, for example, aminoglycosides (33), anti-cancer therapy, (34) and contrast media (77).

The best-studied cell-specific biomarkers in drug discovery are the glutathione S-transferases (GSTs). These provide the opportunity to monitor injury to both the proximal and distal tubules separately and simultaneously.

3.2.1. In Vitro Toxicity Studies

Histologically defined biomarkers can provide sensitive and site-specific information on renal injury from cell cultures. For example, Sonee et al. (88) demonstrated the release of αGST from renal cell cultures in response to cisplatin. Similarly, Vickers et al., using cultured slices of human kidney, could demonstrate cell specificity in vitro by the greater release of αGST than πGST in response to the proximal tubular toxin cisplatin (89). Furthermore, the differential toxicity of different drugs could also be shown by the release of αGST from human tissue slices (90).

3.2.2. Animal Studies

Animal studies offer the opportunity to obtain quantitative information between the release of biomarkers and histological changes. Urinary levels of biomarkers can be compared with the severity and time course of histological changes. For example, urinary αGST levels correlate closely with the extent of renal injury as shown in a study by Kharasch et al. (91) where even a few percent of necrotic renal tubules caused urinary αGST levels to increase several folds (Fig. 6). In both human and animal studies, urinary αGST levels correlated closely with the dose of toxin (92).

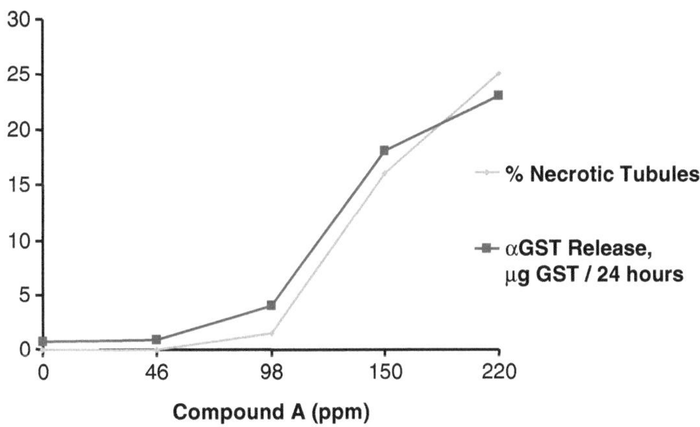

Fig. 6. Proximal tubular injury: compound A nephrotoxicity in rats. Based on Kharasch et al. 1998. Reproduced from reference 37 with permission.

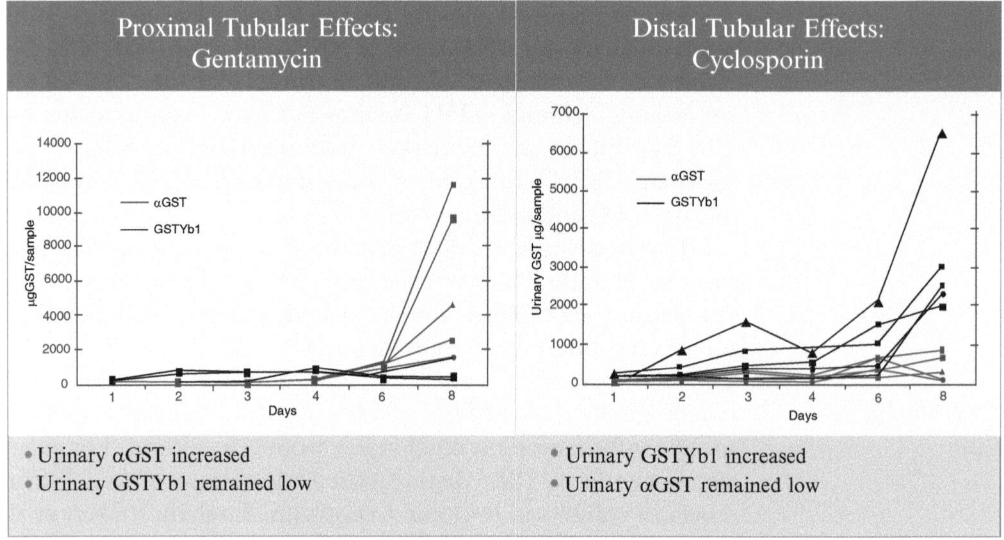

Fig. 7. Urinary GSTs: site of renal injury. Davies et al. 2000. Reproduced from reference 93 with permission.

In neither of these studies did serum creatinine or blood urea nitrogen change. In rats, GSTYb1 is found in the distal tubules (38). Previously, it has been difficult to monitor this important tissue non-invasively, but for example, the toxic effects of cyclosporin on the distal tubules of rats could be detected using urinary GSTYb1 (93).

Combining biomarkers of different parts of the renal tubule can be advantageous. For example, Davies et al. could localise the effects of gentamicin to the proximal tubule and cyclosporin to the distal tubule in rats by simultaneously measuring both urinary αGST and GSTYb1 (Fig. 7). A low response from a sensitive-specific biomarker can be very valuable as it indicates where injury *is not happening*. αGST is a proven biomarker for proximal tubular injury in both humans and rats. πGST in humans and GSTYb1 in rats have similar distributions in the distal tubules enabling comparisons between species.

Using cell-specific biomarkers will make it easier to compare various animal models. It will be possible to see if the sites of renal injury are equivalent in different species. For example, cyclosporin is a proximal tubular toxin in humans but a distal tubular toxin in rats and this can be seen by the types of GST released in humans and rats.

Using enzymes to study different parts of the nephron is difficult due to the absence of simple biomarkers for the distal tubules and renal papilla. However, Obata showed, using electrophoretic separation, that the papillary toxin para amino phenol caused a relatively greater release of the LDH isoforms 4 and 5 (found mainly in the papilla) than the LDH isoforms 1 and 2 (found mainly in the proximal tubules) and the proximal tubule enzymes NAG and ALP (34).

The linking of biomarkers to distinct cell types could reduce the need for expensive traditional histopathology testing and provide better estimates of acute toxicity, particularly regarding toxicokinetics and the "No Observable Effect Level". This will be particularly important in clinical studies where the collection of histological samples may not be possible. Their early elevation in clinical trials, preceding those of traditional biomarkers, will have clear safety advantages in phase I studies.

3.2.3. Human Toxicology Trials

Histological validation of renal injury is seldom possible in clinical trials; therefore, the use of histologically proven biomarkers is valuable in that they can provide a window to look into the kidney and a means of comparing animal and human responses.

Published human studies include antibiotics (20, 79), cyclosporin (94, 95), contrast media (80), and anaesthetics (94, 96, 97). As a general rule, specific biomarkers are released in advance of changes in serum creatinine. Combining biomarkers can also provide information as to the mechanism of renal injury. For example, Amphotericin caused the release of αGST and πGST in human subjects, while urinary NAG rose little (79). This matches the toxicological mechanism of Amphotericin, which is to affect cell membranes, resulting in the release of cytosolic biomarkers, but not lysosomal biomarkers (at the doses given). Furthermore, only male volunteers showed increases in urinary GSTs, which agrees with the greater frequency of reactions in male subjects.

Specific biomarkers also proved the opportunity to compare the renal effect of different dosage regimes. For example, Ahlmén et al. demonstrated that the addition of a calcium blocker to cyclosporin reduced the release of αGST into the urine (95). In humans, where biopsy is seldom possible in trials, the opportunity for development of biomarkers to regions of the nephron apart from the proximal tubule (e.g. πGST for the distal tubule) will be especially valuable.

3.3. Summary

Compounds are failing late in the development process resulting in great expense in time and money. Using more sensitive and specific biomarkers can identify potentially toxic biomarkers early in development, freeing up resources for potentially more successful compounds. Cell-specific biomarkers can simplify cross-species comparisons of toxicology results and can provide quantitative information as to the site and extent of renal injury.

4. Cell Recovery Biomarkers

4.1. Introduction

Recently, renal injury biomarkers have been complemented by the introduction of biomarkers for the next step of the injury process, renal remodelling, and recovery. Being elevated for longer periods than markers of cell death, these biomarkers may make it easier to identify acute renal injury when the time that the injury occurred may not be accurately known or if sample collection is delayed. They will fill in a gap between biomarkers of acute renal injury and biomarkers of decreased renal function, e.g. serum creatinine. The three best-studied biomarkers are clusterin, kidney injury molecule-1 (KIM-1), and neutrophil gelatinase-associated lipocalin (NGAL). The majority of these studies have focused on upregulation and/or induction of gene expression, rather than the appearance of the proteins in the urine. Some of the findings regarding their urinary levels and gene expression are presented below.

4.1.1. Clusterin

Clusterin is a dimer of two 40 kDa units that has been discovered several times, resulting in several names (98), e.g. SP-40,40 (serum protein 40 kDa40 kDa), SGP-2 (sulphated glycoprotein 2), and apolipoprotein J. The name clusterin arises from the property of the protein to cause agglutination of Sertoli cells and is the most commonly used term regarding the study of renal injury.

Clusterin has several known functions that may be relevant to its role in renal recovery, including promoting cell agglutination, possible role in protecting cells on fluid surfaces, protecting against oxidative stress, defence against apoptosis, protection against complement activation, acting as a molecular chaperone and as a heat shock protein (98), and covering denuded epithelia to enable reattachment of cells (99). There is a 55-kDa truncated form of clusterin that, when expressed in cell nuclei indicates that apoptosis is occurring and may, therefore, be pro-apoptotic (98). The protective properties of clusterin are indicated by the increased sensitivity of clusterin-deficient cancer cells to chemotherapy (100) and the protection of renal cell culture against gentamicin toxicity by the addition of clusterin (101).

In a normal kidney, clusterin expression is low, but following renal injury, it is expressed in cells of the thick ascending loop of Henle, distal convoluted tubule, and the S3 section of the proximal convoluted tubule. Cells expressing clusterin in the proximal tubule did not seem to be involved in renal regeneration, but were undergoing apoptosis (102).

4.1.2. Kidney Injury Molecule-1

KIM-1 (also known as Hepatitis A Cellular Receptor 1) is a type I transmembrane glycoprotein expressed on dedifferentiated renal proximal tubule epithelial cells undergoing regeneration. The extracellular domain of KIM-1 is composed of an immunoglobulin-like domain topping a long mucin-like domain, a structure that points to a possible role in cell adhesion by homology to several known adhesion proteins. The shedding of KIM-1 in the kidney undergoing regeneration constitutes an active mechanism allowing dedifferentiated regenerating cells to attach to denuded patches of the basement membrane and reconstitutes a continuous epithelial layer (103).

4.1.3. Neutrophil Gelatinase-Associated Lipocalin

Neutrophil-associated lipocalin (also known as 24p3, SIP24, lipocalin 2, or siderocalin) is a protein that binds iron containing ligands (siderophores). It was originally purified from human neutrophils (104), but in spite of its name, it is not only found associated with neutrophil gelatinase. Its function may be to bind extracellular iron, reducing the risk of oxidative injury and restricting the growth of bacteria, while it may also serve to transport iron to growing cells (105). NGAL protein is rapidly upregulated in regenerating renal tubular cells following acute renal injury (106). The potential protective function of NGAL is indicated by the fact that the systemic administration of NGAL to mice reduced renal ischaemia–reperfusion injury (107).

4.2. Cell Regeneration Biomarkers in Renal Disease

Few studies have been published on the use of cell regeneration biomarkers in drug discovery, but a range of studies have been published on their use in studying different forms of renal injury and renal diseases.

4.2.1. Chronic Renal Disease

Long-term studies on the use of regeneration biomarkers in chronic renal disease do not seem to have been published, but van Timmeren et al. showed increases in urinary KIM-1 in a range of renal diseases. Its expression seemed to be linked to renal inflammation (108). In puromycin-induced proteinuria, renal expression of clusterin mRNA, and the clusterin protein, increased, and these changes preceded the onset of fibrosis and occurred while cell morphology was still apparently normal (109). Elevations of urinary NGAL were seen in children with systemic lupus erythromatosis and renal involvement (110) and in subjects with IgA nephropathy (111). Several studies have shown that urinary clusterin is a biomarker of obstructive renal disease. In models of unilateral obstructive renal disease, clusterin is elevated in urine derived from the injured kidney (112). Expression of clusterin in the injured kidney falls after obstruction is removed, but it does not normalise fully (113).

4.2.2. Renal Transplantation

The expression of cell regeneration biomarkers is increased in transplanted kidneys and increased urinary levels may be found in the event of complications. In the renal graft, NGAL staining is associated with cold ischaemia time (114) and urinary NAGL immediately post-transplant is predictive of delayed graft function (115) indicating that NGAL could be guide to graft function.

Clusterin, KIM-1, and NAGL are all upregulated in kidneys suffering from rejection and other post-transplant complications. Urinary NGAL levels were higher in subjects with clinical tubulitis than those with subclinical tubulitis or stable graft function, but there was a considerable overlap (116). van Timmeren et al. found increased renal levels of KIM-1 in subjects with acute rejection or chronic transplant nephropathy (108) and, similarly, Dvergsten et al. found elevated renal levels of clusterin in transplanted kidneys (117).

4.2.3. Acute Renal Failure

The commonest models to have been studied involve renal ischemia, either directly due to clamping renal blood vessels or indirect injury resulting from major surgery. Cell recovery biomarkers have been shown to be sensitive tests for predicting increases in serum creatinine. Studies have been performed with clusterin (118), KIM-1 (119) and, in particular, with NGAL. In a rat model of renal ischaemia injury, clusterin gene expression rose after 1 day (120) while clusterin appeared in the renal tubules and the urine before serum creatinine rose (118). NGAL was upregulated within 3 hour in the kidneys of mice and rats that have undergone ischaemic injury (121) and elevated urinary NGAL levels in children and adults who had undergone cardiac surgery were predictive of a post-operative increase in serum creatinine (Fig. 8) (122, 123).

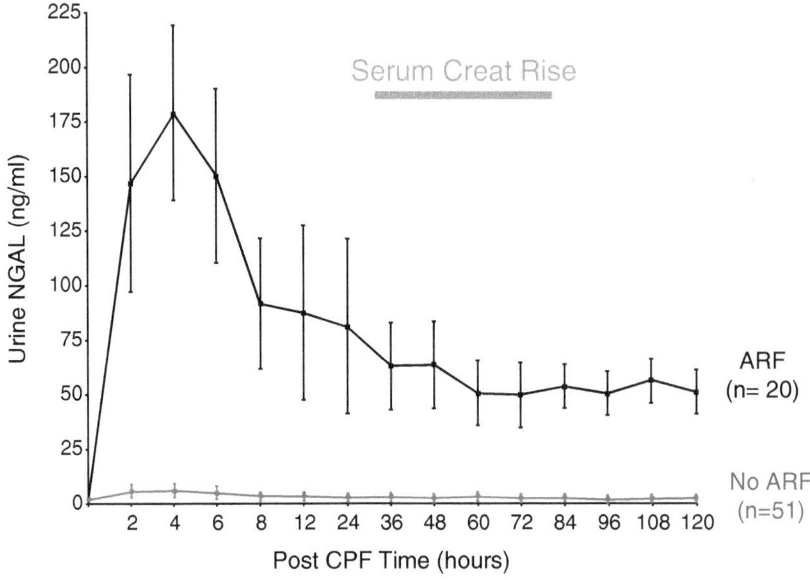

CPB = CardioPulmonary Bypass

ARF = Acute Renal Failure

Fig. 8. Prediction of increased serum creatinine by increased urinary NGAL. Mishra et al. 2005. Reproduced from reference 122 with permission.

In a study on subjects with established urinary KIM-1 rose together with urinary NAG as the APACHE score increased (124). Increasing NAG and KIM-1 revels were associated with an increasing risk for death or dialysis requirement.

4.2.4. Nephrotoxicity

Only a few toxins have been studied using regeneration biomarkers, but they seem to rise earlier and to a greater extent than serum creatinine following administration of test substances. Urinary levels of clusterin, KIM-1, and NGAL all rose in rats given cisplatin (125–127) or gentamicin (128, 129). In a rodent model of chronic gentamicin toxicity, urinary clusterin rose together with NAG but stayed elevated longer (130). Regeneration could possibly be valuable to models of chronic nephrotoxicity.

Regarding other toxins, KIM-1 was elevated in rats with nephrotoxicity following high doses of folic acid and S-(1,1,2,2-tetrafluoroethyl)-L-cysteine (TFEC) (126), both clusterin and KIM-1 gene expression were increased in rats exposed to compound A (131) and KIM-1 was elevated in subjects receiving contrast media (132).

4.3. Summary

Cell regeneration biomarkers show great promise as biomarkers of renal injury and recovery. Results with studies on renal ischaemic injury of different causes are especially encouraging. However, so far, few experimental models or disease conditions have been studied and comparisons with cell-specific acute injury biomarkers are limited. The extended elevation of urinary renal recovery biomarkers following the removal of the cause of injury could make their use in monitoring therapy response or toxicokinetics difficult.

5. "Histomics®", The Use of Immunohistology to Identify Novel Biomarkers

Genomics, proteomics, and the other -omics identify many possible markers and then, through data analysis, identify potentially useful candidates. There is an alternative approach which is to identify target cells and tissues and then develop specific biomarkers for them utilising immunohistological screening – Histomics®. Histomics involves the development of monoclonal antibodies against biomarkers associated with organ or cellular injury and using these as probes to locate their origin. For example, to develop new biomarkers for nephrotoxicity, rats were treated with toxins known to cause injury to defined parts of the renal tubule. Urine samples were collected, the proteins concentrated, and then used to develop monoclonal antibodies. The specific location of the antigen was identified by immunohistological staining of rat kidney tissue sections. Clones expressing antibodies with

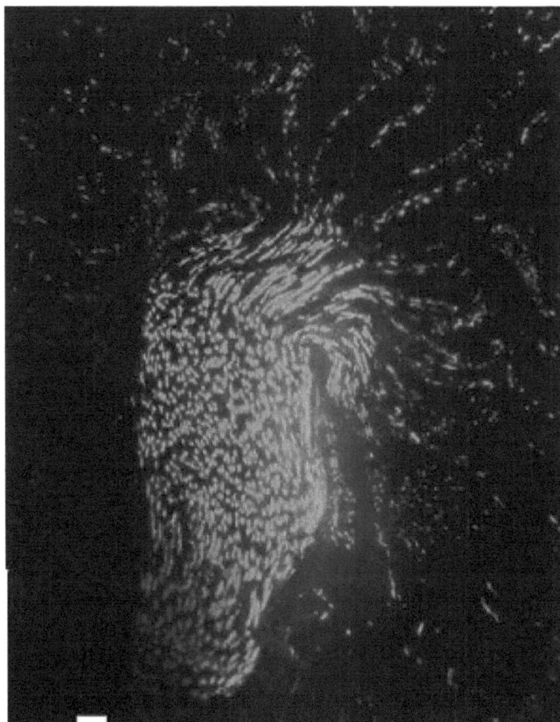

Fig. 9. RPA-1 in the rat kidney. Falkenberg et al. 2005. Reproduced from reference 37 with permission.

specificity to defined renal structures are then cultured further and these antibodies are used to develop immunoassays. Antisera to several defined regions of the nephron have been identified including specific targets in collecting ducts and loop of Henle that have already shown great value in studies of as biomarkers for renal papillary necrosis (Renal Papillary Antigen 1 – RPA-1, Fig. 9) (41, 133, 134). Studies are now continuing utilising urine samples from human subjects with renal pathologies. Eventually, it could be possible to obtain a complete picture of histological injury to the nephron from a urine sample.

6. Comparison of Cell-Specific Biomarkers with Other -Omics Methods

There are many new biomarkers being discovered as a result of the -omics technologies. However, many of them lack a direct link to the standard histopathology scoring systems and they may not directly indicate which part of the kidney is involved. Patterns of biomarkers may be more significant than individual biomarkers. Cell-derived biomarkers are not the only important renal biomarkers and the broad range of proteins identified using -omics platform technologies could identify, for example, inflammatory changes. Genomics studies have revealed information on the kidney's response to injury (135) and identified potential new biomarkers of these processes. For example, as a result of the Health and Environmental Science Institute/International Life Science Institute (HESI/ILSI) study on gene-based markers of renal toxicity, the protein clusterin was identified as probable urinary biomarker for renal recovery following toxic insult (136). Genomics can provide valuable information as to the renal effects of toxins, but the need for biopsy material limits its use in human studies. In the field of toxicology, proteomics has the ability to give a broad picture of changes in protein levels and structure, including changes in protein structure resulting from post-translational modification or enzymatic or chemical changes that can be missed using specific reagents such as antisera. Proteomic techniques have been used to study a variety of models; for example, nephrotoxicity, where changes were found in the renal cortex of rats following gentamicin exposure (137), in the urine of human subjects who had received contrast media (138), and in the urine of human subjects suffering from renal transplant rejection (139). Analysis of data from proteomics can lead to the discovery of new biomarkers (140).

Metabonomics (alternatively metabolonomics) covers a range of techniques where changes in the presence and concentrations of low molecular weight compounds (mostly below 1,000 Da) are studied (141). The compounds detected may be unknown, but the pattern of changes can be analysed to provide information as to the status of the kidneys of the subject that provided the sample (142–144). It has the advantage of indicating not only the effects of the toxic compound on the kidney, but also the knock-on effects of nephrotoxicity on the whole organism.

Each of the "omics" technologies has its advantages and disadvantages and it is not a case of one being superior to any other. It is a case of choosing the appropriate biomarkers to illuminate different aspects of a problem. Cell-specific biomarkers (Histomics) provide information as to the site, extent and kinetics of injury. Other -omics technologies (and routine biochemistry and clinical testing) provide information as to the mechanisms and causes of this injury (145). Different biomarkers have different kinetics and combining them will give more information. For example, detecting preformed cell proteins can give an immediate indication of effect, renal recovery biomarkers can provide an indication as to the renal adaptive response, and genomics and metabonomics can provide further information regarding the genetic and metabolic responses of the organism as a whole to the stress and injury of the toxin (146). The simultaneous use of different biomarkers requires judgement as to their relative importance and this must be based on knowledge of the biochemistry of the substance and the organ affected.

7. Discussion and Conclusions

Cell-specific biomarkers offer the potential to improve both patient care and drug development. The main problem is in their practical introduction. Current clinical procedures and toxicological methods were developed based on, and with regard to, the weaknesses of traditional techniques, especially serum creatinine. Introducing novel biomarkers will necessitate potentially long validation processes in comparison with established techniques to enable comparison with the historical data. Introducing new biomarker systems will involve much more than just changes in chemistry. It could involve changes in patient management, in animal husbandry (use of metabolic cages), sample collection times (earlier), and sample type in urine (instead of serum), for example, to obtain the optimum results. To obtain the maximum benefit of cell-specific biomarkers, panels of biomarkers should be run containing both those expected to be a positive and those expected to be negative if the diagnosis is correct and/or positive if an alternative diagnosis is found. A negative result provides very valuable information. This leads to the need for multiple assays, a problem that is being addressed by the development of multiplex assays.

An important problem with all new biomarkers is education in their use. Clinicians, members of the pharmaceutical industry, and the regulatory authorities are experienced in using established techniques and will need educating and convincing of the value of new biomarkers. This will particularly be the case when there is a difference in interpretation in the results obtained with the established techniques and new biomarkers. The result is that the introduction of new biomarkers into development is proceeding slowly, mainly being used in early drug discovery or when there are potential toxicity issues with a compound.

There is need for rigorous clinical and scientific validation of biomarkers. In the clinical field, this will require controlled prospective studies to compare the patient and cost outcomes of the use of novel biomarkers with traditional monitoring. In the field of drug development, the data obtained from internal company validations of biomarkers is proprietary, valuable and, usually, if any information is released, it is from studies on type toxins rather than true drug candidates. This leads to the repetition of validation efforts and unnecessary expense. The several current initiatives on the validation of biomarkers are of great value here as they will share the cost across the industry and the regulatory authorities are involved in the study design and evaluation. The data will also be in the public domain. The FDA C-path initiative (147) and the HESI/ILSI project on nephrotoxicity (148) are worthy of note. This data will simplify the implementation of new biomarkers in drug development, plus provide supporting evidence for the use of these biomarkers in patient care. Maybe the most important of the "omics" is economics and it is here that the introduction of new biomarkers could add most. The new biomarkers are more expensive than traditional biochemistry testing, but with the costs of chronic renal failure in the billions of dollars every year and the average cost of developing a drug at around a billion dollars, using more expensive, but more accurate, biomarkers can easily be justified if it leads to better decision making.

8. Notes

The current reliance on changes in serum creatinine and clinical signs as indicators of renal effects can lead to suboptimal sampling and loss of data from biomarker studies. The following are some points to note when designing urinary biomarker studies.

1. Base the sampling times on the likely time course of injury or toxicokinetics of the substance being studied.
2. Collect samples starting at time points before any effects are expected. These may reveal unexpected renal effects and they will provide a base line against which to compare later responses.
3. Collect samples from each subject/animal as frequently as is practical. Assaying a subset of these will then reveal which time points seem to be most informative and only these need be studied further. Urinary biomarkers tend to change much faster than serum creatinine and clinical signs.
4. Urinary biomarkers tend to be more variable in the afternoon. If possible, for human subjects, avoid collecting sample then. 24 hour, overnight, or second morning urine collections are mostly used. Timed quantitative urine samples are preferred but a sample for urinary creatinine should be taken to allow for correction for unusual dieresis.
5. The release of urinary biomarkers tends to be constant with time, not with urine volume; therefore, results are best expressed as ng/min or μg/day for example.
6. Not all biomarkers are stable in urine, e.g. β_2 microglobulin; therefore, urine samples may need to be mixed with a stabilising buffer. Furthermore, some urine enzymes lose activity on freezing.
7. Plan for the possibility of assaying biomarkers from parts of the nephron where effects are not expected. They will act as negative controls and may reveal unexpected effects.

8. If possible use each subject or animal as its own control and express changes in terms of the biomarker levels at zero dose or zero time.
9. If possible, compare urinary biomarkers with changes in renal histology as well as the dose of toxin or severity of renal insult. If serum creatinine or clinical picture is being used, compare the biomarker release with the most marked changes in these parameters during the study period. It may take days or weeks between a change in a urinary biomarker and changes in serum creatinine or clinical signs.
10. The greatest value of site-specific biomarkers will be when the degree of renal injury is mild, at low doses, and/or early time points. Nephrons function as units; therefore, injury to one part of the nephron will eventually lead to the release of biomarkers from other parts of the nephron.

References

1. Lysaght, M.J. (2002) Maintenance dialysis population dynamics: Current trends and long term implications. *J. Am Soc. Nephrol.* **13**, S37–S42.
2. Hook, J.B. and Hewitt, W.R. (1980) Toxic responses of the kidney. In The Basic Science of Poisons. 2nd Edition (Doll, J. and Klaasen, C.D. eds). Macmillan, New York. pp 310–329.
3. Price, C.P. and Finney, H. (2000). Developments in the assessment of the glomerular filtration rate. *Clin Chim Acta* **297**, 55–66.
4. Shemesh, O., Golbetz, H., Kriss, J.P. and Myers, B.D. (1985). Limitations of creatinine as a filtration marker in glomerulopathic patients. *Kidney Int* **28(5)**, 830–838.
5. D'Amico, G and Bazzi, C. (2003) Urinary protein and enzyme excretion as markers of tubular damage. *Curr Opin Nephrol Hypertens* **12(6)**, 639–643.
6. Clyne, D.H. and Pollak, V.E. (1981) Renal handling and pathophysiology of Bence Jones proteins. *Contrib Nephrol* **24**, 78–87.
7. Okonogi, H., Nishimura, M., Utsunomiya, Y., Hamaguchi, K., Tsuchida, H., Miura, Y., Suzuki, S., Kawamura, T., Hosoya, T. and Yamada, K. (2001) Urinary type IV collagen excretion reflects renal morphological alterations and type IV collagen expression in patients with type II diabetes mellitus. *Clinical Nephrology* **55**, 357–364.
8. Hardwicke, J. (1965) Estimation of renal permeability to protein on Sephadex G200. *Clin Chim Acta* **12**, 89–96.
9. Hardwicke, J. and Squire, J.R. (1955) The relation between plasma albumin concentration and protein excretion in patients with proteinuria. *Clin Sci* **14(3)**, 509–530.
10. Pesce, A.J., Boreisha, I. and Pollak, V.E. (1972) Rapid differentiation of glomerular and tubular proteinuria by SDS polyacrylamide gel electrophoresis. *Clin Chim Acta* **40(1)**, 27–34.
11. Scherberich, J.E. (1989) Kidney and serum derived proteins in urine of patients suffering from renal diseases or arterial hypertension. *Klinisch Wochenschrift* **67(Suppl XVII)**, 44–47.
12. Grant, G.H. (1935) The proteins of normal urine II. From the urinary tract. *J Clin Pathol* **12**, 510–517.
13. Gillman, G. (1959) Urinary proteins: The appearance of kidney protein in the urine of some cases of severe chronic glomerulonephritis. *J Urol* **34**, 727–731.
14. Antoine, B. and Neveu, T. (1967) L'histurie, debit augmente des macromolecules tissulaires dans certains urine pathologiques. *Mars Med* **104**, 1–13.
15. Hardwicke, J. (1975) Laboratory aspects of proteinuria in human disease. *Clin Nephrol* **3(2)**, 37–41.
16. Scherberich, J.E., Mondorf, A.W., Falkenberg, F.W., Pierard, D. and Schoeppe, W. (1985) Monitoring tubular damage under clinical conditions: use of polyclonal and monoclonal antibodies as a tool. In Advances in Non-Invasive Nephrology (Lubec, G. and Campese, V. eds). Libbey, London. pp 278–285.
17. Schmidt, U. and Guder, W.G. (1976) Sites of enzyme activity along the nephron. *Kidney Int* **9**, 233–242.

18. Scherberich, J.E. (1990) Urinary proteins of tubular origin: Basic immunological and clinical aspects. *Am J Nephrol* **10(Suppl 1)**, 43–51.
19. Harrison, D.J. et al. (1989) Glutathione S-transferase isoenzymes in the human kidney: Basis for possible markers of renal injury. *J Clin Pathol* **42**, 624–629.
20. Price, R.G. (1982) Urinary enzymes, toxicity and renal disease. *Toxicology* **23**, 99–134.
21. http://www.mesoscalediscoveries.com.
22. Stuhler, K. and Meyer, H.E. (2004) MALDI: more than peptide mass fingerprints. *Curr Opin Mol Ther* **6**, 239–248.
23. Nguyen, M.T., Ross, G.F., Dent, C.L. and Devarajan, P. (2005) Early prediction of acute renal injury using urinary proteomics. *Am J Nephrol* **25(4)**, 318–326.
24. http://hesiglobal.og/committees/technical-committees/biomarkers/default.htm.
25. Mattenheimer, H. (1968) The enzymology of renal tissue. In Enzymes in Urine and Kidney (Dubach, U.C. ed). Hans Huber, Switzerland. pp 119–145.
26. Raab, W.P. (1972) Diagnostic value of urinary enzyme determinations. *Clin Chem* **18(1)**, 5–25.
27. Bourbouze, R., Baumann, F.C., Bonvalet, J.P. and Farman, N. (1984) Distribution of *N*-acetyl-β-D-glucosamidase isoenzymes along the rabbit nephron. *Kidney International* **25**, 636–647.
28. Bosomworth, M.P., Aparicio, S.R., Hay, W.H.M. (1989) Urine *N*-acetyl-β-D-glucosamidase – A biomarker of tubular damage? *Nephrol Dial Transplant* **14**, 620–626.
29. Sanchez-Bernal, C., Vlitosi, M., Cabezasi, J.A. and Price, R.G. (1991) Variation in the isoenzymes of *N*-acetyl-β-D-glucosamidase and protein excretion in aminoglycoside nephrotoxicity in the rat. *Cell Biochem Funct* **9**, 209–214.
30. Nouwen, E.J. and de Broe, M.E. (1994) Human intestinal versus tissue-nonspecific alkaline phosphatase as complementary urinary markers for the proximal tubule. *Kidney Int* **46 Suppl** 47, 43–51.
31. Jung, K. and Schulze, G. (1986) Diuresis-dependent excretion of multiple forms of renal brush-border enzymes in urine. *Clin Chim Acta* **156(1)**, 77–83.
32. Kotanko, P., Straunthaler, G. and Pfaller, W.W. (1984) Harnenzyme zur nichtinvasiven Diagnostik von Nierenepithtelschäden im akuten Nierenversagen. *Klin Wochenschr* **96(16)**, 625–629 [Article in German].
33. Ohata, H, Momose, K., Takahashi, A. and Omori, Y. (1987) Urinalysis for detection of chemically induced renal damage (1) – Changes in urinary excretions of enzymes and various components caused by mercuric chloride and Gentamicin. *J Toxicol Sci* **12(4)**, 341–355.
34. Ohata, H, Momose, K, Takahashi, A. and Omori, Y. (1987) Urinalysis for detection of chemically induced renal damage (2) – Changes in urinary excretions of enzymes and various components caused by p-aminophenol, puromycin aminonucleoside and hexadimethrine. *J Toxicol Sci* **12(4)**, 357–372.
35. Maatman, R.G., Van Kuppevelt, T.H. and Veerkamp, J.H. (1991) Two types of fatty acid-binding protein in human kidney. Isolation, characterization and localization. *Biochem J* **273(3)**, 759–766.
36. Kimura, H., Odani, S., Nishi, S., Sato, H., Arakawa, M. and Ono, T. (1991) Primary structure and cellular distribution of two fatty acid-binding proteins in adult rat kidneys. *J Biol Chem* **266(9)**, 5963–5972.
37. Shaw, M. (2005) The use of histologically defined specific biomarkers in drug development with special reference to the glutathione S-transferases. *Cancer Biomark* **1**, 69–74.
38. Rozell, B., Hansson, H.A., Guthenberg, C., Tahir, M.K. and Mannervik, B. (1993) Glutathione transferases of classes alpha, mu and pi show selective expression in different regions of rat kidney. *Xenobiotica* **23(8)**, 835–849.
39. Feinfeld, D.A. and Fuh, Y.L, (1986) Urinary glutathione-S-transferase in cisplatin nephrotoxicity in the rat. *J Clin Chem Clin Biochem* **8**, 529–532.
40. Beckett, G.J. and Hayes, J.D. (1987) Glutathione S-transferase measurements and liver disease in man. *J Clin Biochem Nutr* **2**, 1–24.
41. Falkenberg, F.W., Hildebrand, H., Lutte, L., Schwengberg, S., Henke, B., Greshake, D., Schmidt, B., Friederich, A., Rinke, M., Schlüter, G. and Bomhard, E. (1996) Papillary antigens as markers of papillary toxicity. I. Identification and characterisation of rat papillary antigens with monoclonal antibodies. *Arch Toxicol* **71**, 80–92.
42. Taal, M.W. and Brenner, B.M. (2006) Predicting initiation and progression of chronic kidney disease: Developing renal risk scores. *Kidney Int* **70**, 1694–1705.
43. Abbate, M., Zoja, C. and Remuzzi, G.J. (2006) How does proteinuria cause progressive renal damage? *Am Soc Nephrol* **17(11)**, 2974–2984.
44. Holdt-Lehmann, B., Lehmann, A., Korten, G., Nagel, H., Nizze, H. and Schuff-Werner, P. (2000) Diagnostic value of urinary alanine

aminopeptidase and *N*-acetyl-beta-D-glucosaminidase in comparison to alpha 1-microglobulin as a marker in evaluating tubular dysfunction in glomerulonephritis patients. *Clin Chim Acta* **297(1–2)**, 93–102.

45. Cameron, J.S. (1992) Tubular and interstitial factors in the progression of glomerulonephritis. *Pediatr Nephrol* **6**, 292–303.
46. Bazzi, C., Petrini, C., Rizza, V., Arrigo, G., Napodano, P., Paparella, M. and D'Amico, G. (2002) Urinary *N*-acetyl-β-glucoronidase is a marker of tubular cell dysfunction and a predictor of outcome in primary glomerulonephritis. *Nephrol Dial Transplant* **17**, 1890–1896.
47. Kamijo, A., Sugaya, T., Hikawa, A., Yamanouchi, M., Hirata, Y., Ishimitsu, T., Numabe, A., Takagi, M., Hayakawa, H., Tabei. F., Sugimoto, T., Mise, N. and Kimura, K. (2005) Clinical evaluation of urinary excretion of liver-type fatty acid-binding protein as a marker for the monitoring of chronic kidney disease: a multicenter trial. *J Lab Clin Med* **145(3)**, 125–133.
48. Branten, A.J., Mulder, T.P., Peters, W.H., Assmann, K.J. and Wetzels, J.F.. (2000) Urinary excretion of isoenzymes of glutathione S-transferase alpha and pi in patients with proteinuria. Reflection of the site of tubular injury. *Nephron* **85**, 120–126.
49. Nenov, V.D., Taal, M.W., Sakharova, O.V. and Brenner B.M. (2000) Multi-hit nature of chronic renal disease. *Curr Opin Nephrol Hypertens* **9**, 85–97.
50. Wang, A.-Y. and Lai, K.-N. (2006) The importance of residual renal function in dialysis patients. *Kidney Int* **69**, 1726–1732.
51. Salvesen Blix, H., Kilvik Viktil K., Tron, A.M. and Reikvam, A. (2006) Use of renal risk drugs in hospitalized patients with impaired renal function an underestimated problem? *Nephrol Dial Transplant* **21**, 3164–3171.
52. Uslu, S., Efe, B., Alatas, O., Kebapçi, N., Colak, O., Demirüstü, C. and Yörük, A. (2005) Serum cystatin C and urinary enzymes as screening markers of renal dysfunction in diabetic patients *J Nephrol* **18**, 559–567.
53. Maxwell, P.R., Gordon, D. and Coyle, A. (2004) Differentiation between renal injury and compensation and compensatory responses by the use of specific biomarkers. Poster presented at the 43rd annual meeting of the American Society of Toxicology. Baltimore USA, March 21–25, 2004.
54. Bernard, A., Thielemans, N., Roels, H. and Lauwerys, R. (1995) Association between NAG-B and cadmium in urine with no evidence of a threshold. *Occup Environ Med* **52(3)**, 177–180.
55. Garcon, G., Leleu, B.F., Marez, T., Haguenoer, J.-M., Furon, D. and Shirali, P. (2004) Biologic markers of oxidative stress and nephrotoxicity as studied in biomonitoring of adverse effects of occupational exposure to lead and cadmium. *J Occup Environ Med* **46(11)**, 1180–1186.
56. Green, T., Dow, J., Ong, C.N., Ng, V., Ong, H.Y., Zhuang, Z.X., Yang, X. F. and Bloemen, L. (2005) Biological monitoring of kidney function among workers occupationally exposed to trichloroethylene. *Occup Environ Med* **61**, 312–317.
57. Hariharan, S., Johnson, C.P., Bresnahan, B.A., Taranto, S.E., McIntosh, M.J. and Stablein, D. (2000) Improved graft survival after renal transplantation in the United States, 1988 to 1996. *N Engl J Med* **342**, 605–612.
58. Jung, K., Diego, J., Strobeft, V., Scholz, D. and Schreiber, G. (1986) Diagnostic significance of some urinary enzymes for detecting acute rejection crises in renal-transplant recipients: Alanine aminopeptidase, alkaline phosphatase, γ-glutamyltransferase, *N*-acetyl-β-D-glucosaminidase and lysozyme. *Clin Chem* **32(10)**, 1807–1811.
59. Kotanko, P., Keiler, R., Knabl, L., Aulitzky, W., Margreiter, R., Gstraunthaler, G. and Pfaller, W. (1986) Urinary enzyme analysis in renal allograft transplantation. *Clin Chim Acta* **160(2)**, 137–144.
60. Sundberg, A.G.M., Appelkvist, E.-L., Bäckman, L. and Dallner, G. (1994) Urinary pi-class glutathione transferase as an indicator of tubular damage in the human kidney. *Nephron* **67**, 308–316.
61. Wellwood, J.M., Ellis, B.G., Hall, J.H., Robinson, D.R. and Thompson, A.E. (1973) Early warning of rejection? *Br Med J* **2**, 261–265.
62. Bornstein, B., Arenas, J., Morales, J.M., Praga, M., Rodicio, J.L., Martinez, A. and Valdivieso, L. (1996) Cyclosporine nephrotoxicity and rejection crisis: diagnosis by urinary enzyme excretion. *Nephron* **72(3)**, 402–406.
63. Bäckman, L., Appelkvist, E.-L., Ringden, O. and Dallner, G. (1998) Glutathione transferase in the urine: A marker for post-transplant tubular lesions. *Kidney Int* **33**, 571–577.
64. Iványi, B., Hansen, H.E. and Olsen, S. (1993) Segmental localization and quantitative characteristics of tubulitis from patients undergoing acute rejection. *Transplantation* **56(3)**, 581–585.
65. Falkenberg, F.W., Mai, U., Puppe, P., Riise, P., Herrmann, G., Hecking E., Bremer, K., Mondorf, A.W. and Shapira, Z. (1980)

Kidney-derived urinary antigens assayed with monoclonal antibodies for the detection of renal damage. *Clin Chim Acta* **160**, 171–182.
66. Kootstra, G.K., Kievit, J.K. and Nederstigt, A. (2002) Organ donors, heartbeating and non-heartbeating. *World J Surg* **26**, 181–184.
67. Daeman, J.-W. H.C., Oomen, A.P.A., Janssen, M.A., van de Schoot, L., van Kreel, B.K., Heineman, E. and Kootstra, G.K. (1997) Glutathione S-transferase as predictor of functional outcome in transplantation of machine preserved non-heart-beating donor kidneys. *Transplantation* **63(1)**, 89–93.
68. Carter, J.L., Tompson, C.R.V., Stevens, P.E. and Lamb, E.J. (2006) Does urinary tract infection cause proteinuria or microalbuminuria? A systematic review. *Nephrol Dial Transplant* **21**, 3031–3037.
69. Mengoli, C., Leche, A., Arosio, E., Rizzotti, P., Lechi, C., Corgnati, A., Micciolo, R. and Pancera, P. (1982) Contributions of four markers of tubular proteinuria in detecting upper urinary tract infections. A multivariate analysis. *Nephron* **32**, 234–238.
70. Ivanyi, B., Rumpeldt, H.J. and Thoenes, W. (1988) Acute human pyelonephritis: Leukocytic infiltration of the tubules and localization of bacteria. *Virchows Arch A Pathol Anat* **414**, 29–37.
71. Sundberg, A.G.M. (1997) Glutathione S-transferase. Markers for renal disease. Doctoral dissertation Karolinska Institute, Huddinge, Sweden.
72. Bouissou, F., Birambaux, X. et al. (1998) Urinary glutathione-S-transferase: excretion in normal children and children with pyelonephritis. Presented at the meeting of the French Society of Infectious Diseases in Paediatrics. Limoges, France, May 1998.
73. Westhuyzen, J., Endre, Z.H., Reece, G., Reith, D.M., Saltissi, D. and Morgan, T.J. (2003) Measurement of tubular enzymuria facilitates early detection of acute renal impairment in the intensive care unit. *Nephrol Dial Transplant* **18**, 543–551.
74. Cressey, G., Roberts, D.R. and Snowden, C.P. (2002) Renal tubular injury after infrarenal aortic aneurysm repair. *J Cardiothorac Vasc Anesth* **16(3)**, 290–293.
75. Choudhury, D. and Ahmed, Z. (2006) Drug-associated renal dysfunction and injury. *Nat Clin Pract Nephrol* **2(2)**, 80–91.
76. Hartmann, J.T., Fels, L.M., Franzke, A., Knop, S., Renn, M., Maess, B, Panagiotou, P., Lampe, H., Kanz, L., Stolte, H. and Bokemeyer, C. (2000) Comparative study of the acute nephrotoxicity from standard dose cisplatin +/− ifosfamide and high-dose chemotherapy with carboplatin and ifosfamide. *Anticancer Res* **20(5C)**, 3767–3773.
77. Jakobsen, J.A., Nossen, J.O., Jorgensen, N.P. and Berg, K.J. (1993) Renal tubular effects of diuretics and X-ray contrast media. A comparative study of equimolar doses in healthy volunteers. *Invest Radiol* **4**, 319–324.
78. Ohta, S., Ishimitsu, T., Minami, J., Ono, H. and Matsuoka, H. (2005).Effects of intravascular contrast media on urinary excretion of liver fatty acid-binding protein] *Nippon Jinzo Gakkai Shi* **47(4)**, 437–444 [Article in Japanese].
79. Pai, M.P., Norenberg, J.P., Telepak, R.A., Sidney, D.S. and Yang, S. (2005) Assessment of effective renal plasma flow, enzymuria, and cytokine release in healthy volunteers receiving a single dose of Amphotericin B desoxycholate. *Antimicrob Agents Chemother* **49(9)**, 3784–3788.
80. Behrends-Steins, B., Sorensen, R. and Langer, M. (1991) The evaluation of renal tolerance for roentgen contrast media by determination of urinary kidney-specific proteins. *Wien Med Wochenschr* **141(8)**, 164, 166–70 [Article in German].
81. Sherman, R.A., Feinfeld, D.A., Ohmi, N., Arias I.M. and Levine, S.D. (1984–1985) A prospective study of urinary ligandin in patients at risk of renal tubular injury. *Uremia Invest* **8(2)**, 111–115.
82. Donta, S.T. and Lembke, L.A. (1985) Comparative effects of gentamicin and tobramycin on excretion of *N*-acetyl-β-D-glucosaminidase. *Antimicrob Agents Chemother* **28(4)**, 500–503.
83. Dupond, J.L., Gibey, R., Iehl-Robert, M., Mallet, H., Leconte des Floris, R. and Henry, J.C. (1984) Nephrotoxicity of aminoglycosides and cephalosporins. Value of examining the isoenzyme profile of *N*-acetyl-beta-D-glucosaminidase. *Rev Med Interne* **4**, 321–327 [Article in French].
84. Cittanova, M.L. (2001) Is peri-operative renal dysfunction of no consequence? *Br J Anaesth* **86(2)**, 164–166.
85. Ezzell, C. (2003) The price of pills. *Scientific American* July, 16.
86. Crawford, L.M. Acting Commissioner of the FDA's speech to the Drug Information Association 26 October 2004. http://www.fda.gov/oc/speeches/2004/dia1026.html.
87. Stonard, M.D. (1986) Assessment of nephrotoxicity. In Animal Clinical Chemistry, A Primer for Toxicologists (Evans, G.O. ed). Taylor & Francis, London. pp 87–98.
88. Sonee, M., Hannay, A.K. and Abele, L.G. (2007) Evaluation of kidney toxicity biomarkers in human kidney cells. Presentation at the

46th Society of Toxicology meeting. Charlotte, NC, USA, 25–29 March 2007.

89. Vickers, A.E.M., Alegret, M., Jimenez, R.M., Pflimlin, V., Fisher, R., Spaans, C. and Brendel, K. (1998) Changes in human liver and kidney slice function related to potential side-effects in the presence of biotransformation of 4 cyclosporin derivatives, CSA, IMM. OG and PSC. *In-vitro and Molecular Toxicology* **11(2)**, 119–131.

90. Vickers, A.E.M. et al. (1998) Changes in human liver and kidney slice function related to potential side-effects in the presence of biotransformation of 4 cyclosporin derivatives, CSA, IMM. OG and PSC. *In vitro Mol Toxicol* **11(2)**, 119–131.

91. Kharasch, E.D., Hoffman, G.M., Thorning, D., Hankins, D.C. and Kilty, C.G. (1998) Role of renal cysteine conjugate β-lyase pathway in inhaled compound A nephrotoxicity in rats. *Anesthesiology* **88(6)**, 1624–1633.

92. Goldberg, M.E., Cantillo, J., Gratz, I., Deal, E., Vekeman, D., McDougall, R., Afshar, M., Zafeiridis, A. and Larijani, G. (1999) Dose of compound A, not Sevoflurane, determines changes in the biochemical markers of renal injury in volunteers. *Anesth Analg* **88**, 437–445.

93. Davies, D., Bradshaw, J., Coope, M., Westwood, R., Knight, R. and Murfin, K. (2000) Novel biomarkers for the detection of regional kidney injury in the rat Poster presented at the EMBODY meeting. Cambridge, England, April 3–7, 2000.

94. Tataranni, G., Zavagli, G., Farinelli, R., Malacarne, F., Fiocchi, O., Nunzi, L., Scaramuzzo, P. and Scorrano, R. (1992) Usefulness of the assessment of urinary enzymes and microproteins in monitoring ciclosporin nephrotoxicity. *Nephron* **60(3)**, 314–318.

95. Ahlmén, J., Sundberg, A., Gustavsson, A. and Strömbom, U. (1995) Decreased nephrotoxicity after the use of a microemulsion preparation of cyclosporin A compared to a conventional solution. *Transplant Proc* **27(6)**, 3432–3437.

96. Sekeroglu, M.R., Kati, I., Noyan, T., Dulger, H. and Yalcinkaya, A.S. (2005) Alterations in the biochemical markers of renal function after Sevoflurane anaesthesia. *.Nephrology* **10(6)**, 544–547.

97. Eger II, E.I., Koblin, D.D., Bowland, T., Ionescu, P., Laster, M.J., Fang, Z., Gong, D., Sonner, J. and Weiskopf, R.B. (1997) Nephrotoxicity of Sevoflurane versus Desflurane in volunteers. *Anesthesia and Analgesia* **84**, 160–168.

98. Jones, S.E. and Jomary, C. (2002) Molecules in focus: clusterin. *Int J Biochem Cell Biol* **34(5)**, 427–431.

99. Silkensen, J.R., Skubitz, K.M., Skubitz, A.P.N., Chmielewski, D.H., Manivel, J.C., Dvergsten, J.A. and Rosenberg, M.E. (1995). Clusterin promotes the aggregation and adhesion of renal porcine epithelial cells. *J Clin Invest* **96**, 2646–2653.

100. Zellweger, T., Miyake, H., July, L.V., Akbari, M., Kiyama, S. and Gleave, M.E. (2001) Chemosensitization of human renal cell cancer using antisense oligonucleotides targeting the antiapoptotic gene clusterin. *Neoplasia* **3(4)**, 360–367.

101. Girton, R.A., Sundin, D.P. and Rosenberg, M.E. (2002) Clusterin protects renal tubular epithelial cells from gentamicin-mediated cytotoxicity. *Am J Physiol Renal Physiol* **282(4)**, F703–F709.

102. Witzgall, R., Brown, D., Schwarz, C. and Bonventre, J.V. (1994) Localization of proliferating cell nuclear antigen, vimentin, c-fos, and clusterin in the postischemic kidney: evidence for a heterogenous genetic response among nephron segments, and a large pool of mitotically active and dedifferentiated cells. *J Clin Invest* **93**, 2175–2188.

103. Ichimura, T., Bonventre, J.V., Bailly, V., Wei, H., Hession, C.A., Cate, R.L. and Sanicola, M. (1998) Kidney injury molecule-1 (KIM-1), a putative epithelial cell adhesion molecule containing a novel immunoglobulin domain, is up-regulated in renal cells after injury. *J Biol Chem* **273(7)**, 4135–4142.

104. Kjeldsen, L., Johnsen, A.H., Sengeløv, H. and Borregaard, N. (1993). Isolation and primary structure of NGAL, a novel protein associated with human neutrophil gelatinase. *J Biol Chem* **268(14)**, 10425–10432.

105. Yang, J., Mori, K., Li, J.Y. and Barasch, J. (2003). Iron, lipocalin, and kidney epithelia. *Am J Physiol Renal Physiol* **285**, F9–F18.

106. Devarajan, P. (2006). Update on mechanisms of ischemic acute kidney injury. *Am Soc Nephrol* **17**, 1503–1520.

107. Mishra, J., Mori, K., Ma, Q., Kelly, C., Yang, J., Mitsnefes, M., Barasch, J. and Devarajan, P. (2004). Amelioration of ischemic acute renal injury by neutrophil gelatinase-associated lipocalin. *J Am Soc Nephrol* **15**, 3073–3082.

108. van Timmeren, M.M., van den Heuvel, M.C., Bailly, V., Bakker, S.J., van Goor, H., Stegeman, C.A. (2007) Tubular kidney injury molecule-1 (KIM-1) in human renal disease. *J Pathol* **212(2)**, 209–217.

109. Correa Rotter, R., Ibarra Rubio, M.E., Schwochau, G., Cruz, C., Silkensen, J.R., Chaverri, J.P., Chmielewski, D. and Rosenberg, M.E. (1998). Induction of clusterin in tubules of nephrotic rats. *J Am Soc Nephrol* **9**, 33–37.

110. Brunner, H.I., Mueller, M., Rutherford, C., Passo, M.H., Witte, D., Grom, A., Mishra, J. and Devarajan, P. (2006) Urinary neutrophil gelatinase-associated lipocalin as a biomarker of nephritis in childhood-onset systemic lupus erythematosus. *Arthritis Rheum* **54(8)**, 2577–2584.
111. Ding, H., He, Y., Li, K., Yang, J., Li, X., Lu, R. and Gao, W. (2007) Urinary neutrophil gelatinase-associated lipocalin (NGAL) is an early biomarker for renal tubulointerstitial injury in IgA nephropathy. *Clin Immunol* **123(2)**, 227–234.
112. Ishii, A., Sakai, Y. and Nakamura, A. (2007) Molecular pathological evaluation of clusterin in a rat model of unilateral ureteral obstruction as a possible biomarker of nephrotoxicity. *Toxicol Pathol* **35(3)**, 376–382.
113. Chevalier, R.L., Kim, A., Thornbill, B.A. and Wolstenholme, J.T. (1999) Recovery following relief of unilateral renal obstruction in the neonatal rat. *Kidney Int* **55**, 793–807.
114. Mishra, J., Ma, Q., Kelly, C., Mitsnefes, M., Mori, K., Barasch, J. and Devarajan, P. (2006) Kidney NGAL is a novel early marker of acute injury following transplantation. *Pediatr Nephrol* **21(6)**, 856–863.
115. Parikh, C.R., Jani, A., Mishra, J., Ma, Q., Kelly, C., Barasch, J., Edelstein, C.L. and Devarajan, P. (2006). Urine NGAL and IL-18 are predictive biomarkers for delayed graft function following kidney transplantation. *Am J Transplant* **6(7)**, 1639–1645.
116. Schaub, S., Mayr, M., Hönger, G., Bestland, J., Steiger, J., Regeniter, A., Mihatsch, M.J., Wilkins, J.A., Rush, D. and Nickerson, P. (2007) Detection of subclinical tubular injury after renal transplantation: comparison of urine protein analysis with allograft histopathology. *Transplantation* **84(1)**, 104–112.
117. Dvergsten, J., Manivel, J.C., Correa-Rotter, R. and Rosenberg M.E. (1994) Expression of clusterin in human renal diseases. *Kidney Int* **45(3)**, 828–835.
118. Hidaka, S., Kränzlin, B., Gretz, N. and Witzgall, R. (2002).Urinary clusterin levels in the rat correlate with the severity of tubular damage and may help to differentiate between glomerular and tubular injuries. *Cell Tissue Res* **310**, 289–296.
119. Han, W.K., Bailly, V., Abichandani, R., Thadhani, R. and Bonventre, J.V. (2002) Kidney Injury Molecule 1 (KIM-1): A novel biomarker for human renal proximal tubule injury. *Kidney Int* **62**, 237–244.
120. Yoshida, T., Kurella, M., Beato, F., Min, H., Ingelfinger, J.R., Stears, R.L., Swinford, R.D., Gullans, S.R. and Tang, S.-S. (2002) Monitoring changes in gene expression in renal ischemia-reperfusion in the rat. *Kidney Int* **61**, 1646–1654.
121. Mishra, J., Ma, Q., Prada, A., Mitsnefes, M., Zahedi, K. Yang, J., Barasch, J. and Devarajan, P. (2003) Identification of neutrophil gelatinase-associated lipocalin as a novel early urinary biomarker for ischemic renal injury. *J Am Soc Nephrol* **14**, 2534–2543.
122. Mishra, J., Dent, C., Tarabishi, R., Mitsnefes, M.M., Ma, Q., Kelly, C., Ruff, S.M., Zahedi, K., Shao, M., Bean, J., Mori, K., Barasch, J. and Devarajan, P. (2005) Neutrophil gelatinase-associated lipocalin (NGAL) as a biomarker for acute renal injury after cardiac surgery. *Lancet* **365(9466)**, 1231–1238.
123. Wagener, G., Jan, M., Kim, M., Mori, K., Barasch, J.M., Sladen, R.N. and Lee, H.T. (2006) Association between increases in urinary neutrophil gelatinase-associated lipocalin and acute renal dysfunction after adult cardiac surgery. *Anesthesiology* **105(3)**, 485–491.
124. Perianayagam, M.C., Vaidya, V.S., Han, W.K., Wald, R., Tighiouart, H., MacKinnon, R.W., Li, L., Balakrishnan, V.S., Pereira, B.J., Bonventre, J.V., Jaber, B.L. and Liangos, O. (2007) Urinary *N*-acetyl-beta-(D)-glucosaminidase activity and kidney injury molecule-1 level are associated with adverse outcomes in acute renal failure. *J Am Soc Nephrol* **18(3)**, 904–912.
125. Silkensen, J.R., Agarwal, A., Nath, K.A., Manivel, J.C. and Rosenberg, M.E. (1997) Temporal induction of clusterin in cisplatin nephrotoxicity. *J Am Soc Nephrol* **8**, 302–305.
126. Ichimura, T., Hung, C.C., Yang, S.A., Stevens, J.L. and Bonventre, J.V. (2004) Kidney injury molecule-1: a tissue and urinary biomarker for nephrotoxicant-induced renal injury. *Am J Physiol Renal Physiol* **286**, F552–F563.
127. Mishra, J., Mori, K., Ma, Q., Kelly, C., Barasch, J. and Devarajan, P. (2004) Neutrophil gelatinase-associated lipocalin: a novel early urinary biomarker for cisplatin nephrotoxicity. *Am J Nephrol* **24(3)**, 307–315.
128. Aulitzky, W.K., Schlegel, P.N., Wu, D.F., Cheng, C.Y., Chen, C.L., Li, P.S., Goldstein, M., Reidenberg, M. and Bardin, C.W. (1992) Measurement of urinary clusterin as an index of nephrotoxicity. *Proc Soc Exp Biol Med* **199(1)**, 93–96.
129. Davis, J.W. II., Goodsaid, F.M., Bral, C.M., Obert, L.A., Mandakas, G., Garner, C.E. II., Collins, N.D., Smith, R.J. and Rosenblum, I.Y. (2004) Quantitative gene expression analysis in a nonhuman primate model of

antibiotic-induced nephrotoxicity. *Toxicol Appl Pharmacol* **200(1)**, 16–26.

130. Eti, S., Cheng, C.Y., Marshall, A. and Reidenberg, M.M. (1993) Urinary clusterin in chronic nephrotoxicity in the rat. *Proc Soc Exp Biol Med* **202(4)**, 487–490.
131. Kharasch, E.D., Schroeder, J.L., Bammler, T., Beyer, R. and Srinouanprachanh, S. (2006). Gene expression profiling of nephrotoxicity from the sevoflurane degradation product fluoromethyl-2,2-difluoro-1-(trifluoromethyl) vinyl ether ("compound A") in rats. *Toxicol Sci* **90(2)**, 419–431.
132. Han, W.K., Bailly, V., Abichandani, R., Thadhani, R. and Bonventre, J.V. (2002) Kidney Injury Molecule-1 (KIM-1): A novel biomarker for human renal proximal tubule injury. *Kidney Int* **62**, 237–244.
133. Hildebrand, H., Rinke, M., Schlüter, G., Bomhard, E. and Falkenberg, F.W. (1999) Urinary antigens as markers of papillary toxicity. II Application of monoclonal antibodies for the determination of papillary antigens in rat urine. *Arch Toxicol* **73**, 233–245.
134. Kilty, C., Shaw, M., Falkenberg, F.W., Elliott, G., Betton, G. and Roche, A. (2005) Identification of renal papillary necrosis using an EIA for urinary Renal Papillary Antigen-1 (RPA-1); A new biomarker of collecting duct pathology. Poster presented at the 41st Society of Toxicology meeting, New Orleans, USA, March 7–9 2005.
135. Serkova, N. and Christians, U. (2003) Transplantation: toxicokinetics and mechanisms of toxicity of Cyclosporine and macrolides. *Curr Opin Investig Drugs* **4**, 1287–1296.
136. Amin, R.P. et al. (2004) Identification of putative gene based markers of renal toxicity. *Environ Health Perspect* **112**, 465–479.
137. Charlwood, J., Skehel, J.M., King, N., Camiller, P., Lord, P., Bugelski, P. and Atif, U. (2002) Proteomic analysis of rat kidney cortex following treatment with gentamicin. *J Proteome Res* **1(1)**, 73–82.
138. Hampel, D.J., Sansome, C., Sha, M., Brodsky, S., Lawson, W.E. and Goligorsky, M.S. (2001) Toward proteomics in uroscopy: urinary protein profiles after radiocontrast medium administration. *J Am Soc Nephrol* **12(5)**, 1026–1035.
139. Schaub, S., Rush, D., Wilkins, J., Gibson, I.W., Weiler, T., Sangster, K., Nicolle, L., Karpinski, M., Jeffery, J. and Nickerson, P. (2004) Proteomic-based detection of urine proteins associated with acute renal allograft rejection. *J Am Soc Nephrol* **15**, 219–227.
140. Bandara, L.R., Kelly, M.D., Lock, E.A. and Kennedy, S. (2003) A potential biomarker of kidney damage identified by proteomics: Preliminary findings. *Biomarkers* **8(3–4)**, 272–286.
141. Nicholson, J.K., Connelly, J., Lindon, J.C. and Holmes, E. (2002) Metabonomics, a platform for studying drug toxicity and gene function. *Nat Drug Rev* **1**, 153–160.
142. Gartland, K.P., Bonner, F.W., Timbrell, J.A. and Nicholson, J.K. (1989) Biochemical characterisation of para-aminophenol-induced nephrotoxic lesions in the F344 rat. *Arch Toxicol* **63(2)**, 97–106.
143. Holmes, E., Nicholls, A.W., Lindon, J.C., Ramos, S., Spraul, M., Neidig, P., Connor, S.C., Connelly, J., Damment, S.J., Haselden, J. and Nicholson, J.K. (1998) Development of a model for classification of toxin-induced lesions using 1H NMR spectroscopy of urine combined with pattern recognition. *NMR Biomed.* **11**, 235–244.
144. Holmes, E., Bonner, F.W., Sweatman, B.C., Lindon, J.C., Beddell, C.R., Rahr, E. and Nicholson, J.K. (1992) Nuclear magnetic resonance spectroscopy and pattern recognition analysis of the biochemical processes associated with the progression of and recovery from nephrotoxic lesions in the rat induced by mercury(II) chloride and 2-bromoethanamine. *Mol. Pharmacol.* **42**, 922–930.
145. Okamoto, K., Collette W. III., Mondal, M., de Peyster, A. and Stevens, G.J. (2005) Evaluation of kidney toxicity detection methods in rat urine: GSTs by α and μGST enzyme immunoassays, metabolite profiling by NMR, and protein profiling by SELDI-TOF-MS. Poster presented at the 44th annual meeting of the Society of Toxicology. New Orleans, USA.
146. Nicholson, J.K., Holmes, E., Lindon, J.C. and Wilson, I.D. (2004) The challenges of modelling mammalian complexity. *Nat. Biotechnol.* **22**, 1268–1274.
147. http://www.fda.gov/cder/genomics/default.htm.
148. http:// hesiglobal.og/committees/technicalcommittees/biomarkers/default.htm.

Chapter 17

Proteomic Assays for the Detection of Urothelial Cancer

Kris E. Gaston and H. Barton Grossman

Abstract

Bladder cancer is one of the most expensive cancers from diagnosis to death of the patient due to life-long surveillance involving upper tract imaging, urinary cytology, and cystoscopy. Cytology has been historically used in conjunction with cystoscopy to help detect disease that may be missed by routine cystoscopy (e.g., carcinoma in situ and upper tract disease). Urine cytology is highly cytopathologist dependent and has reasonable sensitivity for detecting high grade disease. However, its sensitivity drops precipitously with regard to well-differentiated low grade cancers. Intensive investigations have been undertaken using proteomics to find an alternative to cystoscopy and cytology. Urine proteomic markers currently evaluated critically in the literature include bladder tumor antigen, nuclear matrix protein 22, BLCA-4, hyaluronic acid, hyaluronidase, cytokeratin 8, cytokeratin 18, cytokeratin 19, tissue polypeptide antigen, and tissue polypeptide-specific antigen. Markers used as alternatives to cystoscopy must be accurate with high sensitivity and specificity, cost effective for life-long surveillance, and minimally invasive to minimize the burden to the patient. To date, no proteomic marker has been developed that can replace cystoscopy for the detection of bladder cancer. However, several urinary markers appear to have higher sensitivity albeit lower specificity than cytology and can be used to supplement cystoscopy. Some of those markers are herein described in this chapter. By defining and characterizing the current state of the art in protein based markers, we are poised to evaluate and benchmark newly discovered protein biomarkers that will be isolated through new proteomics based investigations of urine.

Key words: Bladder tumor antigen, Nuclear matrix protein 22, BLCA-4, Hyaluronic acid, Hyaluronidase, Cytokeratin 8, Cytokeratin 18, Cytokeratin 19, Urine cytology, Tissue polypeptide antigen, Tissue polypeptide-specific antigen

1. Introduction

1.1. Epidemiology

Bladder cancer is the fourth most common cancer and the eighth most common cause of cancer-specific mortality in United States men. In 2007, approximately 67,160 new cases of bladder cancer and 13,750 deaths resulted from this disease (1). According to the Surveillance Epidemiology and End Results (SEER) database between 2000 and 2003, the age-adjusted incidence rate for

Alex J. Rai (ed.), *The Urinary Proteome: Methods and Protocols*, Methods in Molecular Biology, vol. 641,
DOI 10.1007/978-1-60761-711-2_17, © Springer Science+Business Media, LLC 2010

bladder cancer was 20.9 cases/100,000 (2). According to the same registries, men were diagnosed with bladder cancer at a rate of 37 cases/100,000 compared to women diagnosed at 9.3 cases/100,000 (fourfold difference). Caucasian American men have the highest incidence of bladder cancer (40.2 cases/100,000), followed by African American men (19.8 cases/100,000), followed by Caucasian American women (10 cases/100,000) and lastly African American women (7.4 cases/100,000). During the same period, SEER mortality data for bladder cancer shows that African American men die at a slightly lower rate than Caucasian men (5.4 deaths/100,000 versus 7.8 deaths/100,000); however, African American women die at a slightly higher rate than Caucasian American women (2.8 deaths/100,000 versus 2.3 deaths/100,000).

Approximately 70% of bladder tumors present as nonmuscle invasive (NMI) disease, and 10–20% progress to muscle invasion. The term "NMI" represents a diverse spectrum disease that includes Ta, Tis, and T1 lesions. It is well established that tumor grade is an independent predictor of progression. Low grade tumors tend to be NMI with a recurrence rate of 50–70%; however, progression to muscle invasion is low (less than 5% of cases). High grade tumors (including Tis) have a high recurrence rate (approximately 80%) and up to 50% progress in stage. Invasion of the muscular wall of the bladder is associated with a higher risk for metastasis and death.

1.2. Economics of Bladder Cancer Diagnosis and Surveillance

According to SEER data, as of January 1, 2001, there were 505,765 men and women alive who had a history of bladder cancer (372,313 men and 133,452 women) (2). The vast majority of patients present with NMI disease treated by transurethral resection (TUR) with or without intravesical therapy (i.e., bacillus Calmette-Guérin (BCG) or chemotherapy). The current standard is to follow these patients with cystoscopy, urine cytology or other tests, and upper tract imaging throughout their lifetime. Bladder cancer is the fifth most expensive cancer in the United States in Medicare expenditures with an estimated cost in 2001 of $ 3.7 Billion (3). Additionally, bladder cancer has the highest Medicare expenditure per-patient from diagnosis to death due to the long survival with NMI disease combined with life-long surveillance (2001 values, Bladder Cancer $96,000–187,000/patient) (3).

The need for active surveillance should not be underestimated even for low-risk patients. In a study that examined recurrence and progression of 152 patients with Ta grade 1 urothelial carcinoma (the lowest risk bladder tumor category), at a mean follow-up of 76 months, tumor recurrence occurred in 55% of patients, with 27% of those recurrences occurring 2–5 years after the first bladder tumor and 14% occurring after

5 years. Of those that recurred, 37% progressed in grade or stage, including 5 that progressed to muscle invasion (4). In another long-term study of 217 patients with Ta urothelial carcinoma at a mean follow-up of 84 months, 61% of tumors recurred and 30% progressed in stage (5). These findings substantiate the need for long-term surveillance.

1.3. Urine Cytology: The Standard

Urine microscopy dates back to the nineteenth century when Sanders and Dickenson described the presence of abnormal cellular findings in the urine of men that were diagnosed with bladder cancer (6, 7). Papanicolaou and Marshall in the mid-twentieth century used urine cytology to assist in the clinical detection of bladder cancer (8). Urine cytology features suspicious for malignancy include increased cellularity, increased cell size, increased nuclear-to-cytoplasmic ratio, dense hyperchromatic nuclei, and variability in nuclear size. Urine cytology can be an effective adjunct to cystoscopy in the surveillance for new bladder lesions as cell–cell basement membrane cohesiveness is lost in malignant degeneration allowing for collection of exfoliated cancer cells by either voided cytology or bladder-wash cytology. It has been well established that bladder washes will cause these cells to be exfoliated resulting in a greater cellular yield than voided urine specimens. Voided urine cytology is noninvasive but is highly subjective, as demonstrated by a recent international evaluation of urine cytology diagnoses (9). Cytopathologists must be able to make relatively accurate diagnoses sometimes using specimens with low cellular yields. Additionally, cellular yield can be highly variable between voided specimens. In one study of patients with biopsy-proven high grade urothelial cancer, 27% of urine specimens had no tumor cells (10). Studies have also assessed the positive predictive value of urine cytology, which varies between 17 and 100% depending on the histological grade of the lesion. The higher the grade the better the diagnostic yield. In particular, carcinoma in situ of the bladder tends to have the highest diagnostic yield, approaching 95–100%, due to the lack of cohesiveness, its superficial nature, and highly dysmorphic cells. The sensitivity of urine cytology generally ranges from 11% to 76% and its specificity exceeds 90% (11). The positive predictive value of urine cytology drops substantially with low grade lesions as well as with upper tract urothelial carcinoma. Other factors that can influence the interpretation of a urine cytologic specimen are infection/inflammation, intravesical chemotherapy, BCG therapy, or instrumentation. Each of these conditions can cause abnormal exfoliation of the urothelium and can lead to suspicious but spurious results. Urine cytology cannot be the sole monitoring tool for bladder cancer because a high number of low grade cancers will be missed and some tumors will progress in stage and grade.

Another potential method for following patients after being rendered disease-free by TUR with or without intravesical therapy for bladder cancer is urine proteomic testing. The ultimate goal of these techniques is to improve the detection of bladder cancer and eventually decrease the frequency of cystoscopy in patients that have a history of bladder cancer. The immediate challenge with urine proteomic testing is to demonstrate greater sensitivity than urine cytology while maintaining high specificity to avoid false positive results.

2. Review of the Literature

An intensive review of the literature was undertaken for this report. Search terms included bladder cancer, urothelial cancer, bladder tumor antigen, nuclear matrix protein 22, BLCA-4, hyaluronic acid, hyaluronidase, cytokeratin 8, cytokeratin 18, cytokeratin 19, urine cytology, tissue polypeptide antigen, tissue polypeptide-specific antigen, 2-dimensional polyacrylamide gel electrophoresis, matrix-assisted laser desorption ionization time-of-flight mass spectrometry, surface enhanced laser desorption ionization time-of-flight mass spectrometry, liquid chromatography combined with (tandem) mass spectrometry, isotope-coded affinity tags, and isotope tags for relative and absolute quantification. Studies of urine-based clinical tests were clearly identified and commented on if they were used for screening versus surveillance purposes. We made a distinction between proteomic urine-based tests currently approved for patient use in the United States versus those that are not. The utility of the urine proteomic based tests was evaluated in terms of sensitivity and specificity. Sensitivity was defined by the ability of the test to detect cases of urothelial cancer. Specificity was defined by the ability of the test to determine the absence of urothelial cancer.

3. Urine Protein Markers for Bladder Cancer

3.1. U.S. Food and Drug Administration Approved Proteomic Assays

3.1.1. Bladder Tumor Antigen

The bladder tumor antigen (BTA) assay (Bard Diagnostics, Redmond, WA) was originally designed as a latex agglutination assay used to detect a basement membrane antigen in the urine (polypeptides between 16 and 165 kDa). Although originally thought to be sensitive and specific for bladder cancer, the specificity was diminished by any process that caused cellular destruction resulting in the release of basement membrane (e.g., cystitis). Because the BTA test is dependent on the amount of basement membrane released into the urine, it is more sensitive for detection of invasive urothelial cancer than low grade NMI urothelial cancer. These shortcomings led to the development of second-generation BTA tests – BTA Stat (Bion Diagnostic Science Inc., Redmond, WA) and BTA TRAK (Polymedco Inc., Cortlandt Manor, NY). Both of these assays detect complement factor H-related proteins in the urine.

Complement factor H (Complement FH) is a 150-kDa glycosylated plasma protein and the major inhibitor of the alternative complement pathway (12). Complement FH binds to human epithelial and endothelial cells, basement membranes, and some cancer cell surfaces and may have a protective role against complement activation (13). It is theorized that Complement FH is released directly by tumor cells or by leakage from the plasma secondary to hematuria. BTA Stat and BTA TRAK use two monoclonal antibodies (mAb X52.1 as the capture antibody and mAb X13.2 as the detection antibody) that bind to Complement FH and Complement FH-related proteins isolated from urine. Complement FH is composed of 20 short consensus repeat (SCR) domains. mAb X13.2 binds to SCR domain 3 of Complement FH and Complement FH-like protein-1 (Complement FH-like protein-1; 42 kDa, an alternatively spliced product of the Complement FH gene), whereas mAb X52.1 binds to SCR domain 18 of Complement FH and to Complement FH related protein-1 (12). Inflammatory bladder conditions (e.g. intravesical chemotherapy, BCG therapy, urinary tract infections, and stones) can lead to false positive results.

3.1.1.1. BTA Stat

BTA Stat functions as a qualitative point-of-care assay that utilizes a cartridge-form enzyme immunoassay in which five drops of urine are placed in a sample well with a reaction time of 5 min. A positive test is indicated by a red line with a control window to indicate that the test kit is reacting appropriately. Overall sensitivity is about 56% (14). Sensitivity is low for the detection of low grade tumors (13–55%) and increases for higher grade tumors (sensitivity for grade 3 tumors ranges from 63% to 90%). Specificity ranges from 50% to 70%. Intravesical chemotherapy can dramatically impact the specificity of the BTA Stat test, particularly with intravesical mitomycin C, which was found in one study to decrease specificity to 25% (15). In the same study, the effect of BCG on specificity seemed to endure up to 2 years after therapy. Therefore, patients on active intravesical therapy should not be monitored using this assay, and patients that received BCG therapy should not have this assay utilized for an extended length of time secondary to the complement-mediated activity of BCG leading to false reactivity.

3.1.1.2. BTA TRAK

BTA TRAK is a quantitative, sandwich-type immunoassay in 96-well format. Urine samples are added to wells coated with capture mAb X52.1 and incubated for 1 h at 37°C. The alkaline phosphatase-labeled detection mAb X13.2 is then added to the wells and incubated for another hour at 37°C. The plate is then washed and the bound complex is detected by reaction with substrate for 30 min at 37°C. The amount of antigen is determined by comparing the absorbances in the sample wells with those of calibrators assayed simultaneously (16). The sensitivity of BTA TRAK ranges between 57% and 83%, and specificity ranges between 50% and 90% (17, 18). This assay is also hampered by its low sensitivity in detecting low grade NMI lesions and decreased specificity in detecting benign urologic conditions. In healthy individuals, however, the specificity of this assay can exceed 90%.

3.1.1.3. BTA Assays Versus Urine Cytology

Overall, the BTA assays are more sensitive than urine cytology for low grade disease but have less specificity. They should not be used as a replacement for cystoscopy for bladder cancer surveillance but are an alternative to urine cytology in selected populations that do not have conditions likely to produce a false positive result.

3.1.2. Nuclear Matrix Protein

One central finding in the cytopathologic examination of bladder cancer cells is the abnormal nuclear morphology characterized by increased nuclear-to-cytoplasmic ratios, hyperchromatic nuclei, and abnormal nuclear variability between cells. These findings support looking at nuclear abnormalities as an aid in the diagnosis of bladder cancer.

The nuclear matrix is the framework of the nucleus and consists of peripheral lamins, pore complexes, and nucleoli. Nuclear matrix proteins make up less than 10% of all nuclear proteins and function to maintain nuclear architecture, organize deoxyribonucleic acid (DNA) structure, replication, and gene expression. Several NMPs have been identified as potential tumor markers in prostate, breast, and colon cancer (19–21). Four nuclear matrix proteins have been described that appear relevant to bladder cancer; however, only NMP22 is approved by the U.S. Food and Drug Administration (US FDA).

NMP22 (Matritech, Cambridge, MA) is thought to be involved in the distribution of chromatin to daughter cells and is found in the mitotic spindle during mitosis (22). NMP22 is found in all human cells and may be released during cell death. NMP22 is a protein that can be detected in the urine by immunoassay techniques. This protein is found in low levels in the urine of patients without malignancy, whereas high levels can be assayed in patients with invasive bladder cancer.

The first reported clinical experience with NMP22 in patients with bladder cancer was the utilization of this assay as a proteomic marker following TUR of a bladder tumor. Receiver operating characteristic (ROC) curve demonstrated that an NMP22 level of

10 U/ml yielded a sensitivity of 70%, a specificity of 79%, an accuracy of 76%, a positive predictive value of 58%, and a negative predictive value of 86%. With a reference value of 20 U/ml, the positive predictive value increases to 71%, specificity to 91%, and accuracy to 80% (23). A recent multiinstitutional study of 2,871 patients under surveillance for a history of stage Ta or stage T1 bladder cancer with or without carcinoma in situ found that the performance of NMP22 varied by grade and stage (24). The area under the ROC curve for urothelial cancer detection using NMP22 for all urothelial cancers was 0.735. For grade 3 lesions and stage T2 or greater urothelial cancers, the area under the ROC curve increased to 0.806 (95% confidence interval (CI) 0.780–0.831) and 0.864 (95% CI 0.839–0.890), respectively. The overall sensitivity and specificity at the manufacturer's recommended cutoff of 10 U/ml was 57% and 81%, respectively. NMP22 can be measured quantitatively using an immunoassay or qualitatively using a point-of-care assay to give an immediate result in the clinic. A multiinstitutional study using the NMP22 point-of-care proteomic assay for initial detection of bladder cancer in 1,331 people without a history of this disease demonstrated a sensitivity of 55.7% (versus 15.8% for urine cytology) and a specificity of 85.7% (versus 99.2% for urine cytology) (25). A follow-up multiinstitutional study using the point-of-care assay for surveillance in 668 patients with a history of bladder cancer found that combining the biomarker with cystoscopy increased the sensitivity from 91.3% with cystoscopy alone to 99% with NMP22 with cystoscopy (26). In this study, NMP22 found eight additional cancers missed by cystoscopy (seven of eight of which were high grade). Cytology found three additional cancers, increasing the sensitivity of cytology with cystoscopy to 94.2%. For surveillance, NMP22 was found to have a sensitivity of 49.5% and a specificity of 87.3%. Thus, NMP22 was found to have a higher sensitivity and a comparable specificity to cytology. However, it should be noted that urinary tract infection can cause a false positive NMP22 test. An additional advantage of this point-of-care assay is that the results are available to the patient at the time of the patient's office visit.

3.2. Investigational Proteomic Assays Used for Urothelial Cancer Detection

3.2.1. BLCA-4

BLCA-1 to BLCLA-6 are nuclear matrix proteins that were identified at the University of Pittsburgh in 1996. BLCA-4 has shown promise as a potential urinary marker for bladder cancer (27). Specifically, BLCA-4 was isolated from human urothelial carcinoma cell lines but was not found in any normal cellular controls. BLCA-4 shares sequence homology with the ETS family of transcription factors and is expressed early in carcinogenesis (28).

In a study comparing urine from 51 normal individuals and 54 patients with pathologically confirmed bladder cancer, a BLCA-4 urine-based immunoassay detected BLCA-4 in 53 of 55 urine samples from patients with bladder cancer (malignant and normal urothelium) and was negative in all 51 urine samples from nonbladder cancer patients (29). These results demonstrate a 96.4% sensitivity and 100% specificity. Mechanistic studies have been performed looking at BLCA-4 and its effect on normal urothelial cell growth. Normal urothelial cells transfected with BLCA-4 cDNA underwent a fourfold increase in cell proliferation along with upregulation of interleukin (IL)-1α, IL-8, and thrombomodulin genes (30). The clinical utility of BLCA-4 was demonstrated in a follow-up study using urine-based immunoassay with a cut-off of 13 optical density units per microgram of protein in 51 normal controls, 54 patients with urothelial cancer, and 202 patients with a spinal cord injury. The BLCA-4 level was less than the cut-off in all normal controls and above the cut-off in 53 of 55 specimens from patients with bladder cancer (96.4 %) (31). However, 38 of 202 patients with a spinal cord injury (19%) had BLCA-4 levels above the cut-off. Interestingly, there was a direct correlation with higher levels of BLCA-4 in patients with a spinal cord injury who had a greater duration of injury. One established risk factor for the development of bladder cancer in spinal cord injury patients is chronic long term urinary catheterization. Multivariate analysis showed no correlation between BLCA-4 levels and urinary tract infection, smoking, catheterization, or cystitis (31).

3.2.2. Hyaluronic Acid and Hyaluronidase

Hyaluronic acid (HA) is a glycosaminoglycan that is widely distributed in connective tissue, epithelium, and neural tissue. HA is chemically composed of repeating disaccharide units, *N*-acetyl-D-glucosamine, and glucuronic acid. HA was first discovered in 1984 to be elevated in human tumor-cell cultures (32). The elevation was thought to be due to an interaction between malignant cells and cohabitating fibroblasts that increase the production of HA. The function of HA in malignancy may include cell migration, promotion of cellular adhesion, and cellular proliferation. Elevated HA is not specific to bladder cancer, and elevated HA levels have also been demonstrated in lung cancer, colorectal cancer, and breast cancer cell lines (33). Urine HA levels are measured by an ELISA assay, which uses a biotinylated bovine nasal cartilage HA-binding protein and an avidin-biotin detection system. HA in the urine competes with the HA coated on the microtiter wells to bind to the biotinylated HA-binding protein (34). The mean HA concentration is determined by assaying three aliquots of urine and is calculated using a standard graph created by plotting the optical density at 405 nm versus the concentration of human umbilical cord HA standard.

One of the first clinical studies examining HA and its use as a urine marker for bladder cancer involved urine samples from 144 people, with HA measured by immunoassay (35). Patients with malignancy had a four to ninefold increase in HA levels compared to normal controls with a sensitivity of 92% and a specificity of 93%. Most importantly, substantially elevated levels of HA were found in grade 1 tumors yielding a high sensitivity even in the setting of low grade disease.

Hyaluronidases (HAases) are a family of degradative enzymes that cleave HA. By catalyzing the hydrolysis of HA, a major constituent of the interstitial barrier, HAase lowers the viscosity of HA, thereby increasing tissue permeability. As early as the 1950s, researchers began speculating that HAase was associated with malignancy (36). Elevated HAase levels have been demonstrated in bladder, prostate, head and neck, breast, and neural tumors ((35), (37–42)). HAases are assayed using an ELISA assay as described for HA quantification. There are many different isotypes of HAase; however, the first HAase identified from the urine of patients with high grade bladder cancer was HYAL-1 (43). HYAL-1 is neither unique nor specific to bladder cancer, as expression of HYAL-1 has been identified in prostate, bladder, and head and neck tumors (38, 43, 44). Elevated levels of HAase in the urine have been shown to correspond with malignancy. In a study of 513 urine specimens (261 of urothelial cancer, nine of bladder cancer with nontransitional histology, and 243 from controls), an ELISA assay for HAase quantification found a three to sevenfold increase in HAase levels from cancer specimens versus normal controls. Urine HAase was determined to have a sensitivity of 81.5% and a specificity of 83.8% regardless of tumor stage and grade (34). When the HAase test was combined with the HA test (HA-HAase), the sensitivity and specificity was 91.2% and 84.4%, respectively. A subsequent study evaluated HA-HAase test performance for surveillance of 70 patients with a history of bladder cancer and found a sensitivity and specificity of 91% and 70%, respectively (45). In the same study, a screening evaluation was performed with the HA-HAase test on 401 patients from the Department of Energy, and 14% were found to be positive. However, none of the positive cases were diagnosed with bladder cancer at the time of the report (45). In summary, the HA-HAase test shows promise for surveillance of patients with a history of bladder cancer.

3.2.3. Cytokeratins

Cytokeratins are intracytoplasmic fibrous proteins that function as integral components of the cytoskeleton. Cytokeratins can be classified by molecular weight. Low molecular weight cytokeratins are acidic type I cytokeratins. High molecular weight cytokeratins are basic or neutral type II cytokeratins. Type I cytokeratins make up the Type I intermediate filaments of the cytoskeleton. The gene loci for Type I cytokeratins is chromosome 17q. Type I cytokeratins include CKs 9–20 with molecular weights ranging from 40 to 64 kDa. Type II cytokeratins make up the Type II intermediate filaments of the cytoskeleton. The gene loci for

Type II cytokeratins is chromosome 12q. Type II cytokeratins include CKs 1–8 with molecular weights ranging from 52 to 67 kDa. Cytokeratins are essential for dynamic functions such as mitosis, cell migration, and differentiation.

Subsets of epithelial cytokeratins are uniquely expressed by specific epithelium and corresponding epithelial malignancy and can aid in the diagnosis of a variety of tumors using immunohistochemistry. CKs 7, 8, 18, 19, and 20 have been evaluated by urine-based immunoassays to help detect bladder cancer recurrences. The theory behind CK shedding is that neoplastic cells are rapidly dividing cells and that cell rupture occurs frequently, releasing cytokeratins and cytokeratin fragments (46).

3.2.3.1. Tissue Polypeptide Antigen

Tissue polypeptide antigen (TPA) was identified by Bjorklund in 1957 (47). TPA was later described as a complex of 3 cytokeratins: CKs 8, 18, and 19 (48). TPA has been found in normal and malignant epithelial-associated cancer of the breast, colon, lung, head and neck, and bladder (49–54). The current TPA assay available is TPAcyk™ (IDL Biotech, Bromma, Sweden), which is a quantitative monoclonal immunoassay with reactivity against defined epitope structures on cytokeratins 8 and 18. TPAcyk is available in two quantitative formats: TPAcyk ELISA and TPAcyk IRMA.

TPAcyk ELISA is a solid phase sandwich immunochemical assay. Standards, controls, and samples react simultaneously with solid-phase catcher antibodies (6D7 and 3F3) and the horseradish peroxidase-conjugated detector antibody during incubation in the microstrip wells. After washing, the tetramethylbenzidine substrate is added. Subsequently, the reaction is stopped, and the absorbance is quantified. The developed color is directly proportional to the concentration of the analyte (IDL Biotech).

TPAcyk IRMA is a solid phase radiometric sandwich immunochemical assay. Standards, controls, and samples react simultaneously with solid-phase catcher antibodies (6D7 and 3F3) and the ^{125}I-labelled detector antibody during incubation in assigned tubes. After washing, radioactivity is assessed in a gamma counter. The radioactivity is directly proportional to the concentration of the analyte (IDL Biotech). In one study, urine TPA was evaluated in 264 patients with bladder cancer and correlated with local disease response, recurrence, and progression (55). Another study examined urinary TPA as a correlate for invasive versus superficial disease. Urinary TPA was elevated in 74% of patients with invasive bladder cancer compared to only 15% of patients with superficial disease (56). A study examining multiple markers (CYFRA 21-1, UBC, TPA, and NMP22) in 267 patients found that at a cut-off of 760.8 U/L, TPA had a sensitivity of 80.2%; however, TPA had poor specificity, as the false-positive rate was 36% for benign urologic conditions and 52% for bladder screening or bladder cancer surveillance (57).

3.2.3.2. Tissue Polypeptide-Specific Antigen

Tissue polypeptide-specific antigen (TPS) is specific for the M3 epitope structure on CK 18. TPS (IDL Biotech) is a quantitative serum monoclonal immunoassay available in three formats: ELISA, IRMA, and the Immulite system (DPC, Los Angeles, CA).

TPS ELISA is a solid phase sandwich immunochemical assay. Standards, controls, and samples react simultaneously with solid-phase catcher antibodies and the horseradish peroxidase-conjugated detector antibody (M3) during incubation in the microstrip wells. After washing, the tetramethylbenzidine substrate is added. The reaction is subsequently stopped, and the absorbance is read. The developed color is directly proportional to the concentration of the analyte (IDL Biotech).

TPS IRMA is a solid phase radiometric sandwich immunochemical assay. Standards, controls, and samples react simultaneously with solid phase catcher antibodies and the ^{125}I-labelled detector antibody (M3) during incubation in assigned tubes. After washing, the radioactivity is assessed in a gamma counter. The radioactivity is directly proportional to the concentration of the analyte (IDL Biotech).

The TPS Immulite® system is a sequential chemiluminescent enzyme-labeled immunometric assay based on ligand-labeled monoclonal antibodies and separation by an anti-ligand-coated solid phase. The Immulite® System automatically handles sample and reagent additions, the incubation and separation steps, and measurement of the photon output via the temperature-controlled luminometer. It calculates test results for controls and patient samples from the observed signal, using a stored Master Curve (IDL Biotech). A study of urine TPS was performed using the radioimmunoassay on 211 patients (56 patients with bladder cancer, 36 with a history of bladder cancer, 44 with benign urologic disease, and 75 age- and sex-matched controls) (58).Urine TPS was elevated in patients with active bladder tumors but was also elevated in patients with benign urologic disease. However, higher TPS levels correlated with increasing grade and stage. Another study was performed evaluating the performance of urine TPS ELISA in 355 patients divided into five groups: (1) those presenting with microhematuria, (2) those undergoing surveillance for bladder cancer, (3) those treated for bladder cancer with intravesical chemotherapy, (4) those with benign urologic disease, and (5) healthy controls (59). A TPS cut-off of 279 U/L demonstrated a sensitivity of 64% and a specificity of 84%; however, 45% of patients with benign urologic disease also had an elevated TPS.

3.2.3.3. Cytokeratin-19 (CYFRA 21-1)

The concentration of soluble fragments of cytokeratin 19 can be quantitated by CYFRA 21-1 ELISA. CYFRA 21-1 ELISA (CIS Bio International, Gif-Sur Yvette, France), utilizes 2 monoclonal antibodies (BM19-21 and KS19-1) to recognize fragments of cytokeratin 19 (60). The amount of radioactivity bound to the solid phase is directly proportional to the amount of CYFRA 21-1.

In a study using a sandwich ELISA for CYFRA 21-1 to test specimens from patients with ($n=58$) and without bladder cancer ($n=220$), serum and urine CYFRA 21-1 levels were higher in the bladder cancer group than the normal controls (61). However, this study showed no distinction between patients with bladder cancer and those with cystitis.

In a later study testing 325 patients (152 patients presenting with hematuria or irritative voiding symptoms, 107 patients after TUR of a bladder tumor, 46 with nonbladder urinary tract pathology, and 20 healthy participants) at a cut-off of 4 μg/L, CYFRA 21-1 ELISA demonstrated a sensitivity of 79.3% and a specificity of 88.6% in detecting bladder cancer (60). CYFRA 21-1 had a sensitivity of 70% for grade 1 tumors compared to a sensitivity of 23% with urine cytology. One study evaluated CYFRA 21-1 ELISA in benign prostatic hypertrophy and bladder cancer. In both groups, CYFRA 21-1 levels were elevated, demonstrating that CYFRA 21-1 ELISA had a significant false positive potential for benign urinary tract pathology, which may limit its use as a screening tool (62).

CYFRA 21-1 ELISA has also been used to evaluate the serum of patients undergoing local and systemic therapy for bladder cancer (63). Of patients with localized disease, only 7% had an abnormal CYFRA 21-1 level compared to 66% of patients with metastatic disease. Patients with an elevated serum CYFRA 21-1 level also had a lower overall median survival, and reduction of CYFRA 21-1 levels correlated with response to chemotherapy. This correlation with survival and response to therapy has been demonstrated in other studies (64).

3.2.3.4. Urine Bladder Cancer (UBC) Test™ (Cytokeratin 8 and 18)

The urine bladder cancer (UBC) test (IDL Biotech) detects CK 8 and 18 in the urine. There are currently two UBC tests: the qualitative UBC IRMA and the quantitative UBC II ELISA.

UBC™ IRMA

The UBC IRMA uses a chromatographic principle in which antigens in the urine sample react with gold-labeled monoclonal antibodies in the device forming a complex. This complex migrates to the reaction zone where it reacts with a specific antibody to produce a dark line (positive result) (65). Excess gold-labeled antibodies continue to migrate into the second capture zone to form a second dark line (test control) (65). The UBC IRMA was initially evaluated in 267 patients (111 with active bladder cancer, 76 with a history of bladder cancer, 25 with benign urinary tract pathology, 25 with nonbladder malignant conditions, and 30 healthy subjects) (65). UBC IRMA demonstrated a sensitivity for patients with NMI bladder cancer (pTa to pT1) and muscle invasive bladder cancer (pT2 to pT4) of 81.8% and 81.5%, respectively. However, the false positive rates were 33% with urinary tract infection, 50% with nephritis, 33% with benign prostatic hypertrophy, and 33–66% with nonbladder malignant conditions.

UBC II ELISA™

UBC II ELISA is a solid phase, sandwich ELISA assay. Samples, standard, and controls react simultaneously with solid phase antibodies (6D7 and 3GF3) and the horseradish peroxidase-conjugated detector antibody during incubation in the microstrip wells. After washing, the tetramethylbenzidine substrate is added. The reaction is subsequently stopped, and the absorbance is quantified. The developed color is directly proportional to the concentration of the analyte (IDL Biotech). The first study evaluating the use of UBC ELISA for surveillance of bladder cancer recurrence involved 191 patients with a history of superficial NMI bladder cancer (Tis, Ta, and T1 tumors) (66). All patients underwent cystoscopy and TUR or biopsy for histologic confirmation. The overall sensitivity, specificity, and positive and negative predictive values of the UBC test were 20.7%, 84.7%, 35.3%, and 72.6%, respectively. The area under the ROC curve was 0.5, establishing that the test is not useful for detecting bladder cancer recurrences. Another clinical study examining the performance of UBC II ELISA also involved 191 patients (112 patients awaiting TUR of a bladder tumor, 40 patients awaiting second surgical intervention, 29 healthy controls, and 10 women with acute urinary tract infections) (67). All urine samples were tested by UBC IRMA, UBC II ELISA, and urine cytology. The sensitivities of these assays were 64.4%, 46.6%, and 70.5% and the specificities were 63.6%, 86.3%, and 79.5%, respectively. The three tests were compared for accuracy, and both cytokeratin tests were found to be inferior to urine cytology (UBC IRMA 54.4%, UBC II ELISA 64.2%, and Cytology 72.3%). UBC II ELISA was also compared to NMP22 and bladder wash cytology in a study of 90 patients (60 with confirmed urothelial cancer and 30 with benign urologic disease) (68). Urine samples were obtained preoperatively and postoperatively in patients that underwent TUR of confirmed tumors. Sensitivity of detecting recurrence at 3 months was 52% for NMP22, 19% for UBC II ELISA, and 14% for urine cytology. UBC II ELISA had a slightly better sensitivity than urine cytology. However, its overall performance was poor.

4. Urine Proteomic Profiling in Bladder Cancer

The human proteome is composed of greater than 500,000 proteins compared to about 40,000 genes in the human genome (69). With so many proteins circulating in the serum and urine, a major technological hurdle is discriminating rare proteins from more abundant proteins. There are multiple techniques for protein identification and quantification, including 2-dimensional polyacrylamide gel electrophoresis (2D-PAGE), matrix-assisted laser desorption ionization time-of-flight (MALDI-TOF) mass spectrometry, surface enhanced laser desorption ionization time-of-flight (SELDI-TOF) mass spectrometry, liquid chromatography combined with (tandem) mass spectrometry (LC-MS-MS), isotope-coded affinity tags (ICAT™), and isotope tags for relative and absolute quantification (iTRAQ™) (70–73). Proteomic profiling may aid in the diagnosis of cancer since certain cancers may express a relatively specific proteomic signature that can be identified in the urine or serum long before clinical symptoms of cancer appear. Serum SELDI-TOF has been evaluated in the diagnosis of ovarian cancer, breast cancer, prostate cancer, colorectal cancer, liver cancer, renal cell cancer, pancreatic cancer, head and neck cancers, endometrial cancers, and certain lymphomas (74–100). Protein Chip® (Ciphergen, Fremont, CA) array-based SELDI-TOF mass spectrometry utilizes arrays of mixtures of proteins (for example, from urine or serum) and can be resolved into subsets of proteins with common properties. Arrays are washed to remove weakly bound proteins. After which, a solution containing an energy-absorbing molecule is added and allowed to crystallize, thereby embedding the remaining proteins. These arrays are then read by the Protein Chip reader and peaks of interest can be identified. Once identified, the analyte of interest may be further enriched for more detailed analysis (101).

One of the first urine-based SELDI-TOF studies to evaluate the ability to discriminate patients with bladder cancer from those without bladder cancer using protein profiling involved 108 patients (46 with bladder cancer, 32 with benign urinary tract pathology, and 40 healthy controls) and demonstrated a sensitivity of 93.3% and a specificity of 87% (102). Later studies by different groups have yet to confirm these findings. A study evaluating 104 urine samples using SELDI-TOF (from patients included those with pathologically confirmed bladder cancer, normal controls, benign urinary tract pathology, and nonbladder cancer) using a tree analysis pattern found a sensitivity of 71.4% and a specificity of 72.7% (103). In another study evaluating 227 subjects including patients with pathologically confirmed bladder cancer, normal controls, and benign urinary tract pathology, SELDI-TOF demonstrated a sensitivity of 71.7% and a specificity of 62.5% (104). Major issues surrounding this technology include optimizing the discrimination of peaks of interest and reproducibility.

5. Multiple Proteomic Marker Testing in Bladder Cancer

To date there are very few studies in the literature that examine the strategy of combining multiple proteomic markers for the diagnosis and surveillance of bladder cancer. A comparative urine-based proteomic study evaluated 267 patients (111 with active bladder cancer, 76 with a history of bladder cancer, 25 with benign urinary pathology, 25 with other malignant conditions, and 30 healthy subjects) (57). Urine samples were tested with urinary CYFRA 21-1 ELISA, UBC II ELISA, TPAcyk IRMA, and NMP22 ELISA with cut points set at 95% specificity. With respect to NMI disease, CYFRA 21-1, UBC, TPA, and NMP22 demonstrated sensitivities of 83.1%, 76.6%, 66.2%, and 64.9%, respectively. In comparison, with invasive disease (pT2 or greater), CYFRA 21-1, UBC, TPA, and NMP22 demonstrated sensitivities of 88.8%, 70.4%, 81.5%, and 88.8%, respectively. Used together, these four markers yielded sensitivities of 91.9% for bladder cancer detection. Any combination of three markers yielded sensitivities of about 90% for detection of bladder cancer. The lowest sensitivity of any combination of markers was that of UBC II ELISA and TPAcyk IRMA, which was 85.6 %. NMP22 decreased the false positive rate associated with cytokeratin-based assays. The combination of NMP22 and either CYFRA 21-1 or UBC II ELISA appeared to be most effective.

A subsequent report evaluated 187 samples from 112 patients with symptomatic bladder cancer and 75 with benign urologic conditions using CYFRA 21-1, UBC II ELISA, and NMP 22 tests at standard cut points (105). Combining markers increased sensitivity but decreased specificity. Overall accuracy could be increased by combining markers. The best single test was UBC II ELISA with sensitivity, specificity, and accuracy of 69.4%, 91.3%, and 82.2%, respectively. The best combination was NMP22 with either UBC II ELISA or CYFRA 21-1 with sensitivities, specificities, and accuracies of 76.7%, 86.9%, and 83.0%, respectively.

6. Conclusions

No single proteomic marker has emerged with the ability to replace cystoscopy for the detection and surveillance of bladder cancer. However, these proteomic markers are useful supplements to cystoscopy because they can detect bladder cancer that is not visually evident. NMP22 appears to have the best performance of the currently available proteomic tests. The elucidation of cancer-specific protein expression may result in improved tests with higher sensitivities and specificities. The description and characterization of existing protein based assays provide a benchmark by which we can compare newly discovered protein biomarkers from proteomic based investigations.

References

1. Jemal A., Siegel R., Ward E., Murray T., Xu J., Thun M.J. (2007) Cancer statistics, 2007. *CA Cancer J. Clin.* **57**, 43–66.
2. Surveillance, Epidemiology and End Results Database. http://seer.cancer.gov.
3. Botteman M.F., Pashos C.L., Redaelli A., Laskin B., Hauser R. (2003) The health economics of bladder cancer: a comprehensive review of the published literature. *Pharmacoeconomics.* **21**, 1315–30.
4. Leblanc B., Duclos A.J., Benard F., Cote J., Valiquette L., Paquin J.M., Mauffette F., Faucher R., Perreault J.P. (1999) Long-term follow-up of initial Ta grade 1 transitional cell carcinoma of the bladder. *J. Urol.* **162**, 1946–50.
5. Zieger K., Wolf H., Olsen P.R., Hojgaard K. (2000) Long-term follow-up of noninvasive bladder tumours (stage Ta): recurrence and progression. *BJU Int.* **85**, 824–8.
6. Sanders W.R. (1864) Cancer of the bladder: fragments forming urethral plugs discharged in the urine-concentric colloid bodies. *Edinburgh Med. J.* **10**, 273–75.
7. Dickenson W.H. (1869) Portions of cancerous growth passed by urethra. *Trans. Pathol. Soc. London.* **20**, 233–35.
8. Papanicolaou G.N., Marshall V.P. (1945) Urine sediment smears as a diagnostic procedure in cancers of the urinary tract. *Science.* **101**, 511–29.
9. Glatz K., Willi N., Glatz D., Barascud A., Grilli B., Herzog M., Dalquen P., Feichter G., Gasser T.C., Sulser T., Bubendorf L. (2006) An international telecytologic quiz on urinary cytology reveals educational deficits and absence of a commonly used classification system. *Am. J. Clin. Pathol.* **126**, 294–301.
10. Malik S.N., Murphy W.M. (1999) Monitoring patients for bladder neoplasms: what can be expected of urinary cytology consultations in clinical practice. *Urology.* **54**, 62–6.
11. Habuchi T., Marberger M., Droller M.J., Hemstreet G.P. III, Grossman H.B., Schalken J.A., Schmitz-Drager B.J., Murphy W.M., Bono A.V., Goebell P., Getzenberg R.H., Hautmann S.H., Messing E., Fradet Y., Lokeshwar V.B. (2005) Prognostic markers for bladder cancer: International Consensus Panel on bladder tumor markers. *Urology.* **66**, 64–74.
12. Cheng Z.Z., Corey M.J., Parepalo M., Majno S., Hellwage J., Zipfel P.F., Kinders R.J., Raitanen M., Meri S., Jokiranta T.S. (2005) Complement factor H as a marker for detection of bladder cancer. *Clin. Chem.* **51**, 856–63.
13. Junnikkala S., Jokiranta T.S., Friese M.A., Jarva H., Zipfel P.F., Meri S. (2000) Exceptional resistance of human H2 glioblastoma cells to complement-mediated killing by expression and utilization of factor H and factor H-like protein 1. *J. Immunol.* **164**, 6075–81.
14. Sarosdy M.F., Hudson M.A., Ellis W.J., Soloway M.S., deVere White R., Sheinfeld J., Jarowenko M.V., Schellhammer P.F., Schervish E.W., Patel J.V., Chodak G.W., Lamm D.L., Johnson R.D., Henderson M., Adams G., Blumenstein B.A., Thoelke K.R., Pfalzgraf R.D., Murchison H.A., Brunelle S.L. (1997) Improved detection of recurrent bladder cancer using the Bard BTA stat Test. *Urology.* **50**, 349–53.
15. Raitanen M.P., Hellstrom P., Marttila T., Korhonen H., Talja M., Ervasti J., Tammela T.L., Finnbladder Group. (2001) Effect of intravesical instillations on the human complement factor H related protein (BTA stat) test. *Eur. Urol.* **40**, 422–36.
16. Thomas L., Leyh H., Marberger M., Bombardieri E., Bassi P., Pagano F., Pansadoro V., Sternberg C.N., Boccon-Gibod L., Ravery V., Le Guludec D., Meulemans A., Conort P., Ishak L. (1999) Multicenter trial of the quantitative BTA TRAK assay in the detection of bladder cancer. *Clin. Chem.* **45**, 472–77.
17. Giannopoulos A., Manousakas T., Mitropoulos D., Botsoli-Stergiou E., Constantinides C., Giannopoulou M., Choremi-Papadopoulou H. (2001) Comparative evaluation of the BTAstat test, NMP22, and voided urine cytology in the detection of primary and recurrent bladder tumors. *J. Urol.* **166**, 470–75.
18. Ellis W.J., Blumenstein B.A., Ishak L.M., Enfield D.L. (1997) Clinical evaluation of the BTA TRAK assay and comparison to voided urine cytology and the Bard BTA test in patients with recurrent bladder tumors. The Multi Center Study Group. *Urology.* **50**, 882–7.
19. Miller T.E., Beausang L.A., Winchell L.F., Lidgard G.P. (1992) Detection of nuclear matrix proteins in serum from cancer patients. *Cancer Res.* **52**, 422–27.
20. Khanuja P.S., Lehr J.E., Soule H.D., Gehani S.K., Noto A.C., Choudhury S., Chen R., Pienta K.J. (1993) Nuclear matrix proteins in normal and breast cancer cells. *Cancer Res.* **53**, 3394–98.
21. Keesee S.K., Meneghini M.D., Szaro R.P., Wu Y.J. (1994) Nuclear matrix proteins in human colon cancer. *Proc. Natl. Acad. Sci. U.S.A.* **91**, 1913–16.

22. Berezney R., Coffey D.S. Identification of a nuclear protein matrix. (1974) *Biochem. Biophys. Res. Commun.* **60**, 1410–17.
23. Soloway M.S., Briggman V., Carpinito G.A., Chodak G.W., Church P.A., Lamm D.L., Lange P., Messing E., Pasciak R.M., Reservitz G.B., Rukstalis D.B., Sarosdy M.F., Stadler W.M., Thiel R.P., Hayden C.L. (1996) Use of a new tumor marker, urinary NMP22, in the detection of occult or rapidly recurring transitional cell carcinoma of the urinary tract following surgical treatment. *J. Urol.* **156**, 363–67.
24. Shariat S.F., Marberger M.J., Lotan Y., Sanchez-Carbayo M., Zippe C., Ludecke G., Boman H., Sawczuk I., Friedrich M.G., Casella R., Mian C., Eissa S., Akaza H., Serretta V., Huland H., Hedelin H., Raina R., Miyanaga N., Sagalowsky A.I., Roehrborn C.G., Karakiewicz P.I. (2006) Variability in the performance of nuclear matrix protein 22 for the detection of bladder cancer. *J. Urol.* **176**, 919–26.
25. Grossman H.B., Messing E., Soloway M., Tomera K., Katz G., Berger Y., Shen Y. (2005) Detection of bladder cancer using a point-of-care proteomic assay. *JAMA.* **293**, 810–16.
26. Grossman H.B., Soloway M., Messing E., Katz G., Stein B., Kassabian V., Shen Y. (2006) Surveillance for recurrent bladder cancer using a point-of-care proteomic assay. *JAMA.* **295**, 299–305.
27. Getzenberg R.H., Konety B.R., Oeler T.A., Quigley M.M., Hakam A., Becich M.J., Bahnson R.R. (1996) Bladder cancer-associated nuclear matrix proteins. *Cancer Res.* **56**, 1690–94.
28. Van Le T.S., Myers J., Konety B.R., Barder T., Getzenberg R.H. (2004) Functional characterization of the bladder cancer marker, BLCA-4. *Clin. Cancer Res.* **10**, 1384–91.
29. Konety B.R., Nguyen T.S., Dhir R., Day R.S., Becich M..J, Stadler W.M., Getzenberg R.H. (2000) Detection of bladder cancer using a novel nuclear matrix protein, BLCA-4. *Clin. Cancer Res.* **6**, 2618–25.
30. Myers-Irvin J.M., Van Le T.S., Getzenberg R.H. (2005) Mechanistic analysis of the role of BLCA-4 in bladder cancer pathobiology. *Cancer Res.* **65**, 7145–50.
31. Konety B.R., Nguyen T.S., Brenes G., Sholder A., Lewis N., Bastacky S., Potter D.M., Getzenberg R.H. (2000) Clinical usefulness of the novel marker BLCA-4 for the detection of bladder cancer. *J. Urol.* **164**, 634–9.
32. Knudson W., Biswas C., Toole B.P. (1984) Interactions between human tumor cells and fibroblasts stimulate hyaluronate synthesis. *Proc. Natl. Acad. Sci. U.S.A.* **81**, 6767–71.
33. Knudson W. (1996) Tumor-associated hyaluronan. Providing an extracellular matrix that facilitates invasion. *Am. J. Pathol.* **148**, 1721–6.
34. Lokeshwar V.B., Obek C., Pham H.T., Wei D., Young M.J., Duncan R.C., Soloway M.S., Block N.L. (2000) Urinary hyaluronic acid and hyaluronidase: markers for bladder cancer detection and evaluation of grade. *J. Urol.* **163**, 348–56.
35. Lokeshwar V.B., Obek C., Soloway M.S., Block N.L. (1997) Tumor-associated hyaluronic acid: a new sensitive and specific urine marker for bladder cancer. *Cancer Res.* **57**, 773–7.
36. Balazs E.A., Von Euler J. (1952) The hyaluronidase content of necrotic tumor and testis tissue. *Cancer Res.* **12**, 326–29.
37. Pham H.T., Block N.L., Lokeshwar V.B. (1997) Tumor-derived hyaluronidase: a diagnostic urine marker for high-grade bladder cancer. *Cancer Res.* **57**, 778–83.
38. Franzmann E.J., Schroeder L., Goodwin W.J., Weed D.T., Fisher P., Lokeshwar V.B. (2003) Expression of tumor markers hyaluronic acid and hyaluronidase (HYAL1) in head and neck tumors. *Int. J. Cancer.* **106**, 438–45.
39. Posey J.T., Soloway M.S., Ekici S., Sofer M., Civantos F., Duncan R.C., Lokeshwar V.B. (2003) Evaluation of the prognostic potential of hyaluronic acid and hyaluronidase (HYAL1) for prostate cancer. *Cancer Res.* **63**, 2638–44.
40. Lokeshwar V.B., Lokeshwar B.L., Pham H.T., Block N.L. (1996) Association of elevated levels of hyaluronidase, a matrix-degrading enzyme, with prostate cancer progression. *Cancer Res.* **56**, 651–57.
41. Madan A.K., Yu K., Dhurandhar N., Cullinane C., Pang Y., Beech D.J. (1999) Association of hyaluronidase and breast adenocarcinoma invasiveness. *Oncol. Rep.* **6**, 607–9.
42. Bertrand P., Girard N., Duval C., d'Anjou J., Chauzy C., Menard J.F., Delpech B. (1997) Increased hyaluronidase levels in breast tumor metastases. *Int. J. Cancer.* **73**, 327–31.
43. Lokeshwar V.B., Young M.J., Goudarzi G., Iida N., Yudin A.I., Cherr G.N., Selzer M.G. (1999) Identification of bladder tumor-derived hyaluronidase: its similarity to HYAL1. *Cancer Res.* **59**, 4464–70.
44. Lokeshwar V.B., Rubinowicz D., Schroeder G.L., Forgacs E., Minna J.D., Block N.L., Nadji M., Lokeshwar B.L. (2001) Stromal and epithelial expression of tumor markers hyaluronic acid and HYAL1 hyaluronidase in prostate cancer. *J. Biol. Chem.* **276**, 11922–32.

45. Lokeshwar V.B., Schroeder G.L., Selzer M.G., Hautmann S.H., Posey J.T., Duncan R.C., Watson R., Rose L., Markowitz S., Soloway M.S. (2002) Bladder tumor markers for monitoring recurrence and screening comparison of hyaluronic acid-hyaluronidase and BTA-Stat tests. *Cancer.* **95**, 61–72.
46. Sanchez-Carbayo M., Herrero E., Megias J., Mira A., Soria F. (1999) Initial evaluation of the new urinary bladder cancer rapid test in the detection of transitional cell carcinoma of the bladder. *Urology.* **54**, 656–61.
47. Bjorklund B., Bjorklund B. (1957) Antigenicity of pooled human malignant and normal tissues by cyto-immunological technique; presence of an insoluble, heat-labile tumor antigen. *Int. Arch. Allergy Appl. Immunol.* **10**, 153–84.
48. Weber K., Osborn M., Moll R., Wiklund B., Luning B. (1984) Tissue polypeptide antigen (TPA) is related to the non-epidermal keratins 8, 18 and 19 typical of simple and non-squamous epithelia: re-evaluation of a human tumor marker. *EMBO J.* **11**, 2707–14.
49. Gion M., Boracchi P., Dittadi R., Biganzoli E., Peloso L., Gatti C., Paccagnella A., Rosabian A., Vinante O., Meo S. (2000) Quantitative measurement of soluble cytokeratin fragments in tissue cytosol of 599 node negative breast cancer patients: a prognostic marker possibly associated with apoptosis. *Breast Cancer Res. Treat.* **59**, 211–21.
50. Einarsson R., Lindman H., Bergh J. (2000) Use of TPS and CA 15-3 assays for monitoring chemotherapy in metastatic breast cancer patients. *Anticancer Res.* **20**, 5089–93.
51. Plebani M., Basso D., Navaglia F., De Paoli M., Tommasini A., Cipriani A. (1995) Clinical evaluation of seven tumour markers in lung cancer diagnosis: can any combination improve the results? *Br. J. Cancer.* **72**, 170–3.
52. Bennink R., Van Poppel H., Billen J., Decoster M., Baert L., Mortelmans L., Blanckaert N. (1999) Serum tissue polypeptide antigen (TPA): monoclonal or polyclonal radioimmunometric assay for the follow-up of bladder cancer. *Anticancer Res.* **19**, 2609–13.
53. Nicolini A., Caciagli M., Zampieri F., Ciampalini G., Carpi A., Spisni R., Colizzi C. (1995) Usefulness of CEA, TPA, GICA, CA 72.4, and CA 195 in the diagnosis of primary colorectal cancer and at its relapse. *Cancer Detect. Prev.* **19**, 183–95.
54. Rosati G., Riccardi F., Tucci A. (2000) Use of tumor markers in the management of head and neck cancer. *Int. J. Biol. Markers.* **15**, 179–83.
55. Casetta G., Piana P., Cavallini A., Vottero M., Tizzani A. (1993) Urinary levels of tumour associated antigens (CA 19-9, TPA and CEA) in patients with neoplastic and non-neoplastic urothelial abnormalities. *Br. J. Urol.* **72**, 60–64.
56. Carbin B.E., Ekman P., Eneroth P., Nilsson B. (1989) Urine-TPA (tissue polypeptide antigen), flow cytometry and cytology as markers for tumor invasiveness in urinary bladder carcinoma. *Urol. Res.* **17**, 269–72.
57. Sanchez-Carbayo M., Herrero E., Megias J., Mira A., Soria F. (1999) Comparative sensitivity of urinary CYFRA 21-1, urinary bladder cancer antigen, tissue polypeptide antigen, tissue polypeptide antigen and NMP22 to detect bladder cancer. *J. Urol.* **162**, 1951–56.
58. Yao W.J., Chang C.J., Chan S.H., Chow N.H., Cheng H.L., Tzai T.S., Lin S.N. (1995) Significance of urinary tissue polypeptide specific antigen (TPS) determination in patients with urothelial carcinoma. *Anticancer Res.* **15**, 2819–23.
59. Sanchez-Carbayo M., Urrutia M., Silva J.M., Romani R., Garcia J., Alferez F., Gonzalez de Buitrago J.M., Navajo J.A. (2000) Urinary tissue polypeptide-specific antigen for the diagnosis of bladder cancer. *Urology.* **55**, 526–32.
60. Nisman B., Barak V., Shapiro A., Golijanin D., Peretz T., Pode D. (2002) Evaluation of urine CYFRA 21-1 for the detection of primary and recurrent bladder carcinoma. *Cancer.* **94**, 2914–22.
61. Senga Y., Kimura G., Hattori T., Yoshida K. (1996) Clinical evaluation of soluble cytokeratin 19 fragments (CYFRA 21-1) in serum and urine of patients with bladder cancer. *Urology.* **48**, 703–10.
62. Fatela-Cantillo D., Fernandez-Suarez A., Menendez V., Galan J.A., Filella X. (2005) Low utility of CYFRA 21-1 serum levels for diagnosis and follow-up in bladder cancer patients. *J. Clin. Lab. Anal.* **19**, 167–71.
63. Andreadis C., Touloupidis S., Galaktidou G., Kortsaris A.H., Boutis A., Mouratidou D. (2005) Serum CYFRA 21-1 in patients with invasive bladder cancer and its relevance as a tumor marker during chemotherapy. *J. Urol.* **174**, 1771–75.
64. Morita T., Kikuchi T., Hashimoto S., Kobayashi Y., Tokue A. (1997) Cytokeratin-19 fragment (CYFRA 21-1) in bladder cancer. *Eur. Urol.* **32**, 237–44.
65. Sanchez-Carbayo M., Herrero E., Megias J., Mira A., Soria F. (1999) Initial evaluation of the new urinary bladder cancer rapid test in the detection of transitional cell carcinoma of the bladder. *Urology.* **54**, 656–61.

66. Mungan N.A., Vriesema J.L., Thomas C.M., Kiemeney L.A., Witjes J.A. (2000) Urinary bladder cancer test: a new urinary tumor marker in the follow-up of superficial bladder cancer. *Urology.* **56**, 787–92.
67. Hakenberg O.W., Fuessel S., Richter K., Froehner M., Oehlschlaeger S., Rathert P., Meye A., Wirth M.P. (2004) Qualitative and quantitative assessment of urinary cytokeratin 8 and 18 fragments compared with voided urine cytology in diagnosis of bladder carcinoma. *Urology.* **64**, 1121–26.
68. Kibar Y., Goktas S., Kilic S., Yaman H., Onguru O., Peker A.F. (2006) Prognostic value of cytology, nuclear matrix protein 22 (NMP22) test, and urinary bladder cancer II (UBC II) test in early recurrent transitional cell carcinoma of the bladder. *Ann. Clin. Lab. Sci.* **36**, 31–38.
69. Banks R.E., Dunn M.J., Hochstrasser D.F., Sanchez J.C., Blackstock W., Pappin D.J., Selby P.J. (2000) Proteomics: new perspectives, new biomedical opportunities. *Lancet.* **356**, 1749–56.
70. Binz P.A., Hochstrasser D.F., Appel R.D. (2003) Mass spectrometry-based proteomics: current status and potential use in clinical chemistry. *Clin. Chem. Lab. Med.* **12**, 1540–51.
71. Heck A.J., Krijgsveld J. (2004) Mass spectrometry-based quantitative proteomics. *Expert Rev. Proteomics.* **1**, 317–26.
72. Schneider L.V., Hall M.P. (2005) Stable isotope methods for high-precision proteomics. *Drug Discov. Today.* **10**, 353–63.
73. Camacho-Carvajal M.M., Wollscheid B., Aebersold R. Steimle V., Schamel W.W. (2004) Two-dimensional Blue native/SDS gel electrophoresis of multi-protein complexes from whole cellular lysates: a proteomics approach. *Mol. Cell Proteomics.* **3**, 176–82.
74. Ye B., Cramer D.W., Skates S.J., Gygi S.P., Pratomo V., Fu L., Horick N.K., Licklider L.J., Schorge J.O., Berkowitz R.S., Mok S.C. (2003) Haptoglobin-alpha subunit as potential serum biomarker in ovarian cancer: identification and characterization using proteomic profiling and mass spectrometry. *Clin. Cancer Res.* **9**, 2904–11.
75. Kozak K.R., Amneus M.W., Pusey S.M., Su F., Luong M.N., Luong S.A., Reddy S.T., Farias-Eisner R. (2003) Identification of biomarkers for ovarian cancer using strong anion-exchange ProteinChips: potential use in diagnosis and prognosis. *Proc. Natl. Acad. Sci. U.S.A.* **100**, 12343–48.
76. Zhang Z., Bast R.C. Jr, Yu Y., Li J., Sokoll L.J., Rai A.J., Rosenzweig J.M., Cameron B., Wang Y.Y., Meng X.Y., Berchuck A., Van Haaften-Day C., Hacker N.F., de Bruijn H.W., van der Zee A.G., Jacobs I.J., Fung E.T., Chan D.W. (2004) Three biomarkers identified from serum proteomic analysis for the detection of early stage ovarian cancer. *Cancer Res.* **64**, 5882–90.
77. Rai A.J., Zhang Z., Rosenzweig J., Shih I.M., Pham T., Fung E.T., Sokoll L.J., Chan D.W. (2002) Proteomic approaches to tumor marker discovery. *Arch. Pathol. Lab. Med.* **126**, 1518–26.
78. Kozak K.R., Su F., Whitelegge J.P., Faull K., Reddy S., Farias-Eisner R. (2005) Characterization of serum biomarkers for detection of early stage ovarian cancer. *Proteomics.* **5**, 4589–96.
79. Pawlik T.M., Fritsche H., Coombes K.R., Xiao L., Krishnamurthy S., Hunt K.K., Pusztai L., Chen J.N., Clarke C.H., Arun B., Hung M.C., Kuerer H.M. (2005) Significant differences in nipple aspirate fluid protein expression between healthy women and those with breast cancer demonstrated by time-of-flight mass spectrometry. *Breast Cancer Res. Treat.* **89**, 149–57.
80. Li J., Orlandi R., White C.N., Rosenzweig J., Zhao J., Seregni E., Morelli D., Yu Y., Meng X.Y., Zhang Z., Davidson N.E., Fung E.T., Chan D.W. (2005) Independent validation of candidate breast cancer serum biomarkers identified by mass spectrometry. *Clin. Chem.* **51**, 2229–35.
81. Heike Y., Hosokawa M., Osumi S., Fujii D., Aogi K., Takigawa N., Ida M., Tajiri H., Eguchi K., Shiwa M., Wakatabe R., Arikuni H., Takaue Y., Takashima S. (2005) Identification of serum proteins related to adverse effects induced by docetaxel infusion from protein expression profiles of serum using SELDI ProteinChip system. *Anticancer Res.* **25**, 1197–203.
82. Grizzle W.E., Adam B.L., Bigbee W.L., Conrads T.P., Carroll C., Feng Z., Izbicka E., Jendoubi M., Johnsey D., Kagan J., Leach R.J., McCarthy D.B., Semmes O.J., Srivastava S., Srivastava S., Thompson I.M., Thornquist M.D., Verma M., Zhang Z., Zou Z. (2003–2004) Serum protein expression profiling for cancer detection: validation of a SELDI-based approach for prostate cancer. *Dis. Markers.* **19**, 185–95.
83. Malik G., Ward M.D., Gupta S.K., Trosset M.W., Grizzle W.E., Adam B.L., Diaz J.I., Semmes O.J. (2005) Serum levels of an isoform of apolipoprotein A-II as a potential marker for prostate cancer. *Clin. Cancer Res.* **11**, 1073–85.

84. Liotta L.A., Lowenthal M., Mehta A., Conrads T.P., Veenstra T.D., Fishman D.A., Petricoin E.F. III. (2005) Importance of communication between producers and consumers of publicly available experimental data. *J. Natl. Cancer Inst.* **97**, 310–4.
85. Woong-Shick A., Sung-Pil P., Su-Mi B., Joon-Mo L., Sung-Eun N., Gye-Hyun N., Young-Lae C., Ho-Sun C., Heung-Jae J., Chong-Kook K., Young-Wan K., Byoung-Don H., Hyun-Sun J. (2005) Identification of hemoglobin-alpha and -beta subunits as potential serum biomarkers for the diagnosis and prognosis of ovarian cancer. *Cancer Sci.* **96**, 197–201.
86. Moshkovskii S.A., Serebryakova M.V., Kuteykin-Teplyakov K.B., Tikhonova O.V., Goufman E.I., Zgoda V.G., Taranets I.N., Makarov O.V., Archakov A.I. (2005) Ovarian cancer marker of 11.7 kDa detected by proteomics is a serum amyloid A1. *Proteomics.* **5**, 3790–97.
87. Le L., Chi K., Tyldesley S., Flibotte S., Diamond D.L., Kuzyk M.A., Sadar M.D. (2005) Identification of serum amyloid A as a biomarker to distinguish prostate cancer patients with bone lesions. *Clin. Chem.* **51**, 695–707.
88. Shiwa M., Nishimura Y., Wakatabe R., Fukawa A., Arikuni H., Ota H., Kato Y., Yamori T. (2003) Rapid discovery and identification of a tissue-specific tumor biomarker from 39 human cancer cell lines using the SELDI ProteinChip platform. *Biochem. Biophys. Res. Commun.* **309**, 18–25.
89. Paradis V., Degos F., Dargere D., Pham N., Belghiti J., Degott C., Janeau J.L., Bezeaud A., Delforge D., Cubizolles M., Laurendeau I., Bedossa P. (2005) Identification of a new marker of hepatocellular carcinoma by serum protein profiling of patients with chronic liver diseases. *Hepatology.* **41**: 40–47.
90. Tolson J., Bogumil R., Brunst E., Beck H., Elsner R., Humeny A., Kratzin H., Deeg M., Kuczyk M., Mueller G.A., Mueller C.A., Flad T. (2004) Serum protein profiling by SELDI mass spectrometry: detection of multiple variants of serum amyloid alpha in renal cancer patients. *Lab. Invest.* **84**, 845–56.
91. Rosty C., Christa L., Kuzdzal S., Baldwin W.M., Zahurak M.L., Carnot F., Chan D.W., Canto M., Lillemoe K.D., Cameron J.L., Yeo C.J., Hruban R.H., Goggins M. (2002) Identification of hepatocarcinoma-intestine-pancreas/pancreatitis-associated protein I as a biomarker for pancreatic ductal adenocarcinoma by protein biochip technology. *Cancer Res.* **62**, 1868–75.
92. Wadsworth J.T., Somers K.D., Stack B.C. Jr., Cazares L., Malik G., Adam B.L., Wright G.L. Jr., Semmes O.J. (2004) Identification of patients with head and neck cancer using serum protein profiles. *Arch. Otolaryngol. Head Neck Surg.* **130**, 98–104.
93. Melle C., Ernst G., Schimmel B., Bleul A., Thieme H., Kaufmann R., Mothes H., Settmacher U., Claussen U., Halbhuber K.J., Von Eggeling F. (2005) Discovery and identification of alpha-defensins as low abundant, tumor-derived serum markers in colorectal cancer. *Gastroenterology.* **129**, 66–73.
94. Albrethsen J., Bogebo R., Gammeltoft S., Olsen J., Winther B., Raskov H. (2005) Upregulated expression of human neutrophil peptides 1, 2 and 3 (HNP 1-3) in colon cancer serum and tumours: a biomarker study. *B.M.C. Cancer.* **5**, 8–13.
95. Melle C., Ernst G., Schimmel B., Bleul A., Koscielny S., Wiesner A., Bogumil R., Moller U., Osterloh D., Halbhuber K.J., von Eggeling F. (2003) Biomarker discovery and identification in laser microdissected head and neck squamous cell carcinoma with ProteinChip technology, two-dimensional gel electrophoresis, tandem mass spectrometry, and immunohistochemistry. *Mol. Cell Proteomics.* **2**, 443–52.
96. Cho W.C., Yip T.T., Yip C., Yip V., Thulasiraman V., Ngan R.K., Yip T.T., Lau W.H., Au J.S., Law S.C., Cheng W.W., Ma V.W., Lim C.K. (2004) Identification of serum amyloid a protein as a potentially useful biomarker to monitor relapse of nasopharyngeal cancer by serum proteomic profiling. *Clin. Cancer Res.* **10**, 43–52.
97. Yang E.C., Guo J., Diehl G., DeSouza L., Rodrigues M.J., Romaschin A.D., Colgan T.J., Siu K.W. (2004) Protein expression profiling of endometrial malignancies reveals a new tumor marker: chaperonin 10. *J. Proteome Res.* **3**, 636–43.
98. Lin Z., Jenson S.D., Lim M.S., Elenitoba-Johnson K.S. (2004) Application of SELDI-TOF mass spectrometry for the identification of differentially expressed proteins in transformed follicular lymphoma. *Mod. Pathol.* **17**, 670–78.
99. Guo J., Yang E.C., Desouza L., Diehl G., Rodrigues M.J., Romaschin A.D., Colgan T.J., Siu K.W. (2005) A strategy for high-resolution protein identification in surface-enhanced laser desorption/ionization mass spectrometry: calgranulin A and chaperonin 10 as protein markers for endometrial carcinoma. *Proteomics.* **5**, 1953–66.

100. Diamond D.L., Zhang Y., Gaiger A., Smithgall M., Vedvick T.S., Carter D. (2003) Use of ProteinChip array surface enhanced laser desorption/ionization time-of-flight mass spectrometry (SELDI-TOF MS) to identify thymosin beta-4, a differentially secreted protein from lymphoblastoid cell lines. *J. Am. Soc. Mass. Spectrom.* **14**, 760–65.

101. Weinberger S.R., Dalmasso E.A., Fung E.T. (2002) Current achievements using ProteinChip Array technology. *Curr. Opin. Chem. Biol.* **6**, 86–91.

102. Zhang Y.F., Wu D.L., Guan M., Liu W.W., Wu Z., Chen Y.M., Zhang W.Z., Lu Y. (2004) Tree analysis of mass spectral urine profiles discriminates transitional cell carcinoma of the bladder from noncancer patient. *Clin. Biochem.* **37**, 772–79.

103. Liu W., Guan M., Wu D., Zhang Y., Wu Z., Xu M., Lu Y. (2005) Using tree analysis pattern and SELDI-TOF-MS to discriminate transitional cell carcinoma of the bladder cancer from noncancer patients. *Eur. Urol.* **47**, 456–62.

104. Munro N.P., Cairns D.A., Clarke P., Rogers M., Stanley A.J., Barrett J.H., Harnden P., Thompson D., Eardley I., Banks R.E., Knowles M.A. (2006) Urinary biomarker profiling in transitional cell carcinoma. *Int. J. Cancer.* **119**, 2642–50.

105. Sanchez-Carbayo M., Urrutia M., Silva J.M., Romani R., De Buitrago J.M., Navajo J.A. (2001) Comparative predictive values of urinary cytology, urinary bladder cancer antigen, CYFRA 21-1 and NMP22 for evaluating symptomatic patients at risk for bladder cancer. *J. Urol.* **165**, 1462–67.

Chapter 18

Urine Proteomic Analysis: Use of Two-Dimensional Gel Electrophoresis, Isotope Coded Affinity Tags, and Capillary Electrophoresis

Kimia Sobhani

Abstract

The identities and abundance levels of proteins excreted in urine are not only key indicators of diseases associated with renal function but are also indicators of the overall health of individuals. Urine specimens are readily available and provide a noninvasive means to assess and diagnose many disease states. Proteins in urine originate from two sources: the ultrafiltrate of plasma, and those that are shed from the urinary tract. The protein concentration in urine excreted from a normal adult is approximately 150 mg/day, and is typically not greater than 10 mg/100 mL in any single specimen. Following precipitation, concentration, and fractionation methods, proteins of interest from urine samples can be separated, identified, and quantified. One of the most commonly used techniques in the field of urine proteomics is gel electrophoresis followed by identification with mass spectrometry and protein database search algorithms. In this chapter, two-dimensional gel electrophoresis (2-DE) will be discussed, along with less frequently applied techniques, such as isotope coded affinity tags (ICAT) and capillary electrophoresis (CE). Publications discussing the application of these techniques to urine proteomic analyses of healthy individuals and urinary disease biomarker discovery will also be summarized.

Key words: Two-dimensional gel electrophoresis, Isotope coded affinity tags, Capillary electrophoresis, Urinary disease, Biomarker discovery, Polypeptide biomarkers

1. Introduction to Protein Separation Technology

In 2003, sequencing of the human genome was completed. However, two decades prior, sequencing of the approximately three billion nucleotide bases in human DNA was considered to be a nearly impossible task because sequencing methods at the time were prohibitively expensive and slow. By the mid 1990s, the development of robotic analyzers combining automated sequencers with bioinformatics tools allowed the sequencing of

Alex J. Rai (ed.), *The Urinary Proteome: Methods and Protocols*, Methods in Molecular Biology, vol. 641, DOI 10.1007/978-1-60761-711-2_18,

500,000 bases per day at a cost of 10–15 cents/base. This genomic revolution spurred the boom in efforts to analyze the complete protein content of a particular cell or tissue type expressed under a specific set of conditions, termed "proteomics" in 1996 by Wilkins et al. (1). However, due to the inherent complexity of protein expression and modification and the heterogeneous nature of protein biophysical properties, the goal of complete proteome analysis of a particular organism, tissue, or cell has yet to be fully attained.

Currently, two-dimensional gel electrophoresis (2-DE) lies at the forefront of routine proteomic analysis even though it is a decades old technique. 2-DE has become a classic and continuously used technique because it provides good resolution of proteins and is an informative top-down approach that has seen many advances over the years. Nevertheless, more easily automated approaches incorporating multidimensional chromatography or capillary based separations interfaced on-line with mass spectrometric analysis, as well as differential protein tagging approaches for accurate relative quantification of protein content in two or more samples, are becoming increasingly popular.

There are three major steps involved in the comprehensive protein analysis of a cell, tissue, or biological protein source: separation, identification, and quantification. About 40 years ago it was realized that, due to the inherent complexity of an organism's proteome, multidimensional separation techniques would be a straightforward approach for the study of individual protein components or much smaller protein subsets displaying similar characteristics.

In order to perform a multidimensional separation, two or more independent physical properties of proteins are utilized. The most efficient multidimensional separations are achieved when the selected properties of each separation dimension are orthogonal (or independent of one another) such that proteins not resolved in one dimension can potentially be resolved in the other. Separation mechanisms are commonly based on size, charge, hydrophobicity, and affinity or interaction with specific molecules. In the ideal case of completely orthogonal separations, the resolving power or spot capacity of a two-dimensional separation is a product of the peak capacity for each dimension (2).

2. Materials

2.1. 2-DE Urine Analysis

The following materials are from a paper by Pieper et al. (3) and coincide with methodology summarized from this paper in Subheading 3.1: The IgG fraction of antihuman albumin of rabbit origin for immunoaffinity subtraction chromatography was obtained from Sigma-Aldrich (St. Louis, MO). Antiserum to human α-1-acid glycoprotein from goat was obtained from Kent Laboratories (Bellingham, WA). POROS® A 20 and POROS® G 20 resins were purchased from Applied Biosystems (Foster City, CA). Superdex 75 prep grade resin, the low M_r range protein standard mixture for size exclusion chromatography (SEC), and 2-DE carrier ampholytes (pH 8–10.5 range) were acquired from Amersham Biosciences (Piscataway, NJ). Sequencing grade porcine trypsin was purchased from Promega (Madison, WI). Angiotensin II, $ACTH_{18-39}$ (adrenocorticotropic hormone fragment 18-39), and Glu^1-fibrinopeptide B were purchased from Sigma-Aldrich.

2.2. ICAT Urine Analysis

The following materials are from a paper by Cutillas et al. (4) and coincide with methodology summarized from this paper in Subheading 3.2: C_{18} POROS R2 20 resin for reverse phase chromatography desalting of proteins was obtained from Applied Biosystems (Warrington, UK). 18-cm IPG strips for 2-DE were purchased from Amersham Biosciences (Amersham, UK). Further desalting and extraction of polypeptides was carried out using a POROS HS 20, 800 μm ID × 50 mm long SCX column (Applied Biosystems). Heavy and light cleavable isotope-coded affinity tag (ICAT) reagents were purchased from Applied Biosystems (Framingham, MA).

2.3. CE-MS Urine Analysis

The following materials are from a paper by Wittke et al. (5) and coincide with methodology summarized from this paper in Subheading 3.3: Collected urine sample supernatants were desalted on a Pharmacia C2-column (Amersham Biosciences, Buckinghamshire, UK). Eluates were lyophilized in a Christ Speed-Vac RVC 2-18/Alpha 1-2 (Christ, Osterode a.H., Germany). A P/ACE MDQ CE system (Beckman Coulter, Fullerton, CA) was on-line coupled to an ESI-TOF mass spectrometer (Micro-TOF, Bruker-Daltonik, Bremen, Germany) to separate and identify low M_r proteins and peptides.

3. Methods

3.1. Application of 2-DE to Urine Proteomics

In 2004, Pieper et al. (3) published a urine protein analysis methodology utilizing two-dimensional gel electrophoresis (2-DE) with the aim to seek out new disease biomarkers. With such a goal, reproducible urine sample preparation and a 2-DE procedure with high resolution for repeat comparative analyses were needed. Their sample preparation protocol involved the use of size exclusion chromatography to enrich for proteins with molecular weights less than 30 kDa in one fraction, and proteins with molecular weights greater than 30 kDa in another. In order to remove the most highly abundant proteins contained in the >30 kDa, like albumin and immunoglobulin G (IgG), immunoaffinity subtraction chromatography was employed. By removing these highly abundant proteins better resolution of lower abundance proteins could be attained.

The combined fractions were prepared for solubilization in a buffer compatible with 2-DE analysis. After 2-DE separations were completed, proteins were visualized with Coomassie Brilliant Blue and superior protein spot resolution was observed in comparison to urine concentrates that had not been fractionated. High-throughput MALDI-TOF MS as well as ESI-LC-MS/MS analysis were then applied to approximately 1,400 distinct spots which led to the identification of ~30% of the proteins expected to be present in the sample. Not surprisingly, high levels of post-translational modifications exhibited by many urine proteins and proteolytic products reduced the set of 420 spots in which identifications were made to a set of 150 unique proteins.

Since urine is the ultra-filtrate of plasma, it was expected that a high level of classical plasma proteins would be identified; however, only a third of identified proteins fell into this category. In order to demonstrate the potential of this urine proteome analysis methodology for the discovery of new disease biomarkers, urine samples from a patient with renal cell carcinoma were analyzed before and after nephrectomy. The authors made the observation that the intensities of spots in which kininogen was identified were markedly reduced. Kininogen is a thiol protease inhibitor that is a precursor to a protein involved in the mediation of inflammation. The authors cite previous reports linking quantitative changes in kininogen in carcinoma patients to cancer progression. Such prior evidence further bolsters a hypothesis that kininogen might be useful as a biomarker in patients with renal cell carcinoma. Subheadings 3.1.1–3.1.3 summarize the steps taken by Pieper et al. (3) to perform the research discussed above.

3.1.1. Urine Sample Preparation

Protease inhibitor cocktails (Complete Mini; Roche Diagnostics, Mannheim, Germany) were added to urine samples immediately after collection. Samples were then centrifuged and precipitates were removed. Supernatant portions were then transferred to 80 mL Ultrafree® 5 K membrane concentrators (Millipore, Billerica, MA) and centrifuged to reduce sample volumes to 1 mL. Precipitates formed during concentration were again removed. The BCA assay (Pierce, Rockford, IL) was utilized to determine the mass of protein present in each urine concentrate. Finally, concentrated samples were stored at –70°C until the next step.

3.1.2. Fractionation of Samples by Molecular Weight

Columns packed with Superdex 75 resin were used for size exclusion chromatography (SEC) on a fast performance LC system. Separations were monitored and optimized at 280 nm with protein standards, and the cut-off volume for 30 kDa was determined. Proteins were collected in two fraction pools: >30 kDa and <30 kDa. Diluted, fractionated proteins were reconcentrated using Centriplus® 5 K units (Millipore) to 500 μL final volumes.

3.1.3. Immunoaffinity Subtraction Chromatrography and 2-DE

Serum albumin, α-1-acid glycoprotein, and immunoglobulin G (IgG) were among the high abundance proteins that were depleted from the >30 kDa fraction pool. In order to remove these proteins immunoaffinity and protein affinity chromatography were employed. A recyclable 1.7 mL antibody-derivatized immunoaffinity subtraction column (IASC) specific for albumin, IgG, and α-1-acid glycoprotein was used to deplete these proteins from the >30 kDa using a previously described procedure (6). Both SEC fractions were prepared for 2-DE by exchange into a buffer containing 25 mM ammonium bicarbonate, 0.5 mM sodium EDTA, and 0.5 mM benzamidine. 200–300 μL of protein concentrates were transferred to new sample tubes, lyophilized, and solubilized in a 2% CHAPS, 9 M urea, 62.5 mM dithiothreitol (DTT), and 2% pH 8–10.5 carrier ampholytes IEF buffer. 2-DE was performed using the high-throughput ProGEx system (Large Scale Biology Corporation) as previously described (7).

Gels were stained and fixed overnight in Coomassie Brilliant Blue, then scanned with Kepler® software (Large Scale Biology Corporation), which recorded the intensities and positions of spots. Destaining, and finally trypsin digestion of spots were performed with a Tecan Genesis Workstation 200 (Durham, NC). Peptide extracts were then prepared for MALDI-TOF analysis utilizing a 384-format 600 mm AnchorChip MALDI plate (Bruker, Bremen, Germany) and α-cyano-4-hydroxycinnamic acid matrix. Extracts were also prepared for LC-MS/MS analysis by transferring them to 96-well microtiter plates and diluting to 10 μl final volumes. Refer to Pieper et al. (3) for specific MS instrument settings and subsequent data analysis.

3.2. Application of ICAT to the Analysis of Fanconi Syndrome

In 2004, Cutillas et al. (4) published a study discussing differences in the urinary proteome in Fanconi syndrome (FS), which is an impairment in function of the proximal tubule in the kidney. In normal individuals, polypeptides and other small molecules (such as electrolytes) that are still present after glomerular filtration will be reabsorbed via receptor mediated endocytosis in the proximal tubule. In renal FS, many of these polypeptides are not reabsorbed often due to a low presence of renal epithelial cells.

Three proteomic approaches were applied to the analysis of urinary proteins and polypeptides in patients with a form of FS known as Dent's disease (is caused by a mutation in CLC5) in comparison with urine from normal individuals. Three approaches were used to determine whether differences in protein profiles observed with one method could also be observed with another. 2-DE was utilized to compare Dent's patient protein and polypeptide profiles with control subjects that were both healthy and displayed normal renal function. Proteins and polypeptides were then identified with MALDI-TOF MS and/or ESI-LC-MS/MS. The same comparison was conducted using a multidimensional chromatographic approach, where a microcolumn strong cation exchange (SCX) separation was performed followed by reverse phase (RP) microcapillary liquid chromatography (μLC). Collected chromatographic fractions were subsequently digested with trypsin and identified using ESI-LC-MS/MS.

Both 2-DE and μLC approaches demonstrated several partially quantitative differences between normal and Dent's patient urine polypeptide levels. However, for a more quantitative relative comparison, and also to validate the results obtained with the previous two methods, cleavable ICAT reagents were utilized following a protocol essentially the same as that displayed in Fig. 4. In this case, MALDI-TOF MS was employed for protein identification.

Ultimately, the abundance level differences observed in all three methods were very consistent. In Dent's urine, several prosthetic group carrier proteins, apolipoproteins, complement components, and peptides with potential bioactivity were present at higher levels than in comparison to the control urine group. In contrast, proteins originating from the renal tract itself were shown to be proportionately more abundant in control urine.

The authors state that their findings support the hypothesis that one important function of the proximal tubule is the uptake of filtered vitamins, which would normally be bound to carrier proteins, in order to prevent their losses in urine. Furthermore, they contend that the proximal tubule may also be involved in the readsorption of potentially cytotoxic peptides filtered from plasma, preventing them from interacting with distal portions of the tubule. More specific methodology for the research just described is summarized from the same paper by Cutillas et al. (4) in Subheadings 3.2.1 and 3.2.2.

3.2.1. Urine Sample Collection

Four male patients with Dent's disease were selected for study. Collected urine samples from these patients were either used immediately or frozen in liquid nitrogen and stored at –80°C for later use. To prepare samples for analysis, their compositions were adjusted to yield final concentrations of 1 mM EDTA, 1 mg/mL aprotinin, and 0.1% trifluoroacetic acid (TFA). Samples were analyzed using 2-DE, microcapillary HPLC, and ICAT. However, only ICAT methodology will be covered in the next section.

3.2.2. SCX Chromatography, ICAT Labeling, and Protein Analysis

To prepare samples for strong cation exchange (SCX) chromatography proteins were precipitated in 50% acetone and resuspended in a compatible buffer. After SCX, normal urinary proteins were labeled with the light isotopic cleavable ICAT reagent, whereas Dent's disease proteins were labeled with the heavy isotopic reagent per the manufacturer's protocols. As in Fig. 4, labeled protein samples were then mixed and digested with trypsin overnight at 37°C. SCX chromatography was employed again to clean up the samples and remove excess ICAT reagents. Collected SCX fractions were subsequently run over an avidin affinity column to capture ICAT labeled peptides (nonlabeled peptides were not analyzed). The ICAT labeled fractions collected from this step were then cleaved to remove the biotin portions of the tags. Fractions were then lyophilized and resuspended in a buffer appropriate with MS analysis.

ICAT labeled peptides were analyzed by LC-MALDI-MS/MS with a 4700 Proteomics Analyzer (Applied Biosystems). The specific instrument settings used and data analysis procedures followed have been previously described (8–11).

3.3. Application of CE to the Identification of Polypeptide Biomarkers in Renal Diseases

Recently, Wittke et al. (5) used capillary electrophoresis, coupled with mass spectrometry, operating in CZE mode with an 20% v/v acetonitrile 0.25 M formic acid running buffer to identify a series of polypeptides that are biomarkers for renal diseases. In this same paper, human cerebrospinal fluid (CSF) was also analyzed with the same methodology to identify potential biomarkers for schizophrenia and Alzheimer's disease, but will not be further discussed. CE was selected due to its small sample volume requirement, rapid separation capability, high separation efficiency, and better resolving power (in comparison to gel electrophoretic techniques) for low molecular weight analytes such as polypeptides.

580 urine samples (in total) from patients with six diseases: immunoglobulin A (IgA) nephropathy, focal-segmental glomerulosclerosis, membranous glomerulonephritis, minimal-change disease, lupus nephritis, and diabetic nephropathy were analyzed with CE-MS and compared against 300 samples from healthy individuals. Due to the sensitive, high resolution of the CE-MS method that was developed, discriminate and repeatable polypeptide patterns were obtained for each disease and for the normal control population. Polypeptide patterns were determined by examining three-dimensional contour plots of mass-to-charge (m/z) ratios vs. migration times for analytes, with signal intensity plotted on the z-axis. Numerous polypeptide biomarkers were identified for each disease suggesting that their patterns and levels could be used to track progression and the effects of various therapies.

To validate the utility of this method for such a purpose, the polypeptide status of diabetic nephropathy in 24 patients before and after treatment with angiotensin II receptor blocker was monitored with CE-MS. Several polypeptide biomarkers of diabetic nephropathy decreased in response to treatment. These decreases were associated with an improved clinical status for patients indicating the usefulness of tracking biomarker profiles to guide and assess therapeutic strategies. Subheadings 3.3.1–3.3.3 summarize methodology utilized by Wittke et al. (5) to perform the research discussed above.

3.3.1. Urine Sample Collection and Preparation

Collected urine samples were frozen immediately and stored at −20°C. Second urine samples, collected in the morning, were preferentially used when possible because they exhibited less variability. Thawed urine samples were centrifuged, and 1 mL of supernatant from each sample was run over a Pharmacia C2-column to remove urea, salts, and other interfering substances and also to concentrate the peptides present. A buffer containing 50%v/v acetonitrile and 0.5%v/v formic acid was used to elute peptides. Eluted samples were lyophilized and resuspended in HPLC-grade water just prior to use.

3.3.2. CE-MS Analysis

For CE-MS analysis, a P/ACE MDQ CE Instrument (Beckman Coulter) was on-line coupled to an ESI-TOF mass spectrometer (Micro-TOF, Bruker Daltonik) as described previously (12–14). A 90 cm, 50 μm ID, bare fused silica capillary (Beckman) was used for CE separations. A running buffer containing 20% acetonitrile and 0.25 M formic acid was used for CE. Samples were injected using a positive pressure of 1–6 psi for 99 s such that sample plugs of 50–300 nL were separated and analyzed. Separations were performed by the application of +30 kV for 60 min, at a constant capillary temperature of 35°C.

The mass spectrometer was operated in positive electrospray mode with an ESI-TOF sprayer kit (Agilent Technologies). An accumulation of spectra ranging from 350 to 3,000 m/z were collected for 3 s each. In order to sequence relevant peptides, an Ultraflex MALDI-TOF-TOF instrument (Bruker Daltonik) was used. To perform these polypeptide identifications, CE runs were spotted onto a MALDI plate using a Probot microfraction collector (LC Packings) which was configured to deposit one CE fraction every 15 s.

3.3.3. Data Analysis

MosaiquesVisu, a software tool designed to extract information from raw CE-MS datafiles was utilized. This tool characterized each protein/polypeptide by its molecular mass and normalized migration time. Peaklists generated by MosaiquesVisu were added to an Access database to facilitate comparison to peaklists from separate runs and find matching polypeptide patterns. Mosacluster, a software tool that generates a polypeptide model for diseases based on polypeptides that best allowed for discrimination between disease and control (or between different diseases), was also utilized.

4. Two-Dimensional Gel Electrophoresis (2-DE)

Two-dimensional gel electrophoresis (2-DE), often more specifically referred to as two-dimensional SDS polyacrylamide gel electrophoresis (2D-PAGE) first separates proteins by their isoelectric

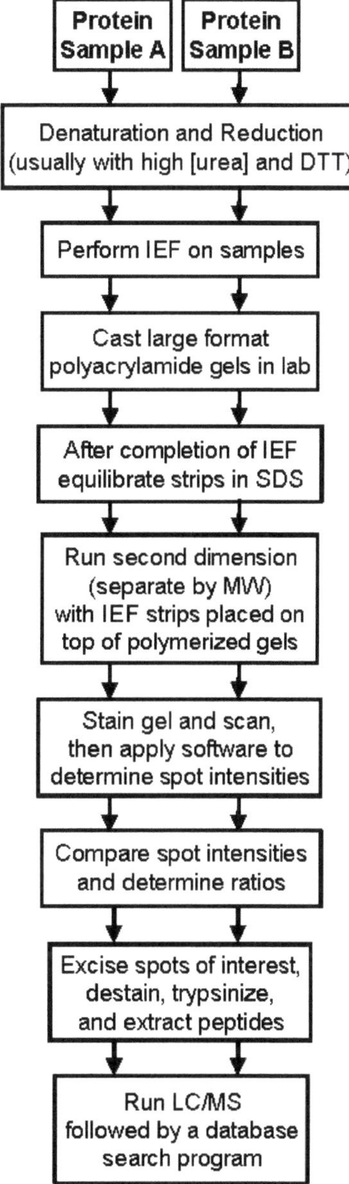

Fig. 1. A work-flow diagram for a comparative 2D polyacrylamide gel electrophoresis experiment.

point (pI), which is the pH at which they are neutral, via liquid or gel based isoelectric focusing (IEF), followed by a size based separation in which denatured proteins are separated by their molecular weight as they migrate through a gel matrix (usually polyacrylamide) under an applied voltage such that the smallest proteins will migrate the fastest (refer to Fig. 1 for a work-flow diagram of a comparative 2-DE experiment). To separate proteins in the second dimension based on differences in their sizes and not differences in their charge-to-size ratios, denatured proteins are coated with the anionic surfactant, sodium dodecyl sulfate (SDS). The SDS molecules coat proteins at a ratio of 1.4 g SDS/g protein. The overwhelming negative charge that SDS imparts to proteins effectively eliminates charge differences for similarly sized proteins; also, rod shaped, SDS coated, denatured proteins all have similar shape profiles, such that the main characteristic proteins will be separated by in the second dimension gel are their varying molecular weights. Visualization of separated proteins in the second dimension gel is most commonly achieved by silver staining.

This now classical proteomic technique was introduced by O'Farrell over 30 years ago (15), and has improved over the years to include immobilized pH gradients in the first dimension, and protein staining and visualization methods with increased sensitivity, which can detect on the order of low nanogram levels of protein. Efficient and highly automated protein identification methods such as matrix-assisted laser desorption/ionization time-of-flight (MALDI-TOF) mass spectrometry, or very commonly used electrospray ionization liquid chromatography mass spectrometry (ESI-LC/MS) have been developed, as well as much more comprehensive protein databases, faster computers, and relatively quick search programs (i.e., SEQUEST and MASCOT) that can output reliable peptide and protein identifications.

Up to a few thousand proteins can routinely be separated on a large format (18 × 20 cm) gel under good conditions, and Inagaki et al. has demonstrated that the separation of up to 11,000 proteins is possible on a 93 cm × 103 cm "cybergel" (16). If it is postulated that a mammalian cell expresses 10,000 genes, it can be estimated that it may have ~20,000 or more modified proteins, so in one of the very best cases reported approximately half of the proteome can hypothetically be resolved.

4.1. Disadvantages of 2-DE

2-DE has several drawbacks, which must be taken into account. For example, 2-DE is a cumbersome, time consuming technique that is difficult to automate. 2-DE is a particularly useful technique for the separation of highly abundant proteins because current, commonly used staining techniques such as Coomassie blue and silver staining have a dynamic range on the order of 3 to 4 orders of magnitude. However, this method is not efficient for the detection of low abundance proteins because protein homogenates from cells or any biological protein source, can have a dynamic range of expression of up to 7 orders of magnitude or more (17). This necessitates the use of preprocessing procedures for depletion of high-abundance proteins and enrichment and detection of lower abundance components for resolution using 2-DE. Other challenges that exist in using 2-DE to resolve proteins are the separation of hydrophobic proteins, which have a tendency to aggregate at their pIs resulting in reduced transfer to the second dimension. Resolution of alkaline proteins is also an issue because most commercially available IEF gel strips do not have immobilized pH gradients above 11. Very large or very small proteins also do not usually get resolved; and finally, the overall, labor intensive, lengthy process of running a 2D gel and identifying proteins from stained spots makes it unfavorable for use in a clinical setting where many samples need to be efficiently run and analyzed.

4.2. Advances in 2-DE Technology

There have been recent advances in 2-DE that continue to keep this method in the forefront of proteomic techniques. Several papers have described enhanced protocols using prefractionation and sequential extraction with new detergents for the enrichment of particular types of proteins (18, 19). Fluorescent dyes have been introduced, such as SYPRO Ruby (a ruthenium-based metal chelate fluorescent stain), which exhibits a somewhat broader dynamic range (20) than silver staining or Coomassie blue; cyanine derived fluorophores "CyDyes" for difference gel electrophoresis (DIGE) are also a relatively recent technical advance. Robots that perform excision of protein spots and automated protein digestion have been developed that can also be interfaced with image analysis software and with MS instrumentation (21, 22); although such automated instrumentation is not commonly used and still requires a good deal of input by the user to ensure that the desired spots are selected and excised. However, such attempts at automation can speed analysis, and also reduces the exposure of samples to contaminants, in particular protein contaminants from researchers themselves.

In recent years, DIGE has become a popular method for comparing two protein sample states in one shot (23). In this approach, proteins from two sample populations are derivatized with two different fluorescent dyes (Cy3 and Cy5) prior to electrophoresis so that only a single gel is required to separate both protein samples and to visualize and quantify differences between them. Currently up to three sample states can be tracked with one gel because there are new forms of CyDyes that have recently been developed with distinct spectral properties. With this method, somewhat higher throughputs can be achieved in comparison to conventional 2-DE and technical variations in gel-to-gel protein migration are eliminated.

4.3. Application of 1D Gel Electrophoresis to Large-Scale Urine Protein Identification

Subheading 3.1 discusses the application of 2-DE methodology to the human urinary proteome. However, the application of 1-D gel electrophoresis to large-scale urine protein discovery will be briefly covered here. Adachi et al. published a set of 1,543 proteins identified from urine samples taken from 10 healthy individuals in late 2006 (24). One-dimensional SDS polyacrylamide gel electrophoresis and reverse phase high-performance liquid chromatography were employed for protein separation and fractionation. One-dimensional gel electrophoresis is essentially equivalent to its two-dimensional counterpart with the isoelectric focusing protocol removed. Fractionated proteins were digested in-gel or in-solution (in-solution digestion was used for proteins separated using reverse phase chromatography), and digests were analyzed with a linear ion trap-Fourier transform (LTQ-FT) mass spectrometer and a linear ion trap-orbitrap (LTQ-Orbitrap) mass spectrometer with accuracy at the parts per million level. A work-flow diagram for the protocol used in this research is shown in Fig. 2. Gene Ontology (GO) analysis revealed that nearly half of the proteins identified were membrane proteins. In particular, when compared with all GO entries, there was enrichment for extracellular, lysosomal, and plasma membrane proteins in urine. It was hypothesized by the authors that plasma membrane proteins are present in urine due, for the most part, to exosomal secretions.

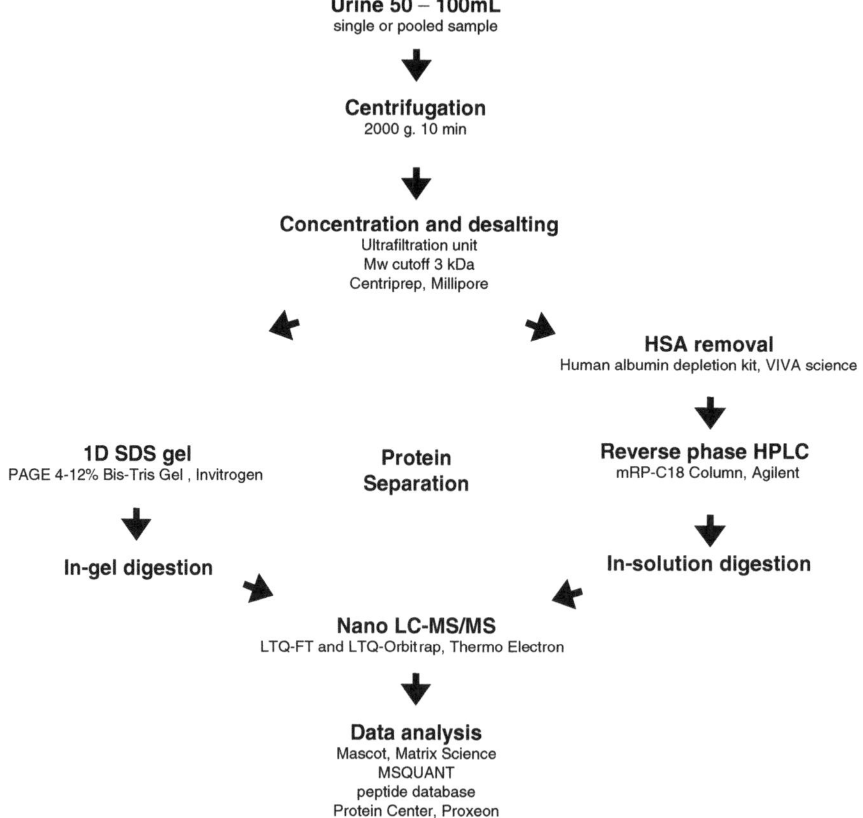

Fig. 2. Workflow diagram for the separation of urine proteins with 1D polyacrylamide gel electrophoresis and reverse phase chromatography followed by identification with LTQ-FT and LTQ-Orbitrap mass spectrometry.

This study is primarily useful for the knowledge it provides on the types of proteins that constitute the urinary proteome and as a reference to compare proteins identified in other studies using different methodologies. Furthermore, the authors speculate that the complexity of the urinary proteome published in this work could potentially give way to the discovery of new disease biomarkers.

5. Quantitative Proteomics: Isotope Coded Affinity Tag (ICAT) Technology

In addition to the goal of identifying as many proteins as possible in a particular proteome, it is often important to quantitatively analyze protein abundance levels. Besides more qualitative approaches like 2-DE and DIGE, there are two other major techniques that exist for the large-scale identification and relative quantitative profiling of proteins: stable isotope labeling of proteins and peptides, and protein expression arrays, both followed by MS analysis. Stable isotope labeling techniques based on ^{2}H, ^{13}C, ^{15}N, and ^{18}O have been developed for methods that incorporate such labels after cell harvest, or in situ (during cell growth in culture). However, only isotope coded affinity tags (ICATs) and the two major forms of this reagent technology are discussed in this section.

In 1999, Gygi et al. published a relative quantitative protein analysis technique discussing the development and utilization of ICAT reagents, which are applied to samples after cells have been harvested and lysed, and protein contents have been denatured and reduced (25). These reagents have three major components: (1) a reactive site that covalently binds to the free sulfhydryl groups of reduced cysteine side chains, (2) a biotin moiety that is used to capture labeled peptides with immobilized avidin, and (3) a side chain containing light or heavy isotope atoms that allow for discrimination between and relative quantification of two samples.

With first generation ICAT reagents, proteins are isolated and selectively alkylated on their cysteine residues with either a heavy isotope reagent that incorporates eight deuteriums (d8-ICAT) or an isotopically normal reagent containing eight hydrogen atoms (d0-ICAT), resulting in a nominal mass difference of 8 Da between the two (see Fig. 3). Following labeling, protein samples are combined, digested with trypsin, and labeled peptides are isolated by passage over an avidin affinity column. Capture of labeled peptides with avidin occurs because ICAT reagents possess the aforementioned biotin linker to facilitate this selective isolation. Relative quantification is deduced when labeled peptides are identified from their MS/MS collision induced dissociation (CID) spectra and their corresponding elution profiles are extracted from the total ion chromatogram to determine

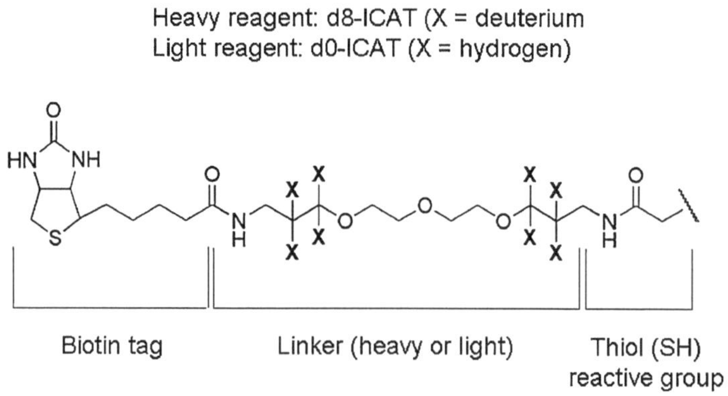

Heavy reagent: d8-ICAT (X = deuterium
Light reagent: d0-ICAT (X = hydrogen)

Biotin tag
Linker (heavy or light)
Thiol (SH) reactive group

Fig. 3. Structure of first generation ICAT reagents. The light reagent contains eight hydrogen atoms and the heavy reagent contains eight deuterium atoms. These reagents covalently bind to the reduced sulfhydryl groups of cysteine residues.

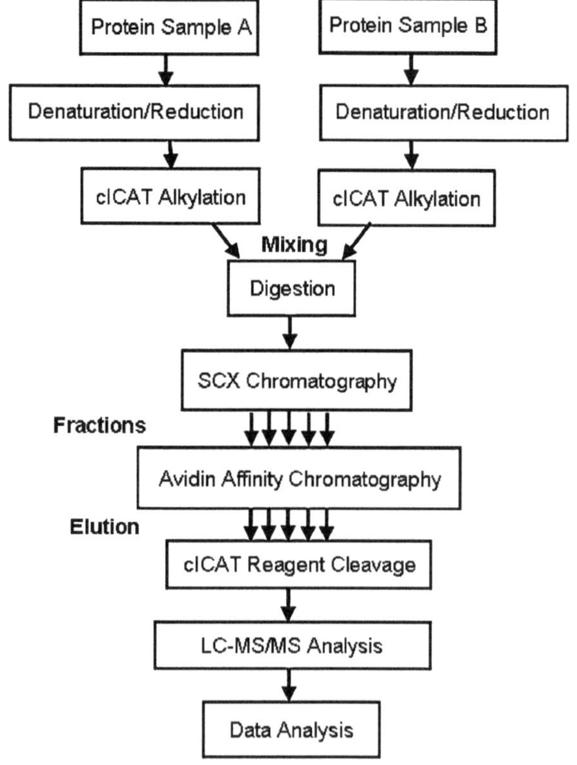

Fig. 4. Work-flow diagram for cleavable ICAT (cICAT) reagents. This protocol is essentially the same as that introduced for the first generation of ICAT reagents, the main difference being that there is no cleavage step with these original reagents.

abundance ratios by integrating the respective light and heavy peptide peak areas. Figure 4 shows a flow diagram of the ICAT protein analysis strategy (this diagram applies, in particular, to cleavable ICAT reagents).

5.1. Advantages of ICAT in Comparison to 2-DE

ICAT offers a number of advantages over 2-DE: It is compatible with proteins harvested from virtually all types of biological sources and growth conditions. Only cysteine containing peptides are targeted, which greatly reduces the complexity of the resultant peptide mixture that is analyzed by MS. ICAT labeled proteins are digested into peptides prior to chromatographic separation steps. Therefore, highly alkaline, acidic, or hydrophobic proteins that do not typically get resolved well in 2-DE methods, are not as problematic when broken down and separated at the peptide level. Furthermore, multidimensional chromatographic approaches (which are typically used after ICAT labeling) are much more easily automated than 2-DE methods, and can also be directly interfaced with MS, thereby significantly increasing the throughput of this comparative and quantitative proteomic technique. Although unrelated to a comparison with 2-DE, another desirable characteristic of ICAT methodology is that the alkylation reaction of cysteine residues with ICAT reagents is highly specific, meaning that side reactions with other moieties are not a concern.

5.2. Some Disadvantages of ICAT and the Introduction of Cleavable ICAT Reagents

A few disadvantages of ICAT methodology are that not all proteins of interest contain cysteine residues, ICAT reagents bound to peptides can complicate MS/MS spectra, and ICAT is relatively expensive. However, recently a second generation of commercially introduced cleavable ICAT reagents based on carbon isotopes were developed (11). The light reagent contains nine ^{12}C atoms and the heavy reagent incorporates nine ^{13}C atoms, resulting in reagents with a mass difference of exactly 9 Da (see Fig. 5).

The utilization of a 9 Da mass difference eliminated potential confusion between doubly d8-ICAT labeled peptides and the 16 Da gain in mass that could commonly be observed in peptides due to oxidation of methionine. Furthermore, cleavable reagents

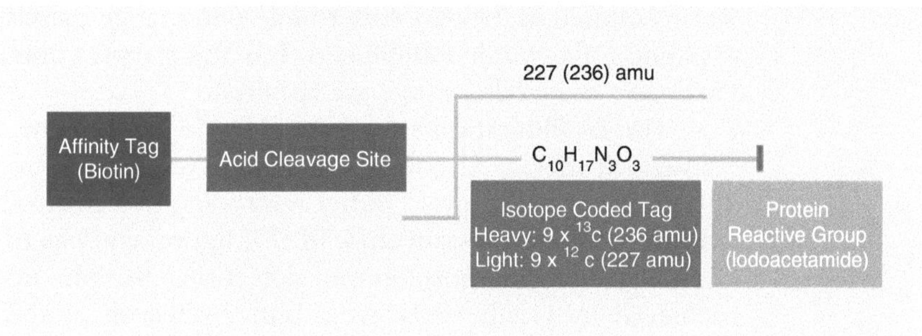

*The formula of cleavable ICAT reagents is proprietary

Fig. 5. The functional components of cleavable ICAT reagents. After cleavage of the biotin portion of the tag, the remainder of the light tag is 227 amu, and the remainder of the heavy tag is 236 amu.

contain biotin moieties that are bound via an acid cleavable linker so that they can be removed prior to MS analysis, which has increased the number of proteins that can be identified and quantified by decreasing the complexity of the resultant spectra due to fragmentation of the biotin tag itself. Also, another drawback of first generation ICAT reagents is that with LC/MS analysis, d0 and d8 modified peptides do not necessarily co-elute during reverse phase chromatography because the d8 reagent is slightly more hydrophilic and therefore tends to elute faster than its nondeuterated counterpart resulting in reduced quantitation accuracy. The $^{12}C/^{13}C$ based reagents do not suffer from this limitation. To date, since its recent introduction in 2003, dozens of papers have been published using the cleavable ICAT reagent technique.

6. Capillary Electrophoresis

Capillary electrophoresis (CE) is the one of the most efficient separation techniques available for the analysis of a large variety of organic and inorganic molecules. One of the main factors that initiated progress to modern CE was the production of inexpensive narrow-bore capillaries for gas chromatography (GC) and the design of highly sensitive on-line detection methods such as those originally developed for high performance liquid chromatography (HPLC). Two significant advantages of CE in comparison to traditional slab gel electrophoresis are (1) the high surface to volume ratio of fused silica capillaries (~200 in comparison to ~2 for a large format slab gel) allows for the use of very high voltages (on the range of 800 V/cm) leading to very short analysis times (on the order of minutes for a 1D separation, and an hour or less for a 2D separation) making it ideal for high throughput analyses (26, 27), and (2) the small dimensions of these capillaries (~20–50 cm in length, and ~10–50 μm inner diameter) yield total column volumes in the nanoliter to microliter range such that buffer volumes of a couple milliliters or less and sample volumes in the nanoliter to picoliter range are required.

The configuration of a basic 1D-CE instrument, which is illustrated in Fig. 6, consists of a high voltage power supply, a fused silica capillary, two buffer reservoirs, two electrodes, and a sample detection system that, in this figure, consists of a sheath flow cuvette for postcolumn detection. Sample injection is achieved by briefly replacing a buffer reservoir at the capillary inlet with a small vial of the sample to be introduced. A specific amount of sample is loaded by manipulating either the injection voltage or the injection pressure. CE separations commonly possess the potential for 50,000–500,000 theoretical plates in a

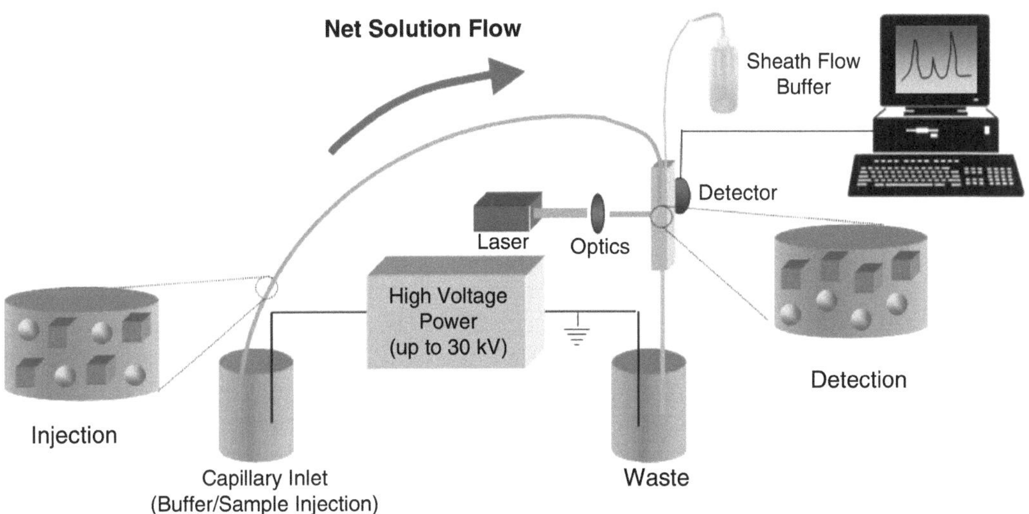

Fig. 6. General configuration of a 1D-CE instrument. The zoomed in views of sections at the beginning and end of the separation indicate how analytes progress from unresolved mixtures to separated discrete zones by the time they are detected, such as in the case of a CZE experiment.

separation capillary ~20 cm in length, which is at least an order-of-magnitude better than competing high performance liquid chromatography (HPLC) methods (28, 29).

CE analyses are typically rapid, can be automated, use miniscule amounts of sample and reagents, and are less expensive than chromatography or classic gel based electrophoresis. While modern CE methods are still developing and evolving, great potential has been established for a broad range of applications, from small molecules like inorganic ions and amino acids, to larger biomolecules such as proteins, nucleic acids, and all the way up to cells.

As highlighted earlier, the architecture of a fused silica capillary leads to an inherently high surface area to volume ratio, especially in comparison to slab gel electrophoresis. The surface area to volume ratio is further enhanced as the inner diameter of the capillary is reduced. Consequently, high voltage drops can be applied in CE; with the current technology, ~30 kV can be applied producing very fast and efficient separations (28, 29). There are two forces that drive capillary electrophoretic separations: the electrophoretic mobility of charged analyte molecules in the capillary under an applied electric field and the electroosmotic movement or flow of the bulk ionic buffer solution (in which the analyte molecules reside) under the force of the electric field. The polarity of the electric field is chosen such that the electroosmotic flow of the bulk buffer solution is in the direction of the detector. The electrophoretic mobility of analyte molecules depends on the sign and magnitude of their charges as well as their sizes (charge-to-size ratios) and can be in the direction of the injection or detection ends of the capillary under a particular electric field polarity. As long as the electroosmotic flow of the buffer solution is sufficiently large, all analyte molecules will eventually be carried toward the detector at differing rates based on their differing electrophoretic mobilities. The apparent velocity at which an analyte molecule migrates toward the detector is the vector sum of the electrophoretic and electroosmotic velocities it experiences.

Avalanche photodiodes (APD) are a highly sensitive and efficient means for detection of photons emitted from fluorescently labeled proteins excited by a laser of the appropriate wavelength. In combination with CE, laser induced fluorescence detection, in comparison to other commonly used methods (such as UV excitation), offers the most sensitivity for the detection of proteins and biogenic amines with a limit of detection (LOD) of 10^{-17} to 10^{-24} moles (30, 31).

6.1. Separation Modes and Principles

CE can be performed under many separation conditions some of the most common of which are capillary zone electrophoresis (CZE), micellar electrokinetic capillary chromatography (MEKC), and capillary sieving electrophoresis (CSE) or capillary gel electrophoresis (CGE). Each of these modes uses a high voltage to achieve a highly efficient separation.

CZE is the simplest mode of CE. In this mode, sample is injected as a narrow zone (band), which is bordered on both sides by the separation buffer. When an electric field is applied, each charged component in the sample zone migrates based on its individual apparent velocity (the sum of the electroosmotic velocity of the buffer and the electrophoretic velocity of the analyte). In an ideal situation, all sample components will eventually separate from each other to form discrete zones containing a single component. In CZE mode, neutral molecules do not separate from one another because they do not have differing electrophoretic mobilities and therefore migrate at the rate of the electroosmotic flow toward the detector. The most efficient separations of charged molecules are achieved when differences in their apparent velocities are maximized while diffusion of the individual zones is minimized.

The development of micellar electrokinetic capillary chromatography (MEKC) has extended CE applications to the separation of both neutral and charged molecules through the use of micelles in the separation buffer. Micelles are spherical aggregates of amphiphilic molecules called surfactants. Surfactants combine two main chemical moieties, a hydrophilic head group and a hydrophobic tail group. The hydrophobic tail often consists of a straight or branched chain of a hydrocarbon, or a steroidal skeleton; the hydrophilic head can take on many forms and charge states such as cationic, anionic, zwitterionic, or nonionic (28, 29).

The aggregative behavior of surfactants is dictated by their critical micelle concentration (CMC), such that when this threshold is achieved or exceeded, individual surfactant molecules are pushed together by the polar solution in which they reside. Spherical micelles form with the hydrophobic tails of surfactant molecules directed toward the inner core of the aggregate and the hydrophilic heads facing the outside solution. SDS is a commonly used surfactant in MEKC separations. SDS is highly anionic and has a CMC of 8 mM in pure water. Neutral molecules are separated in MEKC mode based on differences in their hydrophobicities and tend to switch back and forth between the micellar and solute phases at differing rates depending on the degree of their hydrophobicity. MEKC separations result from the overall effects of the differential partitioning of molecules between the aqueous buffer and the micellar phase, and the varying electrophoretic mobilities of ionic species.

Capillary gel electrophoresis (CGE) or capillary sieving electrophoresis (CSE) are the capillary equivalents of traditional slab gel electrophoresis. CSE is a variation of CGE except that a viscous polymer solution rather than a gel is used as the separation matrix. Just as slab gel electrophoresis is applied to the size based separation of biomolecules such as DNA and proteins, CGE is primarily useful for the same purpose. Noncrosslinked polymers such as linear polyacrylamide, polyethylene glycol, and cellulose derivatives have been applied to CGE separations, as well as crosslinked polymers such as agarose and polyacrylamide, although the latter reagents have a tendency to polymerize imperfectly within the narrow volume of a capillary.

The use of a polymer matrix reduces the analyte diffusion rate and the interaction or adhesion of analyte to the capillary wall, while suppressing electroosmotic flow. These effects allow the use of shorter capillaries due to the increased efficiencies that can be attained. With crosslinked polymers (which are static), the resolution of the capillary can be easily optimized for a given range of molecular weights by varying the total monomer concentration and degree of cross-linking. Yet, noncrosslinked polymers (i.e., polymers of glucose or other carbohydrates) can be easily purged from the capillary when a problem develops (such as a trapped air bubble or clog) and can be reintroduced so that a fresh separation medium can be utilized for each experiment.

In order to separate proteins or molecules based on their size differences alone the electroosmotic flow needs to be suppressed. This suppression can be achieved by coating the inner capillary wall with a polymer (often polyacrylamide based) that prevents interaction of capillary contents with the charged silanol groups of the inner wall. Furthermore, negative polarity is used in CSE experiments because submicellar concentrations of SDS are utilized to complex with denatured proteins such

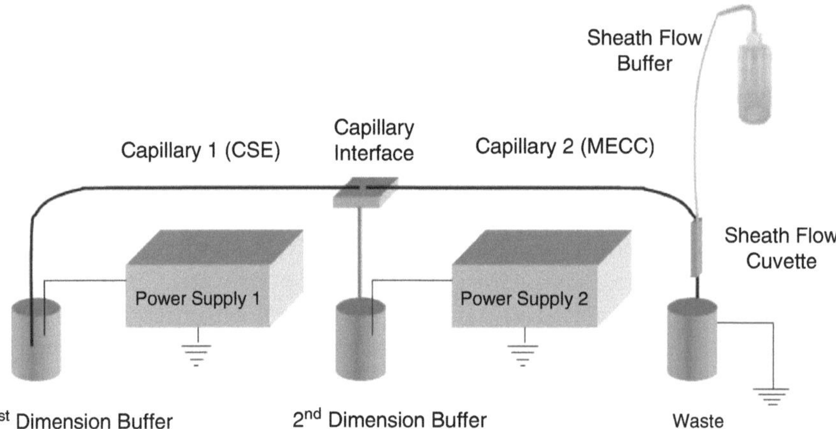

Fig. 7. General configuration of a 2D-CE instrument. The photon detector apparatus is not indicated but would sit behind the sheath flow cuvette and be in alignment with the laser beam and sample stream exiting the capillary tip.

that the size-to-charge ratios for the analytes are similar (due to the negative charge imparted by SDS to all the proteins). Under ideal conditions, proteins are therefore separated only by differences in their sizes as they migrate to the positive pole at varying rates through the sieving or gel matrix.

Two-dimensional capillary electrophoresis (2D-CE) uses a separation principle similar to that of 2-DE, where two separation mechanisms based on different protein properties are combined in sequence to obtain a high spot capacity (refer to Fig. 7 for a diagram of a 2D-CE instrument). The Dovichi group (27) commonly combines CSE (as the first dimension) and MEKC (as the second dimension). The instrumental set-up for a 2D-CE separation is fairly similar to that for a 1D-CE separation; the primary differences being that a third buffer reservoir (which holds the second dimension running buffer), a second high voltage power supply, and a handmade capillary interface (where the two separation dimension capillaries meet) are employed.

7. Concluding Remarks

Testing for the presence and abundance levels of proteins and peptides in urine has enormous potential for the diagnosis and management of urinary tract diseases. In recent years, many studies have been carried out that have founded large urinary proteome datasets. These studies have not only identified proteins and polypeptides present in urine from normal individuals, but also have yielded distinct and useful biomarker profiles that can assist in the diagnosis of particular renal diseases and allow their progression to be assessed and monitored.

As the discussion of 2-DE, ICAT, and CE in this chapter has shown, reproducible and informative proteomic methodologies have been designed that have already yielded advancements in the noninvasive diagnosis and treatment of urinary tract diseases and can be adapted for implementation in clinical settings. In the near future, the use of more high-throughput protein profiling methodologies utilizing CE-MS, or protein expression arrays hold imminent promise for use in clinical laboratories in order to diagnose renal diseases and monitor treatment. However, large-scale protein datasets generated by the application of 2-DE and/or ICAT with multidimensional chromatography will continue to be of use for ongoing detection of urinary disease biomarkers.

References

1. Wilkins, M. R., Sanchez, J. C., Williams, K. L., and Hochstrasser, D. F. (1996) Current challenges and future applications for protein maps and post-translational vector maps in proteome projects. *Electrophoresis* **17**, 830–838.
2. Giddings, J. C. (1987) Transport, space, entropy, diffusion and flow. Elements underlying separation by electrophoresis, chromatography, field-flow fractionation and related methods. *J. Chromatogr.* **395**, 19–32.
3. Pieper, R., Gatlin, C. L., McGrath, A. M., Makusky, A. J., Mondal, M., Seonarain, M., Field, E., Schatz, C. R., Estock, M. A., Ahmed, N., Anderson, N. G., and Steiner, S. (2004) Characterization of the human urinary proteome: A method for high-resolution display of urinary proteins on two-dimensional electrophoresis gels with a yield of nearly 1,400 distinct protein spots. *Proteomics* **4**, 1159–1174.
4. Cutillas, P. R., Chalkley, R. J., Hansen, K. C., Cramer, R., Norden, A. G. W., Waterfield, M. D., Burlingame, A. L., and Unwin, R. J. (2004) The urinary proteome in Fanconi syndrome implies specificity in the reabsorption of proteins by renal proximal tubule cells. *Am. J. Physiol. Renal Physiol.* **287**, F353–F364.
5. Wittke, S., Mischak, H., Walden, M., Kolch, W., Radler, T., and Wiedemann, K. (2005) Discovery of biomarkers in human urine and cerebrospinal fluid by capillary electrophoresis coupled to mass spectrometry: Towards new diagnostic and therapeutic approaches. *Electrophoresis* **26**, 1476–1487.
6. Pieper, R., Su, Q., Gatlin, C. L., Huang, S. T., Anderson, N. L., and Steiner, S. (2003) Multi-component immunoaffinity subtraction chromatography: An innovative step towards a comprehensive survey of the human plasma proteome. *Proteomics* **3**, 422–432.
7. Steiner, S., Gatlin, C. L., Lennon, J. J., McGrath, A. M., Seonarain, M. D., Makusky, A. J., Aponte, A. M., Esquer-Blasco, R., and Anderson, N. L. (2001) Cholesterol biosynthesis regulation and protein changes in rat liver following treatment with fluvastatin. *Toxicol. Lett.* **120**, 369–377.
8. Benvenuti, S., Cramer, R., Quinn, C. C., Bruce, J., Zvelebil, M., Corless, S., Bond, J., Yang, A., Hockfield, S., Burlingame, A. L., Waterfield, M. D., and Jat, P. S. (2002) Differential proteome analysis of replicative senescence in rat embryo fibroblasts. *Mol. Cell Proteomics* **1**, 280–292.
9. Cutillas, P. R., Norden, A. G. W., Cramer, R., Burlingame, A. L., and Unwin, R. J. (2003) Detection and analysis of urinary peptides by on-line liquid chromatography and mass spectrometry: Application to patients with renal Fanconi syndrome. *Clin. Sci.* **104**, 483–490.
10. Gharbi, S., Gaffney, P., Yang, A., Zvelebil, M. J., Cramer, R., Waterfield, M. D., and Timms, J. F. (2002) Evaluation of two-dimensional differential gel electrophoresis for proteomic expression analysis of a model breast cancer cell system. *Mol. Cell Proteomics* **1**, 91–98.
11. Hansen, K. C., Schmitt-Ulms, G., Chalkley, R. J., Hirsch, J., Baldwin, M. A., and Burlingame, A. L. (2003) Mass spectrometric analysis of protein mixtures at low levels using cleavable C-13-isotope-coded affinity tag and multidimensional chromatography. *Mol. Cell Proteomics* **2**, 299–314.
12. Wittke, S., Fliser, D., Haubitz, M., Bartel, S., Krebs, R., Hausadel, F., Hillmann, M., Golovko, I., Koester, P., Haller, H., Kaiser, T., Mischak, H., and Weissinger, E. M. (2003) Determination of peptides and proteins in human urine with capillary electrophoresis-mass spectrometry, a suitable tool for the establishment of new diagnostic markers. *J. Chromatogr. A* **1013**, 173–181.

13. Weissinger, E. M., Wittke, S., Kaiser, T., Haller, H., Bartel, S., Krebs, R., Golovko, I., Rupprecht, H. D., Haubitz, M., Hecker, H., Mischak, H., and Fliser, D. (2004) Proteomic patterns established with capillary electrophoresis and mass spectrometry for diagnostic purposes. *Kidney Int.* **65**, 2426–2434.
14. Kaiser, T., Wittke, S., Just, I., Krebs, R., Bartel, S., Fliser, D., Mischak, H., and Weissinger, E. M. (2004) Capillary electrophoresis coupled to mass spectrometer for automated and robust pofypaptide determination in body fluids for clinical use. *Electrophoresis* **25**, 2044–2055.
15. O'farrell, P. H. (1975) High-resolution two-dimensional electrophoresis of proteins. *J. Biol. Chem.* **250**, 4007–4021.
16. Inagaki, N. (2004) Large gel two-dimensional electrophoresis: Improving recovery of the cellular proteome. *Curr. Proteomics* **1**, 35–39.
17. Anderson, N. L., and Anderson, N. G. (1998) Proteome and proteomics: New technologies, new concepts, and new words. *Electrophoresis* **19**, 1853–1861.
18. Molloy, M. P., Herbert, B. R., Walsh, B. J., Tyler, M. I., Traini, M., Sanchez, J. C., Hochstrasser, D. F., Williams, K. L., and Gooley, A. A. (1998) Extraction of membrane proteins by differential solubilization for separation using two-dimensional gel electrophoresis. *Electrophoresis* **19**, 837–844.
19. Gianazza, E., Wait, R., Begum, S., Eberini, I., Campagnoli, M., Labò, S., Galliano, M. (2007) Mapping the 5–50-kDa fraction of human amniotic fluid proteins by 2-DE and ESI-MS. *Proteomics Clin. Appl.* **1**, 167–175.
20. Lopez, M. F., Berggren, K., Chernokalskaya, E., Lazarev, A., Robinson, M., and Patton, W. F. (2000) A comparison of silver stain and SYPRO Ruby Protein Gel Stain with respect to protein detection in two-dimensional gels and identification by peptide mass profiling. *Electrophoresis* **21**, 3673–3683.
21. Hille, J. M., Freed, A. L., and Watzig, H. (2001) Possibilities to improve automation, speed and precision of proteome analysis: A comparison of two-dimensional electrophoresis and alternatives. *Electrophoresis* **22**, 4035–4052.
22. Dowsey, A. W., Dunn, M. J., and Yang, G. Z. (2004) ProteomeGRID: Towards a high-throughput proteomics pipeline through opportunistic cluster image computing for two-dimensional gel electrophoresis. *Proteomics* **4**, 3800–3812.
23. Unlu, M., Morgan, M. E., and Minden, J. S. (1997) Difference gel electrophoresis: A single gel method for detecting changes in protein extracts. *Electrophoresis* **18**, 2071–2077.
24. Adachi, J., Kumar, C., Zhang, Y. L., Olsen, J. V., and Mann, M. (2006) The human urinary proteome contains more than 1,500 proteins, including a large proportion of membrane proteins. *Genome Biol.* 7, R80.
25. Gygi, S. P., Rist, B., Gerber, S. A., Turecek, F., Gelb, M. H., and Aebersold, R. (1999) Quantitative analysis of complex protein mixtures using isotope-coded affinity tags. *Nat. Biotechnol.* **17**, 994–999.
26. Rocheleau, M. J., and Dovichi, N. J. (1992) Separation of DNA sequencing fragments at 53 bases minute by capillary gel-electrophoresis. *J. Microcolumn Sep.* **4**, 449–453.
27. Kraly, J. R., Jones, M. R., Gomez, D. G., Dickerson, J. A., Harwood, M. M., Eggertson, M., Paulson, T. G., Sanchez, C. A., Odze, R., Feng, Z. D., Reid, B. J., and Dovichi, N. J. (2006) Reproducible two-dimensional capillary electrophoresis analysis of Barrett's esophagus tissues. *Anal. Chem.* **78**, 5977–5986.
28. Baker, D. R. (1995) *Capillary electrophoresis.* Wiley, New York.
29. Landers, J. P. (1997) *Handbook of capillary electrophoresis.* CRC Press, Boca Raton.
30. Pinto, D., Arriaga, E. A., Schoenherr, R. M., Chou, S. S. H., and Dovichi, N. J. (2003) Kinetics and apparent activation energy of the reaction of the fluorogenic reagent 5-furoylquinoline-3-carboxaldehyde with ovalbumin. *J. Chromatogr. B Analyt. Technol. Biomed. Life Sci.* **793**, 107–114.
31. Pinto, D. M., Arriaga, E. A., Craig, D., Angelova, J., Sharma, N., Ahmadzadeh, H., Dovichi, N. J., and Boulet, C. A. (1997) Picomolar assay of native proteins by capillary electrophores is precolumn labeling, submicellar separation, and laser-induced fluorescence detection. *Anal. Chem.* **69**, 3015–3021.

Chapter 19

Proteomic Analysis of Pancreatic Secretory Trypsin Inhibitor/Tumor-Associated Trypsin Inhibitor from Urine of Patients with Pancreatitis or Prostate Cancer

Leena Valmu, Suvi Ravela, and Ulf-Håkan Stenman

Abstract

The development of proteomic methods, especially mass spectrometry, has brought new possibilities to tumor marker research. Pancreatic secretory trypsin inhibitor (PSTI), a common known biomarker for various malignancies, occurs on genetic variants that we are able to detect at the protein level with proteomic techniques using immunoaffinity capture prior to liquid chromatography–mass spectrometry (LC–MS). We also show that PSTI can be detected in urine from cancer patients using a two-step peptide enrichment technique and LC–MS. These results show that tumor-associated peptides can be detected in urine by proteomic techniques.

Key words: Urine, Proteomics, Peptidomics, Pancreatitis, Cancer, Tumor marker, PSTI, TATI, Mass spectrometry

1. Introduction

Determination of proteins and peptides in urine is widely used in clinical diagnostics. Urine contains a selection of proteins and peptides derived from plasma as a result of glomerular filtration and selective tubular reabsorption. The protein concentration in urine is on average 1,000-fold lower than that in plasma while peptides occur at similar or even higher concentrations than in plasma (1). Thus urine is depleted of proteins and enriched in peptides making it a favorable material for studies of the peptidome.

Pancreatic secretory trypsin inhibitor (PSTI) is a potent low molecular weight (6 kDa) trypsin inhibitor. If trypsinogen is prematurely activated in the pancreas, PSTI serves as a first line of

Alex J. Rai (ed.), *The Urinary Proteome: Methods and Protocols*, Methods in Molecular Biology, vol. 641, DOI 10.1007/978-1-60761-711-2_19, © Springer Science+Business Media, LLC 2010

defense preventing further activation of other pancreatic enzymes, which could eventually lead to development of pancreatitis (2).

An association between PSTI mutations, especially the asparagine-to-serine mutation at amino acid 34 (N34S), and chronic pancreatitis has been demonstrated (3, 4), but since the mutation is relatively common also in the general population (5), it is unlikely that it alone initiates the development of chronic pancreatitis, but it may rather act as a disease modifier (6). Another mutation, a proline-to-serine mutation at amino acid 55 (P55S), in PSTI has also been identified, but it is not significantly associated with pancreatitis (5, 7). We have shown that each of these variants of PSTI can be analyzed in urine by proteomic techniques (8).

PSTI is also associated with numerous malignancies, particularly with ovarian and prostatic cancer (9, 10) and has therefore also been called tumor-associated trypsin inhibitor (TATI) (11). Because of its small size, TATI is readily detectable by peptidomics methodology. Because invasive tumor digest the protein network around it, the resulting peptides can be expected to occur in various body fluids and may be characteristic of cancer (12). A large number of plasma-derived peptides can be detected in urine (1) but clearly tumor-associated peptides have not been detected by proteomic techniques. We have therefore studied whether TATI can be detected in urine from cancer patients by peptidomics technology.

2. Materials

2.1. Determination of Urine Density

1. Use a digital urine refractometer (UG-1, Atago, Tokyo, Japan).

2.2. Immunoaffinity Capture

1. Protein G-coated magnetic beads (Dynal Biotech, Invitrogen, Carlsbad, CA), stored in phosphate buffered saline (PBS), pH 7.4, containing 0.1% Tween-20 and 0.02% sodium azide (NaN_3). Stored at +4°C.
2. An antibody against PSTI (we have used an in-house antibody MAb 11B3 (8)).
3. PBS: phosphate buffered (10 mM) saline (0.15 M NaCl) pH 7.4. Stored at room temperature.
4. 0.01% Tween-20 (Fluka, Sigma, St. Louis, MO) in PBS. Stored at room temperature.
5. MagneSphere 12-position magnetic separation stand (Promega, Madison, WI).
6. 0.1% TFA: 0.1% (v/v) trifluoroacetic acid. Stored at room temperature.

7. Millex-HV 0.45 μm 4 mm low protein binding Durapore (PVDF) membrane for clarification of aqueous and mild organic solutions, nonsterile (Millipore, Billerica, MA).

2.3. Peptide Enrichment

1. PBS: phosphate buffered (10 mM) saline (0.15 M NaCl) pH 7.4. Stored at +4°C.
2. PD-10 size exclusion column (GE Healthcare, Piscataway, NJ). Stored at +4°C. For one-time use only.
3. Low-protein binding Eppendorf tubes (Eppendorf, Westbury, NY).
4. $ZipTip_{C18}$ (P10, Millipore). If possible, use the same batch for all samples analyzed in the same series (see Note 1).
5. 0.1% TFA: 0.1% (v/v) trifluoroacetic acid. Stored at room temperature.
6. ACN: acetonitrile. Harmful, to be stored in a conditioned space at room temperature. Used at various dilutions in the sample enrichment procedure, dilute for one-time use.
7. 0.1% FA: 0.1% (v/v) formic acid. Stored at room temperature.

2.4. Liquid Chromatography–Mass Spectrometry

1. Trapping column: C18 column (Atlantis dC18, NanoEase Trap Column 5 μm, Waters, Milford, MA).
2. Analytical column: 0.075 × 150 mm C18 column (Atlantis dC18, 300 Å, 3.5 μm, Waters).
3. ACN: acetonitrile. Harmful, to be stored in a conditioned space at room temperature. Dilute for one-time use.
4. 0.1% FA: 0.1% (v/v) formic acid. Stored at room temperature.
5. Nanoflow needle: PicoTip Emitter, Silica Tip, coated, diameter 10 μm, (New Objective, Stock#FS360-20-10-D-20-C7, coating 1P-4P).
6. Vials for LC autosampler, Polypropylene plastic screw-top vial, 300 μl (Waters).

3. Methods

Although analysis of genetic variants traditionally is performed with molecular genetic techniques, the development of proteomic technology has facilitated analysis of genetic variants at the protein level. This method furthermore provides information about posttranslational modifications and differences in protein expression levels. Mass spectrometry can be used to characterize variants of PSTI in urine from patients with pancreatitis and cancer. For this purpose a rapid small-scale immunoaffinity capture procedure to isolate and analyze PSTI from small volumes of urine can be employed. PSTI can also be detected by enrichment of endogenous peptides from urine of cancer patients. Here we describe both methods.

3.1. Pretreatment of Urine Samples

1. Collect urine samples from cancer patients and healthy controls of the same age and gender into identical containers (see Note 1).
2. Remove cells by centrifugation (3,000 × *g*, 10 min), and store the supernatant at -80°C.
3. After thawing, centrifuge the urine samples at 3,000 × *g* for 10 min to remove precipitates. Measure specific gravity of urine by refractometry (*).
4. Aliquot samples and store at –20°C. The number of freeze-thaw cycles has to be constant; in this case two (see Note 2).

3.2. Immunoaffinity Capture of PSTI from Urine

1. Couple an antibody against PSTI (i.e., MAb 11B3) covalently to Protein G-coated magnetic beads covalently according to the instructions of the manufacturer (Dynal Bead).
2. Incubate an aliquot of urine containing 0.05–0.5 μg of PSTI with 20 μL of antibody-coated beads for 2 h at room temperature with constant agitation. Dilute the urine to 50 μl with 0.01% Tween-20 in PBS to prevent aggregation of the beads.
3. Wash the beads three times with the same buffer, then three times with PBS, and once with water to remove detergent (see Note 3), which may interfere with the MS analysis. Use a magnetic separation stand to separate the beads.
4. To elute the bound PSTI, incubate the magnetic beads in 10 μL 0.1% TFA for 30 min at room temperature with constant agitation before magnetic separation. After elution, clarify the sample by filtration through a 0.45 μm pore size Millex low protein binding Durapore membrane (Millipore) to ensure that there are no magnetic beads in the sample. Place the filter in a low-protein binding Eppendorf tube or LC-vial. Add 10 μl 0.1% FA to the filter and centrifuge at 3,000 × *g* for 2 min. Add the sample to the filter, and centrifuge. Add 10 μl 0.1% FA, centrifuge and add 2.8 μl 50% ACN in 0.1% FA and centrifuge. The final sample volume to be subjected for LC–MS analyses is thus 30 μl. The efficiency of the immunoaffinity capture for PSTI is shown in Fig. 1.
5. For further use, the beads are transferred after elution into PBS and washed until the pH is neutral. To prevent aggregation, add 0.1% Tween 20 to the storage buffer (PBS) (see Note 4).

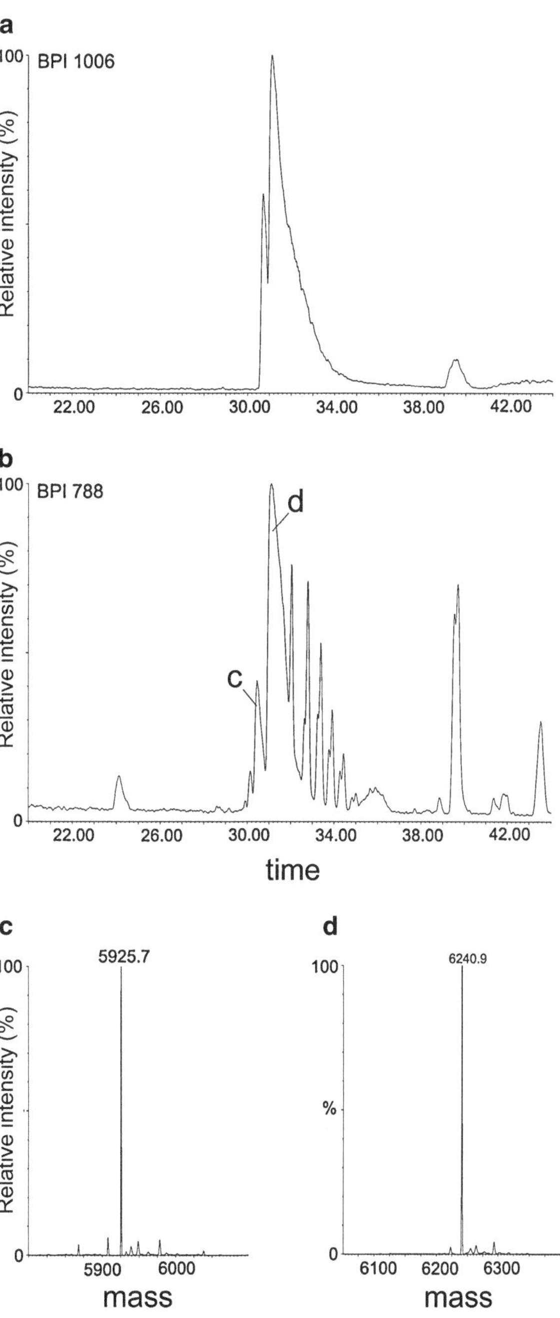

Fig. 1. LC-MS analysis of PSTI purified from urine by immunoaffinity capture. Panel B shows a C18 RP-chromatogram of PSTI purified with MAb-coated magnetic beads from a urine sample containing 10 pmol of PSTI. The chromatographic pattern of 10 pmol of purified PSTI is shown for comparison (**a**). In both chromatograms, the base peak intensity (BPI) is shown and the relative intensity value is indicated in the left hand corner of the chromatogram. To identify PSTI in the peaks, the mass was derived by deconvolution (**c**, **d**) of the mass spectra of peaks annotated (**c**) and (**d**) in panel B. The deconvoluted mass corresponds to the theoretical mass of PSTI (6241.0 Da), which is the main component detected by LC–MS. Minor interfering components are caused by detergents remaining after magnetic bead capture. (Reproduced from ref. (8) with permission from American Association for Clinical Chemistry, AACC).

3.3. Solid-Phase Enrichment of Peptides from Urine (see Note 5)

1. Normalize the amount of urine applied for peptidomic analysis according to a specific gravity of 1.015 (see Note 6). From urine with this gravity a volume of 1 ml is used for analysis. The urine volume subjected for analysis is calculated as follows: If the sample has a specific gravity of 1.0075, the double volume is selected. If the sample has a specific gravity of 1.030, half of the volume is selected. Before size exclusion chromatography always adjust the sample volume to 2.5 ml with PBS. Should the applied sample volume be over 2.5 ml (with very dilute samples) use vacuum dryer to reduce the sample volume. Adjust then the final volume to 2.5 ml with water.
2. Equilibrate a PD-10 size exclusion column with 25 ml PBS, and apply the urine sample to the column. Discard the flowthrough fraction. Elute the column with 4 ml PBS collecting 1 ml fractions into low-protein binding Eppendorf tubes (see Note 7).
3. Take 50 µl aliquots from fractions 2 and 3 and combine them. The combined fraction contains urinary proteins together with urinary peptides. Most urinary metabolites that suppress ionization of peptides in mass spectrometry are removed by this method (see Note 8). The rest of fractions 2 and 3 can be stored at –20°C.
4. Apply the peptide-containing fraction (100 µl) to a C18 ZipTip column ($ZipTip_{C18}$, P10, Millipore) according to the manufacturer's instructions. Elute the peptides with 3 µl 40% ACN in 0.1% TFA. Prior the LC–MS analysis, dilute the samples with 0.1% FA to reduce the ACN concentration to below 5%.

3.4. Liquid Chromatography–Electrospray Mass Spectrometry

1. The peptide sample (see Note 9) is concentrated with a flow rate of 20 µl/min onto a trapping column, from which it is transferred into the analytical 0.075 × 150 mm C18 column using nanoscale HPLC (CapLC, Waters) at a flow rate of 0.3 µL/min. Peptides are eluted with a linear gradient of ACN in 0.1% formic acid (5–50% in 30 min). The eluent is transferred into the mass spectrometer via a fused-silica capillary with an i.d. of 25 µm.
2. Use a nanoflow needle to electrospray the sample into a quadrupole – time-of-flight mass spectrometer (e.g., Q-TOF micro, Waters) using a capillary voltage of 2,300 V and a cone voltage of 45 V.
3. Acquire data for the m/z range 400–1,700 using a Q-TOF instrument. The time range of data acquisition is dependent on the LC–MS setup. The mass spectrometric calibration used for the analysis is dependent on size and adopted charge states of the relevant molecules (see Note 10). An example of MS spectra of different PSTI variants is shown in Fig. 2.
4. The sequence of PSTI/TATI variants is determined by fragmentation of tryptic peptides by tandem mass spectrometry

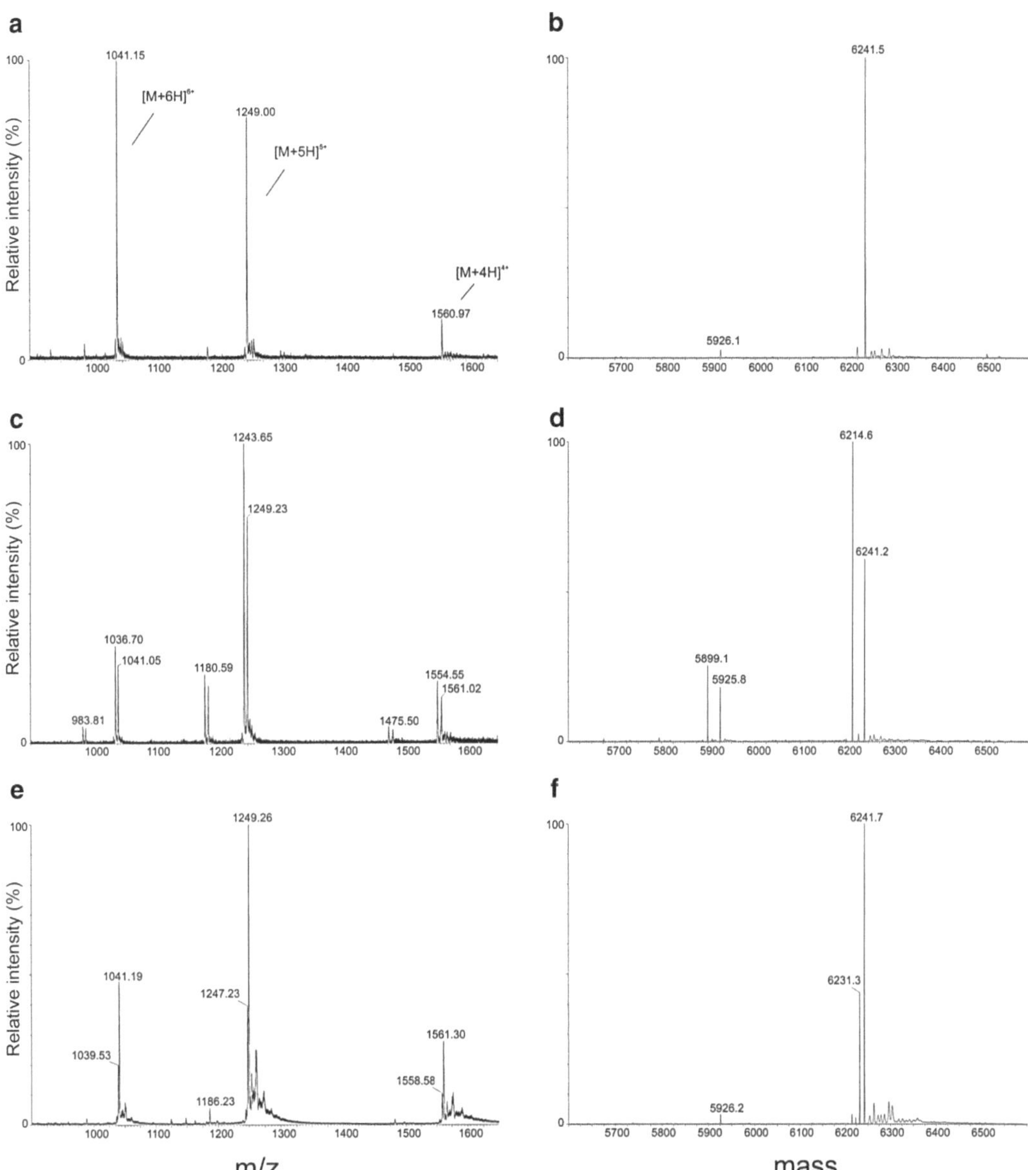

Fig. 2. Mass spectrometric analysis of immunoaffinity captured PSTI variants from patients with pancreatitis. Mass spectra of PSTI purified from the urine of pancreatitis patient with either no mutation (**a**), the N34S (**c**) or P55S mutations (heterozygote) (**e**) showing peaks mainly corresponding to $(M+4H)^{4+}$, $(M+5H)^{5+}$, and $(M+6H)^{6+}$ as indicated. The mass of PSTI was deconvoluted from these mass spectra with MassEnt1 software (**b**, **d**, **f**). Based on the deconvoluted spectra the mass of PSTI WT is 6241.5 (**b**), that of the N34S variant 6214.6 (**d**) and of the P55S variant 6231.3 (**f**). Each of the masses correspond to the theoretical masses calculated for WT PSTI (6241.0 Da), N34S mutated PSTI (6214.0 Da), and P55S mutated PSTI (6231.0 Da) with all putative disulfide bridges present. A smaller proteolytic fragment corresponding to PSTI lacking 3 N-terminal amino acids was also detected (Fig. 1**b**, **d**, **f**). (Reproduced from ref. (8) with permission from AACC).

(MSMS). MSMS fragmentation spectra of the peptides are acquired by colliding the precursor ions with argon collision gas using accelerating collision energies of 30–70 V (see Note 11).

3.5. Differential Data Analysis

1. Analyze the mass spectra collected during LC–MS separation with MassLynx software (Waters) and of PAWS proteomic analysis software (Proteometrics, Rockefeller University, New York, NY). Deconvolute different charge states into actual masses with MaxEnt1 software (Waters).
2. Comparison of peptide profiles from various urine samples can be performed with DeCyder MS software (GE Healthcare). Convert the acquired LC-MS data into ASCII text files using the DataBridge of the MassLynx software (Waters).
3. The profiles can be visualized with DeCyder MS as two-dimensional (2D) and three-dimensional (3D) graphs. Perform semi quantitative differential analysis of the peptides by integrating the ion counts over the spot areas and comparing the integrated ion counts between samples. To calculate total intensity of the deconvulated masses, take into account the ion counts of all different charge states of the same peptide. Compare profiles in different samples using the DeCyderMatch module of the software. You can transfer the overall intensities of deconvoluted peptides to MS Excel (Microsoft) for further comparisons. Analysis of TATI in urine from ten prostate cancer patients and ten healthy patients is shown in Fig. 3. Normalization of peptidomic data can be done using internal endogenous peptides (see Note 12).

4. Notes

1. Serum peptidomics studies have shown that both the type of the sample collection tubes and use of different batches for SPE enrichment affect peptide recovery and quality of the peptide pattern (13).
2. Many freeze-thaw cycles of the samples prior MS analysis have been shown to reduce peptide recovery in serum proteomics (13).
3. Without detergent the magnetic beads are more difficult to handle, but the detergent must be removed before MS-analysis to avoid ion suppression effects.
4. Although some detachment of antibody at each cycle of immunoaffinity capture can be detected, the same batch of antibody-linked magnetic beads can be used at least 100 times.
5. The mass spectrometric analysis is very sensitive to impurities and especially to detergents. Therefore, all the glassware and plastics used have to be free of impurities. Use only new tubes and do not aliquot liquids (e.g., ACN) into detergent-washed

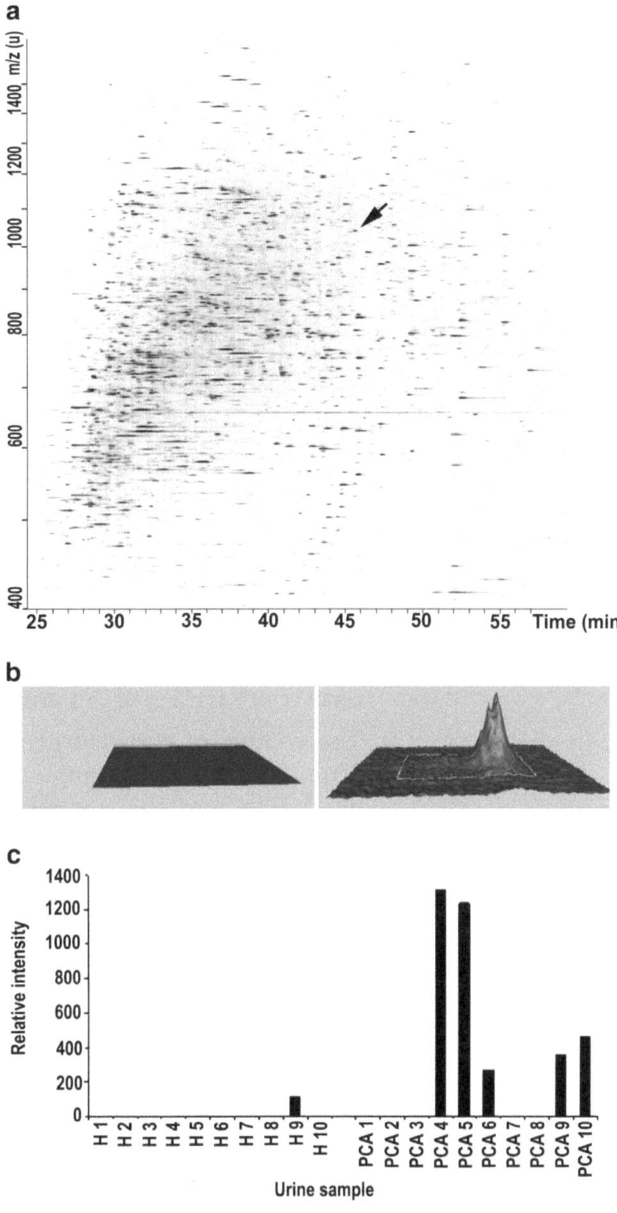

Fig. 3. Peptidomic detection of TATI in urine from prostate cancer (PCa) patients. Peptide profile of the urine of a PCa patient (patient number 4) visualized with DeCyder MS as a 2D-graph (**a**). Each spot represents one peak (certain m/z of a peptide) in the mass spectrum. The arrow indicates the peak of the main charge state of TATI ($(M+6H)^{6+}$). The urine was enriched by size exclusion chromatography-solid phase enrichment and analyzed by LC–MS as described in Methods. (**b**) shows a 3D-graph of TATI at charge state of $(M+6H)^{6+}$ in the urine of a prostate cancer patient (*right*). No peak is detected in the urine of a healthy individual (*left*). (**c**) shows the total intensity of TATI (the summarized intensity of all charge states) in urine from ten healthy subjects and ten cancer patients normalized according to endogenous internal standard (see Note 12).

glassware. We routinely use only dedicated and very thoroughly rinsed (acid, water and alcohol) glassware and low-protein binding plastic ware. Use powder-free gloves and avoid working in a space where other proteins are handled.

6. The concentration of urine varies greatly and therefore some normalization method needs to be applied. Three different normalization methods are generally used; normalization against urine protein concentration, normalization against urine specific gravity, and normalization against urine creatinine concentration. For urine peptidomics, we have shown that normalization against urine specific gravity is the best method (Ravela et al., in preparation).
7. Peptides and proteins adhere to the walls of tubes. To avoid nonspecific loss of peptides we use low-protein binding Eppendorf tubes.
8. Urine contains many endogenous and nonendogenous metabolites that interfere with MS-analysis of peptides. The physical and chemical properties of these metabolites are similar to peptides, and they are difficult to remove from the sample without affecting peptide recovery. The combined use of size-exclusion chromatography (PD-10) and solid phase extraction removes metabolites with little effect on peptide recovery (Ravela et al., in preparation).
9. The volume of enriched peptide sample for LC-MS run is 10 μl. The volume of immunoaffinity captured PSTI varies depending on the sample concentration, inject about 10–50 pmol of PSTI.
10. For analysis of intact PSTI, the Q-TOF is calibrated using the m/z envelope of a small protein (i.e., myoglobin, 400 nM). For urinary peptidomics analysis, the mass spectrometer is calibrated using 2 μM glufibrinogen peptide B fragments.
11. MSMS fragmentation of intact PSTI produces spectra that are not easily interpretable due to the charge state distribution, e.g., $(M+4H)^{4+}$, $(M+5H)^{5+}$, and $(M+6H)^{6+}$. MSMS spectra of good quality are obtained by fragmentation of doubly and triply charges precursor peptides. To obtain these, tryptic digestion of immunoaffinity-purified PSTI can be used (8).
12. Due to external factors affecting ESI-MS analyses (i.e., needle position, partial needle blockage etc), the intensities of the acquired m/z peaks can vary. To limit variation, internal normalization of the acquired data is often needed. Multiple peptides in serum can be used for this purpose (13). We have identified a urinary collagen fragment with a mass of 3457.5 that can be used as an internal standard (Ravela et al., in preparation). This peptide has been identified as alpha-1 type-III collagen fragment comprising amino acids 910–948.

Acknowledgments

The authors would like to thank Ms Maarit Leinimaa and Ms Helena Taskinen for skillful technical assistance. The financial support of the Finnish Academy of Sciences and EU (P-Mark, contract no: LSHC-CT-2004-503011) is gratefully acknowledged.

References

1. Hortin, G. L., and Sviridov, D. (2007) Diagnostic potential for urinary proteomics. *Pharmacogenomics* **8**, 237–255
2. Rinderknecht, H. (1986) Activation of pancreatic zymogens. Normal activation, premature intrapancreatic activation, protective mechanisms against inappropriate activation. *Dig. Dis. Sci.* **31**, 314–321
3. Gomez-Lira, M., Bonamini, D., Castellani, C., Unis, L., Cavallini, G., Assael, B. M., and Pignatti, P. F. (2003) Mutations in the SPINK1 gene in idiopathic pancreatitis italian patients. *Eur. J. Hum. Genet.* **11**, 543–546
4. Threadgold, J., Greenhalf, W., Ellis, I., Howes, N., Lerch, M. M., Simon, P., Jansen, J., Charnley, R., Laugier, R., Frulloni, L., Olah, A., Delhaye, M., Ihse, I., Schaffalitzky de Muckadell, O. B., Andren-Sandberg, A., Imrie, C. W., Martinek, J., Gress, T. M., Mountford, R., Whitcomb, D., and Neoptolemos, J. P. (2002) The N34S mutation of SPINK1 (PSTI) is associated with a familial pattern of idiopathic chronic pancreatitis but does not cause the disease. *Gut* **50**, 675–681
5. Lempinen, M., Paju, A., Kemppainen, E., Smura, T., Kylanpaa, M. L., Nevanlinna, H., Stenman, J., and Stenman, U. H. (2005) Mutations N34S and P55S of the SPINK1 gene in patients with chronic pancreatitis or pancreatic cancer and in healthy subjects: A report from finland. *Scand. J. Gastroenterol.* **40**, 225–230
6. Schneider, A. (2005) Serine protease inhibitor kazal type 1 mutations and pancreatitis. *Clin. Lab. Med.* **25**, 61–78
7. Tukiainen, E., Kylanpaa, M. L., Kemppainen, E., Nevanlinna, H., Paju, A., Repo, H., Stenman, U. H., and Puolakkainen, P. (2005) Pancreatic secretory trypsin inhibitor (SPINK1) gene mutations in patients with acute pancreatitis. *Pancreas* **30**, 239–242
8. Valmu, L., Paju, A., Lempinen, M., Kemppainen, E., and Stenman, U. H. (2006) Application of proteomic technology in identifying pancreatic secretory trypsin inhibitor variants in urine of patients with pancreatitis. *Clin. Chem.* **52**, 73–81
9. Stenman, U. H. (2002) Tumor-associated trypsin inhibitor. *Clin. Chem.* **48**, 1206–1209
10. Paju, A. and Stenman, U. H. (2006) Biochemistry and clinical role of trypsinogens and pancreatic secretory trypsin inhibitor. *Crit. Rev. Clin. Lab. Sci.* **43**, 103–142
11. Huhtala, M. L., Pesonen, K., Kalkkinen, N., and Stenman, U. H. (1982) Purification and characterization of a tumor-associated trypsin inhibitor from the urine of a patient with ovarian cancer. *J. Biol. Chem.* **257**, 13713–13716
12. Villanueva, J., Shaffer, D. R., Philip, J., Chaparro, C. A., Erdjument-Bromage, H., Olshen, A. B., Fleisher, M., Lilja, H., Brogi, E., Boyd, J., Sanchez-Carbayo, M., Holland, E. C., Cordon-Cardo, C., Scher, H. I., and Tempst, P. (2006) Differential exoprotease activities confer tumor-specific serum peptidome patterns. *J. Clin. Invest.* **116**, 271–284
13. Villanueva, J., Philip, J., Chaparro, C. A., Li, Y., Toledo-Crow, R., DeNoyer, L., Fleisher, M., Robbins, R. J., and Tempst, P. (2005) Correcting common errors in identifying cancer-specific serum peptide signatures. *J. Proteome Res.* **4**, 1060–1072

Index

Alex J. Rai (ed.), *The Urinary Proteome: Methods and Protocols*, Methods in Molecular Biology, vol. 641, DOI 10.1007/978-1-60761-711-2, © Springer Science+Business Media, LLC 2010

T

U

V

MIX
Papier aus verantwortungsvollen Quellen
Paper from responsible sources
FSC® C105338

If you have any concerns about our products,
you can contact us on
ProductSafety@springernature.com

In case Publisher is established outside the EU,
the EU authorized representative is:
Springer Nature Customer Service Center GmbH
Europaplatz 3, 69115 Heidelberg, Germany

Printed by Libri Plureos GmbH
in Hamburg, Germany